61

D1288331

# SEMIGROUP ALGEBRAS

# PURE AND APPLIED MATHEMATICS

*A Program of Monographs, Textbooks, and Lecture Notes*

# MONOGRAPHS AND TEXTBOOKS IN
# PURE AND APPLIED MATHEMATICS

*Other Volumes in Preparation*

# SEMIGROUP ALGEBRAS

Jan Okniński

Warsaw University
Warsaw, Poland

Marcel Dekker, Inc. • New York • Basel • Hong Kong

**Library of Congress Cataloging-in-Publication Data**

Okniński, Jan.
  Semigroup algebras / Jan Okniński.
    p.  cm.-- (Monographs and textbooks in pure and applied mathematics;
  138).
  Includes bibliographical references and indexes.
  ISBN 0-8247-8356-5
  1. Semigroup algebras. 2. Semigroup rings. 3. Noncommutative rings.
  II. Series.
  QA251.5.O38   1990
  512'.2--dc20                                                    90-3879
                                                                  CIP

This book is printed on acid-free paper

MARCEL DEKKER, INC.
270 Madison Avenue, New York, New York 10016

Current printing (last digit):
10 9 8 7 6 5 4 3 2 1

PRINTED IN THE UNITED STATES OF AMERICA

# Preface

This book is intended as the first attempt to gather and unify the results of the theory of noncommutative semigroup rings. Most of the material comes from the literature of the past 10 years, and several new results are included. We follow the line initiated by *Infinite Group Rings* by D. S. Passman and his later monograph *The Algebraic Structure of Group Rings*, and R. Gilmer's *Commutative Semigroup Rings*. The group ring results are basically the starting point for most of the topics considered, while the case of commutative semigroup rings is a base and a motivation for developing the theory of $PI$-semigroup-algebras. Since problems on semigroup rings $R[S]$ over an arbitrary ring $R$ often are easily settled once we handle the case where $R$ is a field, or they lead to specific difficult problems on tensor products, only semigroup rings with coefficients in a field (referred to as semigroup algebras) will be considered.

The theory of semigroup rings, extensively studied in recent years, has been stimulated by the role of this class of rings in the general ring theory. The theory of linear representations of semigroups leads, as in the group ring case, to the class of semigroup rings satisfying polynomial identities. On the other hand, the modern representation theory of finite dimensional algebras exploits some specific classes of semigroup rings (e.g., path algebras). Moreover, it has recently been shown that every finite dimensional algebra of finite representation type (that is, with finitely many isomorphism classes of indecomposable finite dimensional modules) over an algebraically closed field is, in fact, a contracted semigroup algebra of a very special type (compare with Ref. [8]). Another specific type of semigroup algebras arises when developing the theory of growth and Gelfand–Kirillov dimension of arbitrary algebras. Last but not least, the semigroup ring construction has been successfully used to build several important examples. This exploits the fact that algebras defined by generators and relations of a special type are contracted semigroup algebras in which calculations are reduced to some combinatorial problems and often become relatively easy to handle.

The aim is to present, in a unified form, these aspects of the theory where decisive results have been obtained. The book does not cover all the material on semigroup rings available from the literature. We restrict ourselves to the ring theoretical properties for which a systematic treatment is possible at this stage. These particularly include the classical finiteness conditions—strong enough to provide definite general results, and *PI*-algebras—extending the well-developed theory of commutative semigroup rings. We purport to give an account of the current state of knowledge in these topics and we extract the most interesting open problems, to which a special section is devoted. No special class of semigroups (except cancellative ones) is considered for its own sake. Thus, for example, some interesting special results and techniques in semigroup rings of inverse semigroups, due mainly to Munn, are not included.

This book should be of particular interest both to ring theorists and semigroup theorists. While it starts with the fundamental definitions and leads to open research problems, it is addressed to research workers in these fields as well as to students. The reader is basically assumed to be familiar with the fundamental results in ring theory. The necessary background is available from the textbooks and monographs on ring theory and semigroup theory to which precise references for the proofs of the used results are provided. A few deep theorems, such as Gromov's theorem on groups of polynomial growth and the theorem of Braun on the radical of finitely generated $PI$-algebras, are the only exceptions.

Part I is of a preparatory character. It presents the basic definitions, techniques, and results on the classes of semigroups essential as tools for the main body of the book. It consists mostly of published results, but some new results are also included. While one of the main techniques developed in the book consists of looking at (skew) linear representations of a given semigroup algebra, linear semigroups are discussed. Particularly, a general class of semigroups containing the most essential classes of semisimple and periodic semigroups is discussed with this respect in Chapter 3. Some stronger finiteness conditions of semigroups (minimal conditions, local finiteness, nilpotency) are also presented. Fundamental facts and techniques in semigroup algebras are given in Chapters 4 and 5. Some useful methods coming from the theory of graded semigroups and algebras are discussed in Chapter 6.

Semigroup algebras of cancellative semigroups, studied in Part II, are (beside their general importance as a class close to group algebras and polynomial algebras) essential for some of the general theory developed in Part IV. In fact, many questions on $PI$-semigroup rings $K[S]$ may be successfully attacked through semigroup algebras of some cancellative semigroups derived from the linear representations of $S$. For this, relations between the properties of a semigroup algebra of a cancellative semigroup $S$ with a group of fractions $G$, and the group algebra $K[G]$ are studied in Chapter 7. Cancellative semigroups of polynomial growth are described in Chapter 8. An analog of the powerful $\Delta$-methods in group algebras is developed for arbitrary cancellative semigroups in Chapter 9. Unique product (u.p.) semigroups are considered in Chapter 10, with an emphasis on the distinction with the case of groups. Finally, the class of (semigroup) subalgebras of the important class of (noetherian) group algebras of polycyclic-by-finite groups is discussed.

In Part III we study a chain of related ring theoretical finiteness conditions for a semigroup algebra. The main aim is to characterize rings of these types in terms of the underlying semigroups and to describe relations between the considered classes of rings. The results in any of Chapters 12–16 are heavily dependent on the material presented in the foregoing sections of Part III. Chapter 17 presents, without proofs, results on important finiteness conditions of other types (and thus requiring methods different from those developed in the book).

One of the main objectives in Part IV is to find necessary and sufficient conditions on a semigroup $S$ in order that the semigroup algebra $K[S]$ would satisfy a polynomial identity. In Chapter 19 and 20 we present two possible ways of attacking this problem: through the permutational property of semigroups (which is an immediate consequence of the $PI$-property of the corresponding semigroup algebra) and through a reduction to the principal factors of $S$ (and to some derived group algebras consequently). The subsequent chapters are concerned with some important tools in $PI$-algebras and some special classes of $PI$-semigroup-algebras (e.g., Azumaya, prime $PI$). For example, Chapter 21 discusses a description and the properties of the Jacobson radical, while Chapter 23 gives formulas for computing the (classical Krull, Krull, and Gelfand–Kirillov) dimensions of $K[S]$ by means of some invariants of $S$.

The book concludes with a list of open problems related to the presented material. We hope to stimulate interest and further progress in the theory of semigroup algebras in this way.

Each chapter is completed with bibliographical notes and comments on related results appearing in the literature. Beyond this we refer to the survey papers of Ponizovskii [221] and Okniński [185] for an additional bibliography on some other topics in the theory of semigroup rings.

The first draft of the main chapters of the book was written in Warsaw. The work was completed during the author's stay at North Carolina State University in the Fall of 1988 and in 1989.

Jan Okniński

# Contents

# Part I

# Semigroups and Their Algebras

In Part I, we state the most important semigroup theoretic results used throughout the text, and we present some basic facts and methods of the theory of semigroup rings. While one of the main techniques consists in studying a given semigroup "locally," through its principal factors, we start in Chapter 1 with completely 0-simple semigroups. This is then illustrated with the important example of full linear semigroups. Investigating semigroups via their principal factors is especially fruitful in the case of semigroups satisfying certain finiteness conditions. Semigroups of this type often appear in the context of semigroup algebras. We discuss them in Chapter 2. Results on a class of semigroups satisfying a fairly weak finiteness condition (and containing all classes considered in Chapter 2), called weakly periodic semigroups, are presented in Chapter 3. In many cases, problems involving these semigroups can be reduced to the classes of groups and nilpotent semigroups. This is especially apparent from the crucial result on linear semigroups of this type.

The remaining part of this chapter is devoted to algebras. Semigroup algebras, their basic properties, and some general techniques are introduced in Chapter 4. Most of this material extends standard facts on group algebras. A fundamental class of algebras arising from completely 0-simple semigroups is discussed in Chapter 5. We show how the properties of these algebras are determined, and may be studied, by the properties of the group algebras of the maximal subgroups. We close this part with a very useful setting of algebras (and semigroups) graded by semigroups. Our attention is restricted to the two opposite cases: group gradings and semilattice gradings. They arise in several contexts in ring theory and semigroup theory. In particular, we present some aspects of the powerful technique of smash products for group-graded algebras.

# 1

# Completely 0-Simple and Linear Semigroups

In this chapter, we establish some basic notation, and we recall the definitions and fundamental results of the theory of semigroups required in the sequel.

By $S$ we denote an arbitrary semigroup. The operation in $S$ will always be written multiplicatively. If $S$ has a zero element, then it is denoted by $\theta_S$, or $\theta$ if unambiguous. Whether or not $S$ has a zero element, by $S^0$ we mean the semigroup obtained by adjoining a zero element to $S$. Similarly, $S^1$ stands for the monoid obtained by adjoining an identity element to $S$. The set of all idempotents of $S$ is denoted by $E(S)$, while the set of all (two-sided) units of a monoid $S$ is denoted by $U(S)$. If $A$ is a nonempty subset of $S$, then by $\langle A \rangle$ we mean the subsemigroup of $S$ generated by $A$. Further, $C_S(A)$ is the centralizer of $A$ in $S$, and $Z(S) = C_S(S)$ is the center of $S$.

An equivalence relation $\rho$ on $S$ is called a right congruence if for any $s, t, x \in S$ we have $(sx, tx) \in \rho$ whenever $(s, t) \in \rho$. We write $S/\rho$ for the set of $\rho$-classes in $S$. The set of all right congruences on $S$ is a lattice under the natural operations $\wedge$, $\vee$, where $\rho_1 \wedge \rho_2$ is the intersection and $\rho_1 \vee \rho_2$ is the smallest right congruence on $S$ containing $\rho_1$ and $\rho_2$. This lattice is denoted by $\mathcal{R}(S)$. Similarly, $\mathcal{L}(S)$ stands for the lattice of left congruences on $S$, (defined dually), and $\mathcal{T}(S) = \mathcal{R}(S) \cap \mathcal{L}(S)$, for the lattice of congruences on $S$. If $\rho \in \mathcal{T}(S)$, then $S/\rho$ is a semigroup with the natural structure induced by that of $S$.

Let $I$ be an ideal of $S$. Then $S/I$ is the Rees factor semigroup, that is, $S/I$ may be identified with the set $(S \setminus I) \cup \{\theta\}$ subject to the multiplication $\circ$ defined by the formula

$$s \circ t = \begin{cases} st & \text{if } st \notin I \\ \theta & \text{if } st \in I \end{cases}$$

In other words, $S/I \cong S/\rho$, where $\rho$ is the congruence on $S$ given by $(s,t) \in \rho$ if $s = t$ or $s, t \in I$. If $I$ is an empty set, then it is convenient to put $S/I = S$.

Let $s \in S$. The principal ideal $S^1 s S^1$ of $S$ generated by $s$ is denoted by $J_s$, while the subset of $J_s$ consisting of nongenerators of $J_s$ as an ideal of $S$ is denoted by $I_s$. Thus, $I_s = \varnothing$ if and only if $J_s$ is a minimal ideal of $S$, and if it is not the case, then $I_s$ is an ideal of $S$. The factor $S_s = J_s/I_s$ is called the principal factor of $S$ determined by the element $s$. The set $J_s \setminus I_s$ is called the $\mathcal{J}$-class of $s$ in $S$. In fact, this is a class of the equivalence relation $\mathcal{J}$ defined by $(s,t) \in \mathcal{J}$ if $S^1 s S^1 = S^1 t S^1$. Similarly, one defines the relations $\mathcal{R}$ and $\mathcal{L}$, and the corresponding $\mathcal{R}$- and $\mathcal{L}$-classes of $S$, by $(s,t) \in \mathcal{R}$ if $sS^1 = tS^1$, $(s,t) \in \mathcal{L}$ if $S^1 s = S^1 t$.

A semigroup $T$ (not necessarily having a zero element) is 0-simple if it has no nonzero proper ideals and is not a semigroup with zero multiplication of cardinality 2. It is well known that any principal factor $S_s$, $s \in S$, of $S$ is either 0-simple or it is a semigroup with zero multiplication; see [26], Lemma 2.39.

The class of 0-simple semigroups contains an important subclass of completely 0-simple semigroups. These are 0-simple semigroups that have a primitive idempotent, that is, a nonzero idempotent that is minimal with respect to the natural order in the set of (nonzero) idempotents given by $e \leq f$ if and only if $e = ef = fe$ for $e, f \in E(S)$. A trivial semigroup also will be included in the class of completely 0-simple semigroups. We will often use the following well-known result; see [26], Theorem 2.54, Corollary 2.56.

**Lemma 1** Let $S$ be a 0-simple semigroup. If $S$ has a nonzero idempotent, then $S$ is completely 0-simple if and only if $S$ has no subsemigroup isomorphic to the bicyclic monoid $B(x,y)$ defined by generators $x$, $y$ subject to the relation $xy = 1$. In particular, this is the case if $S$ is a nonzero periodic semigroup, that is, $\langle s \rangle$ is a finite semigroup for every $s \in S$.

Principal factors of a given semigroup $S$ are used to study the properties of $S$ through its "local" structure. This motivates the following classical definition. $S$ is called semisimple if all its principal factors are 0-simple, and $S$ is called completely semisimple if all its principal factors are completely 0-simple. We recall two important special classes of semisimple semigroups. $S$ is regular if it satisfies the von Neumann regularity condition: for any $s \in S$, there exists $x \in S$ such that $sxs = s$. $S$ is inverse if it is regular and the idempotents of $S$ commute; or, equivalently, for every $s \in S$, there exists a unique $x \in S$ with $sxs = s$, $xsx = x$; see [26].

While, in general, the principal factors of a homomorphic image of $S$ may not be homomorphic images of the corresponding principal factors of $S$, even more holds in the following special case.

**Lemma 2**   Let $I$ be an ideal of a semigroup $S$, and let $\phi : S \to S/I$ be the natural homomorphism. Then, for any $s \in S \setminus I$, $\phi$ induces an isomorphism of the principal factor of $s$ in $S$ onto the principal factor of $\phi(s)$ in $S/I$.

*Proof.*   Clearly, $\phi(S^1 s S^1) = (S/I)^1 \phi(s)(S/I)^1$. If $t \in S$ is such that $S^1 s S^1 = S^1 t S^1$, then $t \notin I$, and so $\phi(t) \neq \phi(s)$ if $t \neq s$. On the other hand, if $(S/I)^1 \phi(u)(S/I)^1 = (S/I)^1 \phi(s)(S/I)^1$ for some $u \in S$, then $u \notin I$, and so $S^1 u S^1 = S^1 s S^1$. Thus, $\phi$ establishes a one-to-one correspondence between the elements of the $\mathcal{J}$-classes of $s$ in $S$ and of $\phi(s)$ in $S/I$, respectively. The assertion is now a direct consequence of the definition of a principal factor.

Let $G^0$ be a group with a zero adjoined, and let $I$, $M$ be nonzero sets. Further, let $P = (p_{mi})_{m \in M, i \in I}$ be a generalized $M \times I$ matrix over $G^0$, that is, every $p_{mi}$ lies in $G^0$. By $\mathfrak{M}^0(G^0, I, M, P)$ we mean the set of all generalized $I \times M$ matrices over $G^0$ with at most one nonzero entry, and with the multiplication defined by the rule $AB = A \circ P \circ B$ where $\circ$ stands for the usual multiplication of matrices. $\mathfrak{M}^0(G^0, I, M, P)$ becomes a semigroup subject to this operation. Any nonzero element of $\mathfrak{M}^0(G^0, I, M, P)$ is uniquely determined by its nonzero entry, and so it may be denoted by $(g, i, m)$, where $g \in G$, $i \in I$, $m \in M$. Therefore, $\mathfrak{M}^0(G^0, I, M, P)$ may be treated as the set of all triples $(g, i, m)$, $g \in G^0$, $i \in I$, $m \in M$, with the multiplication given by

$$(g, i, m)(h, j, n) = (g p_{mj} h, i, n) \qquad \text{for } g, h \in G^0, i, j \in I, m, n \in M$$

where all triples $(\theta_G, i, m)$ are identified with the zero element of $\mathfrak{M}^0(G^0, I, M, P)$.

If $P$ is an $M \times I$ matrix over $G$, similarly, $T = \mathfrak{M}(G,I,M,P)$ usually stands for the semigroup of all $I \times M$ matrices over $G^0$ with exactly one nonzero entry, subject to the multiplication $AB = A \circ P \circ B$. Thus, $\mathfrak{M}(G,I,M,P) = \{(g,i,m) \mid g \in G, i \in I, m \in M\}$. Clearly, $T \subset T^0 \cong \mathfrak{M}^0(G^0,I,M,P)$ in this case. Since we will often deal with semigroups of the above two types, arising as the principal factors $S_s$, $s \in S$, of a given semigroup $S$, it is very convenient not to make a notational distinction between the cases where $S_s$ has, or does not have, a zero element. Thus, we write $\mathfrak{M}^0(G,I,M,P)$ for both $\mathfrak{M}^0(G^0,I,M,P)$ and $\mathfrak{M}(G,I,M,P)$. In other words, $\mathfrak{M}^0(G,I,M,P)$ will always stand for a semigroup of any of the above types, which will not produce any ambiguity. We refer to $\mathfrak{M}^0(G,I,M,P)$ as a semigroup of matrix type over the group $G$ with the sandwich matrix $P$.

If the set $I$ is finite, then we identify it with the set $\{1,2,\ldots,i\}$ where $i$ is the cardinality of $I$, and we write $\mathfrak{M}^0(G,i,M,P)$ for the semigroup of matrix type $\mathfrak{M}^0(G,I,M,P)$. Similarly, the notation $\mathfrak{M}^0(G,I,m,P)$ is used if $|M| = m < \infty$.

The importance of semigroups of matrix type comes from the following fundamental result; see [26], Chapter 3.

**Theorem 3**  Let $S$ be a semigroup. Then $S$ is completely 0-simple if and only if $S$ is isomorphic to a semigroup of matrix type $\mathfrak{M}^0(G,I,M,P)$ such that, for every $i \in I$, there exists $m \in M$ with $p_{mi} \neq \theta$, and, for every $n \in M$, there exists $j \in I$ with $p_{nj} \neq \theta$. In this case, $S$ is a regular semigroup.

Let $S = \mathfrak{M}^0(G,I,M,P)$ be a semigroup of matrix type. If $i \in I$, then the set $\{(g,i,m) \in S \mid g \in G^0, m \in M\}$ is denoted by $S_{(i)}$, and it is called the $i$th row of $S$. Similarly, for any $m \in M$, the set $S^{(m)} = \{(g,i,m) \in S \mid g \in G^0, i \in I\}$ is called the $m$th column of $S$. We put $S^{(m)}_{(i)} = S_{(i)} \cap S^{(m)}$. More generally, for any nonempty subsets $J \subseteq I$, $N \subseteq M$, define

$$S_{(J)} = \bigcup_{j \in J} S_{(j)}, \quad S^{(N)} = \bigcup_{n \in N} S^{(n)}, \quad S^{(N)}_{(J)} = S_{(J)} \cap S^{(N)} = \bigcup_{j \in J, n \in N} S^{(n)}_{(j)}$$

The following lemma summarizes the basic properties of $S$.

**Lemma 4**  Let $S = \mathfrak{M}^0(G,I,M,P)$ be a semigroup of matrix type. Then,

(i)  For any subsets $J \subseteq I$, $N \subseteq M$, $S_{(J)}$ is a right ideal of $S$, and $S^{(N)}$ a left ideal of $S$.

(ii)    For any subsets $J \subseteq I$, $N \subseteq M$, $S_{(J)}^{(N)}$ is a semigroup isomorphic to a semigroup of matrix type $\mathfrak{M}^0(G, J, N, P_{NJ})$, where $P_{NJ} = (p'_{nj})$ is the $N \times J$ submatrix of $P$ defined by $p'_{nj} = p_{nj}$ for $j \in J$, $n \in N$.

(iii)   $\{(g,i,m) \in S \mid p_{mi} \neq \theta_G, g = p_{mi}^{-1}\}$ is the set of nonzero idempotents of $S$.

(iv)    For any $i \in I$, $m \in M$, such that $S_{(i)}^{(m)}$ is not a semigroup with zero multiplication, $S_{(i)}^{(m)}$ is a group (possibly with zero) isomorphic to $G$ (to $G^0$, respectively) via the mapping $(x,i,m) \to xp_{mi}$. In this case, $sSs = S_{(i)}^{(m)}$ for any $\theta \neq s \in S_{(i)}^{(m)}$. Moreover, every maximal subgroup of $S$ is of the form $S_{(i)}^{(m)}$ (or $S_{(i)}^{(m)} \setminus \theta$) for some $i \in I$, $m \in M$.

(v)     For every $s,t \in S$, $s \neq \theta$, any of the conditions $s = st$, $s = ts$ implies that $t$ is an idempotent.

(vi)    If $S$ is completely 0-simple, then, for every $\theta \neq s \in S$, there exist idempotents $e,f \in S$ such that $s = es = sf$. Moreover, in this case, $Ss = S^{(m)}$, where $m \in M$ is such that $s \in S^{(m)}$, and $sS = S_{(i)}$, where $i \in I$ is such that $s \in S_{(i)}$. Further, every nonzero right (left) ideal of $S$ is of the form $S_{(J)}$ for a subset $J$ of $I$ ($S^{(N)}$ for a subset $N$ of $M$, respectively).

*Proof.*   Let $s = (g,i,m)$, $t = (h,j,n) \in S$. Then, $st = (g,i,m)(h,j,n) = (gp_{mj}h,i,n) \in S_{(i)}^{(n)}$, from which (i) follows. Since, when elements of $S_{(J)}^{(N)}$ are multiplied, only the entries of $P$ lying in $P_{NJ}$ are involved, (ii) is clear.

If $s \neq \theta$, then the equality $s = s^2 = (g,i,m)(g,i,m) = (gp_{mi}g,i,m)$ is equivalent to the fact that $gp_{mi}g = g$. Since $g \neq \theta_G$, this holds if and only if $g = p_{mi}^{-1}$ in $G$. This proves (iii).

Assume that $p_{mi} \neq \theta_G$. Then the map $\phi : S_{(i)}^{(m)} \to G^0$ defined by $\phi((x,i,m)) = xp_{mi}$ is a semigroup homomorphism. Clearly, $\phi$ is an embedding and $\phi$ maps $S_{(i)}^{(m)}$ onto $G^0$, or $G$, depending on the existence of a zero element in $S$. Consequently, if $\theta \neq s \in S_{(i)}^{(m)}$, then, by (i) we must have $sSs = S_{(i)}^{(m)}$. Let $H$ be any subgroup of $S$, and let $e = (f,k,r)$ be the identity of $H$. Then, $H \subseteq eSe \subseteq S_{(k)}^{(r)}$. Thus, (iv) follows.

If $st = s \neq \theta$, then $gp_{mj}h = g \in G$ and, as above, $h = p_{mj}^{-1}$ implies that $t$ is an idempotent. The latter case of (v) is verified similarly.

Assume that $S$ is completely 0-simple. By Theorem 3 there exists $k \in M$ such that $p_{ki} \neq \theta_G$. Then, $(p_{ki}^{-1},i,k)(g,i,m) = (p_{ki}^{-1}p_{ki}g,i,m) = (g,i,m)$, so that $es = s$, where $e = (p_{ki}^{-1},i,k)$ is an idempotent by (iii). Moreover, $Ss = S^{(m)}$ if $s \neq \theta$, because for any $(x,l,m) \in S^{(m)}$ we have $(x,l,m) =$

$(xg^{-1}p_{ki}^{-1}, l, k)(g,i,m) \in Ss$. It follows that any nonzero left ideal $L$ of $S$ is of the form $S^{(N)}$, where $N = \{m \in M \mid S^{(m)} \cap L \neq \varnothing, \theta\}$. The assertion on right ideals of $S$ is verified similarly.

We state a useful observation on certain homomorphisms of semigroups of matrix type.

**Lemma 5**   Let $S = \mathfrak{M}^0(G,I,M,P)$ be a semigroup of matrix type over a group $G$. If $H$ is a normal subgroup of $G$, then $\phi_H((g,i,m)) = (gH,i,m)$ defines a homomorphism of semigroups $\phi_H : S \to S_H$, where $S_H$ is the semigroup of the form $\mathfrak{M}^0(G/H,I,M,P_H)$ with the $M \times I$ matrix $P_H = (p_{mi}^{(H)})$ defined by $p_{mi}^{(H)} = p_{mi}H$ for any $m \in M$, $i \in I$.

*Proof.*   For $(g,i,m),(h,j,n) \in S$, we have

$$\begin{aligned}
\phi_H((g,i,m)(h,j,n)) &= \phi_H((gp_{mj}h,i,n)) = (gp_{mj}hH,i,n) \\
&= (gHp_{mj}HhH,i,n) \\
&= (gH,i,m)(hH,j,n) = \phi_H((g,i,m))\phi_H((h,j,n))
\end{aligned}$$

Clearly, $\phi$ maps $S$ onto $S_H$.

The congruence on $S$ corresponding to the homomorphism $\phi_H$ described above will be denoted by $\rho_H$.

For any $i,j \in I$, let $P^{(i)}$, $P^{(j)}$ denote the $i$th and $j$th columns of the matrix $P$. If $p$ is a prime number or $p = 0$, then we say that $P^{(i)}$, $P^{(j)}$ are $p$-equivalent if, for every $m \in M$, we have $p_{mi}H = p_{mj}H$, where $H$ is the maximal normal $p$-subgroup of $G$. (Here, by a $p$-subgroup with $p = 0$, we mean the trivial group.) In other words, this is the case if and only if the $i$th and $j$th columns of the matrix $P_H$ described in Lemma 5 are identical. Similarly, one defines $p$-equivalence of rows $P_{(m)}$, $m \in M$, of $P$. Clearly, if $p = 0$, then two columns (rows) of $P$ are $p$-equivalent if and only if they are equal.

We now consider an important example of completely 0-simple semigroups arising from skew linear semigroups. Let $K$ be a field, and let $D$ be a division $K$-algebra. For a fixed integer $n \geq 1$, the algebra $M_n(D)$ of $n \times n$ matrices over $D$ may be treated as the ring of homomorphisms of the left $D$-module $D^n$ with fixed standard basis. If $a \in M_n(D)$, then we define the rank $\rho(a)$ of $a$ as the dimension of the subspace $(D^n)a$ of $D^n$ over $D$. Put $I_j = \{a \in M_n(D) \mid \rho(a) \leq j\}$ for $j = 0, 1, \ldots, n$. It is clear that every $I_j$ is an ideal of $M_n(D)$ treated as the semigroup under multiplication. We will often exploit the following fundamental result; see [208], Section I.

**Theorem 6** For any division algebra $D$ and any integer $n \geq 1$, $0 = I_0 \subset I_1 \subset \cdots \subset I_n = M_n(D)$ are the only ideals of the multiplicative semigroup of the algebra $M_n(D)$. Moreover, every Rees factor $I_j/I_{j-1}$, $j = 1, \ldots, n$, is a completely 0-simple semigroup, the maximal subgroups of which are isomorphic to the full skew linear groups of the corresponding algebras $M_j(D)$.

*Proof.* Let $a, b \in M_n(D)$ be such that $\rho(a) = \rho(b) = j \geq 1$. Then $(D^n)a$, $(D^n)b$ are isomorphic as left $D$-linear spaces, so that there exists an invertible matrix $x \in M_n(D)$ such that $(D^n)ax = (D^n)b$. Since $ax$ maps $D^n$ onto $(D^n)b$, then, for every element $e_i$ of the standard basis $e_1, \ldots, e_n$ of $D^n$, there exists $f_i \in D^n$ such that $(f_i)ax = (e_i)b$. Then $yax = b$, where $y \in M_n(D)$ is such that $(e_i)y = f_i$ for any $i = 1, \ldots, n$. Hence, $b \in M_n(D)aM_n(D)$. Since $a, b$ are arbitrary elements of $I_j \setminus I_{j-1}$, this shows that $I_j \setminus I_{j-1}$ is a $\mathcal{J}$-class of the multiplicative semigroup of $M_n(D)$. Thus, $I_j/I_{j-1}$ is a 0-simple semigroup because it has a nonzero idempotent (take any projection of $D^n$ onto $D^j$), and it has a primitive idempotent since $M_n(D)$ has no infinite chain of idempotents. Hence, $I_j/I_{j-1}$ is a completely 0-simple semigroup. Moreover, for $e = e^2 \in I_j \setminus I_{j-1}$, $eM_n(D)e$ is the endomorphism algebra of the left $D$-space $(D^n)e$, so that we have an isomorphism of algebras $eM_n(D)e \cong M_j(D)$. Now, every invertible element $c$ of $eM_n(D)e$ is of the form $c = e(ece)e$, where $ece \in I_j \setminus I_{j-1}$. Since $e(I_j/I_{j-1})e$ is a maximal subgroup of $I_j/I_{j-1}$ with zero adjoined, by Lemma 4 (iv), then $e(I_j/I_{j-1})e$ is isomorphic to the group of invertible matrices in $M_j(D)$ with zero adjoined. The result follows.

It is easy to see that the rank defined for $a \in M_n(D)$ via homomorphisms of the right $D$-module $D^n$ coincides with $\rho(a)$ (for example, use Theorem 6 and its right-left symmetric analog to diagonal idempotents of $M_n(D)$).

Theorem 6 may be used to derive the following useful observation.

**Corollary 7** Let $D$ be a division algebra, and let $a, b, e \in M_n(D)$, $n \geq 1$. Then,

(i) If $ab = e = e^2$ and $\rho(a) = \rho(e)$, then $a = ea$.
(ii) If $\rho(a^k) = \rho(a^{k+1})$ for some $k \geq 1$, $\rho(a^k) = \rho(a^i)$ for every $i \geq k$, and $a^k$ lies in a subgroup of $M_n(D)$ in this case.

*Proof.* (i) Let $j = \rho(a)$. Clearly, we may assume that $j \geq 1$. We have $a(be) = e$, and so $\rho(be) \geq \rho(e) = j$. Thus, $\rho(be) = j$. Therefore, the equality $a(be) = e$ may be considered in the factor semigroup $I_j/I_{j-1}$. From Lemma 4 (vi) it follows that $eI_j = aI_j$, so that $a = ea$. (ii) Treating $a$ as a linear transformation of the left $D$-space $D^n$ (in the fixed standard basis), we have $(D^n)a^k = ((D^n)a^k)a$. Thus, $(D^n)a^k = (D^n)a^i$ for every $i \geq k$, showing that $\rho(a^k) = \rho(a^i)$. In particular, $a^{2k}, a^k$ are in the same $\mathcal{R}$- and $\mathcal{L}$-classes of the principal factor $I_j/I_{j-1}, j = \rho(a^k)$, the intersection of which is a subgroup of $I_j/I_{j-1}$; see Lemma 4.

Our next result, providing a crucial reduction step in some of the considerations of Part IV, deals with an important class of skew linear semigroups. Here, by a 0-cancellative semigroup we mean a semigroup that is either cancellative or isomorphic to a semigroup of the form $S^0$, where $S$ is a cancellative semigroup. Further, if $T$ is a subsemigroup of a matrix algebra $M_n(D)$ over a division algebra $D$, and $Z(D)$ is the center of $D$, then $T$ is called irreducible if $M_n(D)$ is the $Z(D)$-subspace (and hence subalgebra) of $M_n(D)$ generated by $T$. Recall also that $S$ is a nil semigroup if it has a zero element and every $s \in S$ is nilpotent.

**Proposition 8** Let $S$ be a subsemigroup of the multiplicative semigroup of the matrix algebra $M_n(D)$, $n \geq 1$, over a division algebra $D$. Assume that $M_n(D)$ is the classical ring of right quotients of the $Z(D)$-subalgebra generated by $S$. Then,

(i)   If $I$ is a nil ideal of $S$, either $n = 1$ (so that $I = \{0\}$ or $S = \{1\}$), or $\theta_I = 0$ and $I = \{0\}$.

(ii)  There exists $s \in S$ such that $sSs$ is a 0-cancellative semigroup and $sM_n(D)s = eM_n(D)e$ for some $0 \neq e = e^2 \in M_n(D)$ with $\rho(s) = \rho(e)$. Moreover, if $S$ is an irreducible subsemigroup of $M_n(D)$, then $sSs$ is an irreducible subsemigroup of $eM_n(D)e \cong M_{\rho(s)}(D)$.

*Proof.* (i) Clearly $\theta_I = \theta_S$. Then $\theta$ is a central idempotent in the $Z(D)$-subalgebra $Z(D)\{S\}$ generated by $S$. The hypothesis implies that $\theta$ is central in $M_n(D)$. If $\theta = 1$, then $S = I = \{1\}$, and $n = 1$. Assume that $\theta = 0$. It is known that $I$ must be nilpotent; see [55], 17.19. Therefore, the $Z(D)$-subalgebra $Z(D)\{I\}$ of $M_n(D)$ generated by $I$ is a nilpotent ideal of $Z(D)\{S\}$. Since $Z(D)\{S\}$ is an order in $M_n(D)$, then, from [86], Theorem 7.2.3, it follows that $Z(D)\{S\}$ is a prime algebra. This shows that $I = \{0\}$. (ii) Let $T$ be the set of elements of $S$ of the least nonzero rank $j$ as matrices in $M_n(D)$, with the zero matrix adjoined if it is in $S$. Then $T$ is an ideal of $S$, and from (i) it follows that there exists $s \in T$, which is not nilpotent as a matrix in $M_n(D)$. It is clear that $T = S \cap I_j$ and that $T$ embeds into the completely 0-simple semigroup $I_j/I_{j-1}$ defined for $M_n(D)$ as in Theorem 6. Then $sSs = T \cap sSs = I_j \cap sSs$ embeds into $s(I_j/I_{j-1})s$ (here $s$ is identified with its image in $I_j/I_{j-1}$). Since $s$ is not nilpotent, then from Lemma 4 (iv) it follows that $s(I_j/I_{j-1})s$ is a group, possibly with zero. Hence, $sSs \subseteq s(I_j/I_{j-1})s$ is a 0-cancellative semigroup. Moreover, if $e \in M_n(D)$ is an element that maps onto the identity of $s(I_j/I_{j-1})s$, then there exists $t \in sI_js$ such that $st = ts = e$. Then $sM_n(D)s \supseteq stM_n(D)ts = eM_n(D)e$, and the equality $sM_n(D)s = eM_n(D)e$ follows. As in the proof of Theorem 6, we see that $eM_n(D)e \cong M_j(D)$. Finally, if $S$ is an irreducible subsemigroup of $M_n(D)$, that is, $Z(D)\{S\} = M_n(D)$, then $Z(D)\{sSs\} = s(Z(D)\{S\})s = sM_n(D)s$. This completes the proof.

**Remark 9**   In the situation, and with the notation, of Proposition 8, $S \cap I_j$ embeds into the completely 0-simple semigroup $I_j/I_{j-1}$. If $H$ is the maximal subgroup of $I_j/I_{j-1}$ containing $s$ and $e$, then $S \cap H \supseteq sSs \setminus \{\theta\}$ is a cancellative subsemigroup of $S$. Further, $S \cap H$ is irreducible in $eM_n(D)e$ if $S$ is irreducible in $M_n(D)$.

We close this chapter with an example showing that, in contrast to the assertion of Lemma 4 (iv), a cancellative subsemigroup of a completely 0-simple semigroup $S$ need not be contained in a maximal subgroup of $S$.

**Example 10**   Let $X$ be a free semigroup with free generators $x, y$. Consider the completely simple semigroup $S = \mathfrak{M}(X, 1, 2, P)$, where $p_{11} = p_{21} = 1$. Set $T_1 = \{(w, 1, 1) \mid y \text{ is the terminal letter in } w\}$, $T_2 = \{(w, 1, 2) \mid x \text{ is the terminal letter in } w\}$. It is clear that $T = T_1 \cup T_2$ is a subsemigroup of $S$, which is not contained in any subgroup of $S$ by Lemma 4 (iv). If $(w, 1, i)(z, 1, k) = (v, 1, j)(z, 1, k)$ for some $(w, 1, i), (z, 1, k), (v, 1, j) \in T$, then $wz = vz$, and so $w = v$. Consequently, $i = j$ by the definition of $T$, which shows that $T$ is right cancellative. It is easy to see that $S$ is left cancellative. Hence, $T$ is a cancellative semigroup.

## Comments on Chapter 1

The general material on semigroups presented in this chapter is very well known, and beyond our main reference [26], it may also be found in other textbooks on semigroups; see [90] and [209]. A deeper analysis of the full linear transformation semigroups of not necessarily finite dimensional spaces over a division algebra is given in [208]. As will be apparent from the subsequent chapters, the properties of sandwich matrices arising from the completely 0-simple principal factors of a given semigroup $S$ often correspond to ring theoretical properties of the semigroup algebra $K[S]$ of $S$. However, problems involving matrices over group algebras are, in general, difficult to handle. Recently, Okniński and Putcha obtained some decisive results on the sandwich matrices of full matrix semigroups $M_n(F)$ over a field $F$ [192].

# 2

# Semigroups with Finiteness Conditions

A semigroup $S$ is called locally finite if all its finitely generated subsemigroups are finite. Clearly, any locally finite semigroup is periodic, but finitely generated infinite periodic semigroups abound. The first example of a nil semigroup of this type is attributed to Morse and Hedlund (see [132], Section 10.5), whereas a group with this property was first constructed by Golod; see [203], Theorem 10.1.12. For a discussion of the Burnside-type problems, that is, problems on local finiteness of certain periodic semigroups, we refer to [132]. Here, we state some results on local finiteness of semigroups and its connection with some other finiteness conditions, which will prove useful in subsequent chapters.

We start by showing that the class of locally finite semigroups is closed under ideal extensions.

**Lemma 1** Let $S$ be a finitely generated semigroup. If $T$ is a subsemigroup of $S$ such that $S \setminus T$ is finite, then $T$ is finitely generated.

*Proof.* We may assume that a finite set $X$ of generators of $S$ contains the set $S \setminus T$. Put $X' = (X \cap T) \cup (X^2 \cap T) \cup (X^3 \cap T)$. Clearly, $X' \subseteq T$. Let $t \in T$, and let $t = x_1 \ldots x_n$ be a presentation of the shortest length of $t$ in the generators from $X$. Then $x_i x_{i+1} \ldots x_k \in T$ for all $i, k$ such that $1 \leq i < k \leq n$ since, otherwise, $x_i x_{i+1} \ldots x_k \in S \setminus T \subseteq X$, contradicting the minimality of $n$. If $n \leq 3$, then $t \in X' \subseteq \langle X' \rangle$ by the definition of $X'$. Assume that $n > 3$. Then $t = t_1 t_2$, where $t_1 = x_1 x_2$, $t_2 = x_3 \ldots x_n$. Since we know that $t_1, t_2 \in T$ and $t_1, t_2$ have a shorter presentation in the generators from $X$ than $t$, then an easy induction on $n$ shows that $t \in \langle X' \rangle$. This means that $T = \langle X' \rangle$ and proves the assertion because $X'$ is finite.

**Proposition 2**  Let $J$ be an ideal of a semigroup $S$. Then $S$ is locally finite if and only if the semigroups $J$, $S/J$ are locally finite.

*Proof.*  It is clear that the class of locally finite semigroups is closed on subsemigroups and on homomorphic images. Thus, the necessity follows.

Assume that $J$, $S/J$ are locally finite, and let $T$ be a finitely generated subsemigroup of $S$. Then $T/(J \cap T)$ is a finitely generated subsemigroup of $S/J$, so that it must be finite. Hence, $T \setminus (J \cap T)$ is finite, and it follows from Lemma 1 that $J \cap T$ is a finitely generated subsemigroup of $J$. Thus, $J \cap T$ is finite, and so $T = (T \setminus (J \cap T)) \cup (J \cap T)$ is also finite.

We first show that, in some important cases, local finiteness of $S$ may be verified by checking the subgroups of $S$ only.

**Lemma 3**  Let $S$ be a completely 0-simple semigroup. Then $S$ is locally finite if and only if a maximal subgroup of $S$ is locally finite.

*Proof.*  From Chapter 1 we know that $S$ may be identified with a semigroup of matrix type $\mathfrak{M}^0(G,I,M,P)$, where $G$ is isomorphic to any maximal subgroup of $S$. Assume that $G$ is locally finite. Let $X = \{(g_j, i_j, m_j) \mid g_j \in G^0, i_j \in I, m_j \in M, j = 1, \ldots n\}$ be a finite subset of $S$. Let $H$ denote the subgroup of $G$ generated by the set $[\{g_1, \ldots, g_n\} \cup \{p_{mi} \mid m \in M', i \in I'\}] \setminus \{\theta\}$, where $M' = \{m_1, \ldots, m_n\}$, $I' = \{i_1 \ldots, i_n\}$. Then $H$ is a finitely generated subgroup of $G$, and so it must be finite. From the multiplication rule in $\mathfrak{M}^0(G,I,M,P)$, it follows that $\langle X \rangle$ is contained in the semigroup of matrix type $\mathfrak{M}^0(H,I',M',P')$ where $P' = (p'_{mi})$ is the $M' \times I'$ matrix defined by $p'_{mi} = p_{mi}$ for $i \in I'$, $m \in M'$; see Lemma 4 in Chapter 1. Since the latter semigroup is finite, then $\langle X \rangle$ is also finite. This shows that $S$ is locally finite. The converse is clear.

Our next aim is to extend the assertion of Lemma 3 to a wider class of semigroups. As in [26], §.6.6, the descending chain conditions (d.c.c.) on principal right, left, and two-sided ideals in the class of semigroups will be denoted by $M_R$, $M_L$, and $M_J$, respectively. Clearly, any completely 0-simple semigroup satisfies $M_J$, and it satisfies $M_R$, $M_L$ by Lemma 4 in Chapter 1.

**Lemma 4** Let $J$ be an ideal of $S$. Assume that $sJ^1 \subseteq tJ^1$ and $sS^1 = tS^1$ for some $s, t \in S$, $s \neq t$. Then $sJ^1 = tJ^1 = sJ = tJ$.

*Proof.* Since $s \neq t$, then $s = ty$ for some $y \in J$. Moreover, there exists $x \in S$ such that $t = sx$. Thus, $t = sx = tyx = sxyx \in sJ$. Then $tJ^1 \subseteq sJ \subseteq sJ^1 \subseteq tJ \subseteq tJ^1$, and the assertion follows.

**Proposition 5** Let $J$ be an ideal of $S$. Then $S$ satisfies the condition $M_R$ if and only if the semigroups $J$, $S/J$ satisfy $M_R$.

*Proof.* Assume that $S$ satisfies $M_R$. Suppose that $s_1 J^1 \supsetneq s_2 J^1 \supsetneq \cdots$ is a strictly descending chain for some $s_1, s_2 \ldots \in J$. Then $s_{i+1} \in s_i J^1$, so that $s_{i+1} S^1 \subseteq s_i S^1$ for any $i \geq 1$. Thus, there exists $n$ such that $s_{n+1} S^1 = s_n S^1$. From Lemma 4 it follows that $s_{n+1} J^1 = s_n J^1$, contradicting the supposition. Hence, $J$ satisfies $M_R$. The assertion on $S/J$ follows from the easy-to-verify, more general fact that any homomorphic image of $S$ must satisfy $M_R$.

Assume now that the semigroups $J$, $S/J$ satisfy $M_R$, and consider a chain $t_1 S^1 \supseteq t_2 S^1 \supseteq \cdots$, $t_1, t_2, \ldots \in S$. Then $t_{i+1} = t_i x_i$ for some $x_i \in S^1$, and so $t_{i+1} = t_1 x_1 \ldots x_i$ for any $i \geq 1$. Consider two cases:

(i) There exists $i$ such that $x_i x_{i+1} \ldots x_j \notin J$ for every $j \geq i$. Then the fact that the chain $x_i(S/J)^1 \supseteq x_i x_{i+1}(S/J)^1 \supseteq \cdots$ stabilizes by the assumption on $S/J$ implies that there exist $m \geq 1$ and $u \in S^1$ such that $x_i \ldots x_m = x_i \ldots x_m x_{m+1} u$. Consequently, $t_{m+1} = t_i x_i \ldots x_m = t_i x_i \ldots x_{m+1} u = t_{m+2} u$, showing that $t_{m+1} S^1 = t_{m+2} S^1$.

(ii) For every $i \geq 1$, there exists $j \geq i$ such that $x_i x_{i+1} \ldots x_j \in J$. Then we may find a sequence $j_1 < j_2 < \ldots$ such that $x_{j_i} \ldots x_{j_{i+1}-1} \in J$ for $i \geq 1$. Since $t_{j_{i+1}} = t_{j_i} x_{j_i} \ldots x_{j_{i+1}-1} \in t_{j_i} J$, and $t_{j_1} J^1 \supseteq t_{j_2} J^1 \supseteq \cdots$, then the assumption on $J$ implies that $t_{j_k} J^1 = t_{j_{k+1}} J^1$ for some $k > 1$. Therefore, $t_{j_k} S^1 = t_{j_{k+1}} S^1$.

It follows that $S$ satisfies $M_R$.

Whereas a symmetric proof shows that the assertion of Proposition 5 holds with the property $M_R$ replaced by $M_L$, this is not the case with respect to $M_J$; see [240a]. The only true, and obvious, implication is that $S/J$ has $M_J$ whenever $S$ has $M_J$.

An application of the reasoning of the Konig lemma (see [232], Theorem 1.4.9) allows us to prove the following result.

**Lemma 6**  If $S$ is a semigroup satisfying any of the conditions $M_R$, $M_L$, or $M_J$, then $S$ satisfies the descending chain condition (d.c.c.) on finitely generated right, respectively, left, or two-sided ideals.

*Proof.*  Suppose that $A_1, A_2, \ldots$ are finite subsets of $S$ such that the chain of right ideals $I_1 = \bigcup_{s \in A_1} sS^1 \supset I_2 = \bigcup_{s \in A_2} sS^1 \supset \cdots$ does not stabilize. For any $s \in A_1$, put $I_1^{(s)} = sS^1$ and, for $n \geq 2$, let $B_n^{(s)} = I_{n-1}^{(s)} \cap A_n$, $I_n^{(s)} = B_n^{(s)} S^1$ if $B_n^{(s)} \neq \varnothing$ and $I_n^{(s)} = \varnothing$ otherwise. Clearly, $A_2 = \bigcup_{s \in A_1} B_2^{(s)}$ and, inductively, we see that

$$A_n = A_{n-1} S^1 \cap A_n = (\bigcup_{s \in A_1} B_{n-1}^{(s)}) S^1 \cap A_n = \bigcup_{s \in A_1} (B_{n-1}^{(s)} S^1 \cap A_n)$$

$$= \bigcup_{s \in A_1} (I_{n-1}^{(s)} \cap A_n) = \bigcup_{s \in A_1} B_n^{(s)}$$

Moreover,

(a)                                         $$I_2^{(s)} \supseteq I_3^{(s)} \supseteq \cdots$$

and every $I_n^{(s)}$ is a finitely generated right ideal of $S$ or an empty set. Suppose that, for all $s \in A_1$, the chain (a) stabilizes. Since $\bigcup_{s \in A_1} I_n^{(s)} = \bigcup_{s \in A_1} B_n^{(s)} S^1 = A_n S^1 = I_n$, $n \geq 2$, and $A_1$ is a finite set, then the chain $I_1 \supseteq I_2 \supseteq \cdots$ also stabilizes, which is impossible. Therefore, there exists $s \in A_1$ such that the chain (a) does not stabilize. Hence, omitting some of its elements if necessary, we get a strictly descending chain $sS^1 \supset I_{i_1}^{(s)} \supset I_{i_2}^{(s)} \supset \cdots$ of finitely generated right ideals. Iterating this procedure (with respect to the chain $I_{i_1} \supset I_{i_2} \supset \cdots$ in place of the chain $I_1 \supset I_2 \supset \cdots$) leads to an infinite strictly descending chain of principal right ideals. This shows that the condition $M_R$ implies the d.c.c. on finitely generated right ideals.

The proofs of the respective assertions on the minimal conditions for left and two-sided ideals are similar and will be omitted.

**Theorem 7**  Let $S$ be a semigroup such that every subgroup of $S$ is locally finite. Assume that all 0-simple principal factors of $S$ are completely 0-simple and that $S$ satisfies any of the conditions $M_R$, $M_L$, $M_J$. Then $S$ is locally finite.

*Proof.*  Assume first that $S$ satisfies $M_R$ or that $S$ satisfies $M_J$. Let $t_1, \ldots, t_k \in S$, and let $H = \langle t_1, \ldots, t_k \rangle$. Put $Y_1 = \bigcup_{i=1}^{k} S^1 t_i S^1$, and choose a minimal subset $\{s_1, \ldots, s_n\}$ of $\{t_1, \ldots, t_k\}$ such that $Y_1 = \bigcup_{i=1}^{n} S^1 s_i S^1$. If $s_1$ is a zero in $S$, then $Y_1 = \theta$, and so $H \subseteq Y_1$ is finite. Otherwise, define

$X_1 = \bigcup_{i=2}^{n} S^1 s_i S^1 \cup J$, where $J \subseteq S^1 s_1 S^1$ is the ideal of nongenerators of $S^1 s_1 S^1$, and put $Z_1 = \bigcup_{i=1}^{n} s_i S^1$. Then, $J$ is an ideal of $S$, and $Y_1/X_1 \cong (S^1 s_1 S^1)/J = S_{s_1}$ is the principal factor determined by $s_1$ in $S$. If $S_{s_1}$ is 0-simple, then it is locally finite by the hypothesis and Lemma 3. Otherwise, $S_{s_1}$ is nilpotent and so, in any case, $S_{s_1}$ is locally finite. Put $F_1 = H \cap X_1$, $H_1 = H \setminus F_1$. If $F_1 \neq \varnothing$, then $F_1$ is an ideal of $H$, and $H/F_1$ embeds into $Y_1/X_1$. Since $H/F_1$ is finitely generated, then it must be finite. Thus, $H_1$ is finite and, by Lemma 1, $F_1$ is finitely generated. If $Z_2$ is the right ideal of $S$ defined by $Z_2 = \bigcup_{s \in F_1} s S^1$, then $Z_2 \subseteq X_1$, $Z_2 \subset Z_1$, and $s_1 \in Z_1 \setminus X_1$. Put $Y_2 = \bigcup_{s \in F_1} S^1 s S^1$. Then $Y_2 \subseteq X_1$ and, as above, we may construct a maximal ideal $X_2$ in $Y_2$ and put $F_2 = F_1 \cap X_2$, $H_2 = F_1 \setminus F_2$, $Z_3 = \bigcup_{s \in F_2} s S^1$, $Y_3 = \bigcup_{s \in F_2} S^1 s S^1$. As above, $H_2$ is finite and $F_2$ is finitely generated. Proceeding this way, if $F_i \neq \varnothing$ for any $i$, we get a chain of ideals $Y_1 \supset X_1 \supseteq Y_2 \supset X_2 \supseteq \cdots$ and a chain of right ideals $Z_1 \supset Z_2 \supset \cdots$ such that $Z_{i+1} \subseteq X_i$, $Z_{i+1} \nsubseteq X_{i+1}$ for $i \geq 1$. Since all $Y_i$ and $Z_i$ are finitely generated as two-sided, respectively right, ideals of $S$, this contradicts the assumption on $S$ in view of Lemma 6. Therefore, $F_m = \varnothing$ for some $m \geq 1$. This means that $H = H_1 \cup F_1 = H_1 \cup H_2 \cup F_2 = \cdots = H_1 \cup H_2 \cup \cdots \cup H_m$. Since any $H_i$ is a finite set, then $H$ is a finite semigroup. Thus, $S$ is locally finite.

If $S$ satisfies $M_L$, then the same reasoning, with right ideals $Z_i$ replaced by the corresponding left ideals, establishes the assertion.

We note that it was shown by Hotzel in [89] that if $S$ is finitely generated with no infinite subgroups and satisfies $M_R$, then $S$ must be finite.

**Example 8**  Let $S$ be a Baer–Levi semigroup; see [26], §.8.1. It is known that $S$ is right simple, so that it satisfies $M_R$ and $M_J$, and $S$ has no idempotents. Clearly, $S$ is not locally finite. This shows that the hypothesis on the principal factors is essential in Theorem 7. Specifically, $S$ may be chosen as the semigroup of all one-to-one transformations $\phi$ of an infinite denumerable set $X$ such that $X \setminus \phi(X)$ is a finite set.

Observe that, in view of Lemma 1 in Chapter 1, Theorem 7 may be reformulated in the following way: $S$ is locally finite whenever all principal factors of $S$ are locally finite, provided that $S$ satisfies any of the conditions $M_R$, $M_L$, $M_J$.

**Corollary 9**  Assume that $S$ is a periodic semigroup satisfying any of the conditions $M_R$, $M_L$, $M_J$. If every subgroup of $S$ is locally finite, then $S$ is locally finite.

*Proof.* This is now a direct consequence of the fact that any periodic 0-simple semigroup must be completely 0-simple; see Lemma 1 of Chapter 1.

Finally, we consider the condition $M_R$ for the class of nil semigroups.

**Lemma 10**   Let $S$ be a nil semigroup. Then $S$ satisfies $M_R$ if and only if, for any $s_1, s_2, \ldots \in S$, there exists $n \geq 1$ such that $s_1 s_2 \ldots s_n = \theta$.

*Proof.*   Assume that $t_1 S^1 \supset t_2 S^1 \supset \cdots$ for some $t_i \in S$. Then there exist $s_i \in S$ such that $t_{i+1} = t_i s_i$, and so $t_{i+1} = t_i s_1 \ldots s_i$ for $i \geq 1$. Thus, $s_1 \ldots s_i \neq \theta$ because, otherwise, $t_{i+1} S^1 = t_{i+2} S^1 = \theta$, a contradiction.

On the other hand, if, for some $s_1, s_2, \ldots \in S$, we have $s_1 \ldots s_n \neq \theta$ for all $n \geq 1$, then $s_1 S^1 \supset s_1 s_2 S^1 \supset \cdots s_1 s_2 s_3 S^1 \supset \cdots$. In fact, otherwise, there exists $n \geq 1$ such that $s_1 \ldots s_n = s_1 \ldots s_{n+1} x$ for some $x \in S^1$. Then, for any $j \geq 1$, $s_1 \ldots s_n = s_1 \ldots s_n (s_{n+1} x)^j$, and so $s_1 \ldots s_n = \theta$ because $s_{n+1} x \in S$ is a nilpotent element.

The semigroups characterized by the equivalent conditions of Lemma 10 are called left $T$-nilpotent. The analog of Lemma 10 holds for right $T$-nilpotent semigroups defined symmetrically. We note that the classes of right and left $T$-nilpotent semigroups do not coincide. For example, let $S$ be the semigroup defined by generators $x_0, x_1, x_2, \ldots$, subject to the relations $x_i x_j = x_0, x_i x_0 = x_0 = x_0 x_i$ for $j \geq i \geq 0$. Then $x_0$ is the zero element of $S$, and $x_n x_{n-1} \ldots x_1 \neq x_0$ for every $n \geq 1$, but $x_{i_1} x_{i_2} \ldots x_{i_r} = x_0$ whenever $r > i_1$.

We continue with a useful result on a distinction between the classes of nilpotent and $T$-nilpotent semigroups. For a further convenience, we state it with respect to the right $T$-nilpotency.

**Lemma 11**   Let $S$ be a nil semigroup. Then,

(i)   If $SX \supseteq X$ or a finite subset $X$ of $S$, then $X = \theta$.
(ii)   Assume that $S$ is not nilpotent. Then, for any $m, n \geq 1$, there exists $s \in S$ such that $|S^m s| \geq n$.
(iii)   Assume that $S$ is right $T$-nilpotent but not nilpotent. Then, for any $n \geq 1$, there exists $s \in S$ such that $|S^n s| \geq n + 1$ and $S^{2n} s = \theta$.

*Proof.*   (i) Take $x_1 \in X$. Then $x_1 = y_1 x_2$ for some $y_1 \in S, x_2 \in X$. Similarly, for any $k > 1$, $x_k = y_k x_{k+1}$, where $y_k \in S$, $x_{k+1} \in X$. Since $X$ is finite, then there exist integers $i < j$ such that $x_i = x_j$. Now $x_i = y_i x_{i+1} = \cdots = y_i y_{i+1} \ldots y_{j-1} x_j$. If $z = y_i y_{i+1} \ldots y_{j-1}$, then $x_i = z x_i = z^2 x_i = \cdots$ and, since $z \in S$ is nilpotent, then $x_i = \theta$. Consequently, $x_{i-1} = \theta, \ldots, x_1 = \theta$, which shows that $X = \theta$.

(ii) Suppose that there exist $m, n \geq 1$ such that $|S^m s| < n$ for any $s \in S$. If $S^m s \neq \theta$, then, from (i), it follows that $SS^m s \subset S^m s$, and so $|S^{m+1} s| < n - 1$. Repeating this argument at most $n - 2$ times, we come to $S^{m+n-2} s = \theta$. Since $s \in S$ is arbitrary, then $S^{m+n-1} = \theta$, contradicting the hypothesis on $S$ and proving (ii).

(iii) From (ii) we know that, for any fixed integer $n \geq 1$, there exists $s \in S$ such that $|S^n s| \geq n + 1$. Since $S$ satisfies $M_L$, by the hypothesis, we may assume that $Ss \cup \{s\}$ is minimal among the left ideals $Sx \cup \{x\}, x \in S$, such that $|S^n x| \geq n + 1$. Then, for every $y \in S$, $s \notin Sys \cup \{ys\}$ because $S$ is a nil semigroup. The minimality of $Ss \cup \{s\}$ implies that $|S^n ys| \leq n$. Therefore, as in (ii), we conclude that $S^{2n-1} ys = \theta$. Consequently, $S^{2n} s = \theta$.

From Corollary 9, it may be derived that any right or left $T$-nilpotent semigroup $S$ is locally nilpotent; that is, any finitely generated subsemigroup of $S$ must be nilpotent. In fact, more may be proved.

**Corollary 12**   Assume that $S$ is a right $T$-nilpotent semigroup. Then, for any finite subsets $A_1, A_2, \ldots$ of $S$, there exists $n \geq 1$ such that $A_n A_{n-1} \ldots A_1 = \theta$.

*Proof.*   From Lemma 6 it follows that there exists $n \geq 1$ such that $S^1 A_n \ldots A_1 = S^1 A_{n+1} \ldots A_1$. Since $S^1 A_{n+1}$ is a nil semigroup and $A_n \ldots A_1$ is a finite set, then Lemma 11 implies that $A_n \ldots A_1 = \theta$.

We close this chapter with an important result on the nilpotency of nil semigroups with certain finiteness conditions. The proof can be found in [55], 17.20, 17.22.

**Proposition 13**   Let $S$ be a nil semigroup. Then,

(i)   If $S$ has the ascending chain condition (a.c.c.) on right and left annihilator ideals, then $S$ is nilpotent.

(ii)   If $S$ is a multiplicative subsemigroup of a ring $R$ of finite right Goldie dimension, and $S$ satisfies the a.c.c. on right annihilator ideals, then $S$ is nilpotent.

(iii)   If $S \subseteq M_n(D)$ for a division algebra $D$, then $S^n = 0$.

**Remark 14**   Let $R$ be a ring with unity, and let $S$ be a subsemigroup of the multiplicative semigroup of $R$. Assume that $I$ is a nil ideal of $S$ with zero $e$. As we have seen in Proposition 8 in Chapter 1, in some important cases, we must have $e = 0$. In general, since $es = se = e$ for all $s \in S$, then $\phi : S \rightarrow R' = (1-e)R(1-e)$ defined by $\phi(s) = (1-e)s(1-e)$ is a semigroup embedding. Moreover, $\phi(e) = 0$. In particular, this can be applied to the case of a skew linear semigroup $S \subseteq R = M_n(D)$ over a division algebra $D$. Then $S$ can be treated as a subsemigroup of $M_k(D) \cong R'$, $k = \rho(1 - e)$, so that the zero of $S$ is that of $M_k(D)$.

Finally, we note that Proposition 13 can be applied to multiplicative subsemigroups of right noetherian rings because the a.c.c. on annihilator ideals is hereditary on subsemigroups.

## Comments on Chapter 2

The study of the connections between the minimal conditions $M_J$, $M_R$, $M_L$ was begun by Munn in [155]; see also [26], § 6.6. In [89], Hotzel proved that the condition $M_J$ implies any of the conditions $M_R$, $M_L$, provided that all 0-simple principal factors of $S$ are completely 0-simple (even less may be assumed: all 0-simple principal factors of $S$ have 0-minimal right, respectively left, ideals). We note this in conjunction with our Theorem 7, which is an extension of a result formerly proved in [178] for the class of completely semisimple semigroups. Minimal conditions on one-sided congruences were studied by Kozhukhov in [117], [120]. Results analogous to those of Lemmas 6 and 10 are also known for one-sided ideals of associative rings. They appeared when characterizing perfect rings; see [55], 22.29, and [232], Theorem 6.2.5. Lemma 11 is taken from the paper of Kljushin and Kozhukhov [115], while Corollary 13 was originally established by Patterson in [206]. Deep results on the structure of semigroups with maximal conditions on one-sided ideals or congruences were obtained by Hotzel in [88], [89].

# 3

# Weakly Periodic Semigroups

In this chapter, we consider an important class of semigroups defined through a relatively weak finiteness condition. We say that $S$ is a weakly periodic semigroup if, for every $s \in S$, there exists $n \geq 1$ such that $S^1 s^n S^1$ is an idempotent ideal of $S$, that is, $(S^1 s^n S^1)^2 = S^1 s^n S^1$. This condition is equivalent to saying that a power of every element of $S$ determines a 0-simple principal factor of $S$. It is clear that any semisimple semigroup is weakly periodic. On the other hand, periodic semigroups form another important subclass of the class of weakly periodic semigroups. We will also be concerned with some auxiliary intermediate classes of semigroups. $S$ is called $\pi$-regular if, for every $s \in S$, there exists $n \geq 1$ such that $s^n$ is a regular element, that is, $s^n x s^n = s^n$ for some $x \in S$. $S$ is called strongly $\pi$-regular if, for every $s \in S$, there exists $n \geq 1$ such that $s^n$ lies in a subgroup of $S$. Further, $S$ is said to be strongly $\pi$-regular of bounded index if there exists $n \geq 1$ such that $s^n$ lies in a subgroup of $S$ for every $s \in S$.

**Lemma 1** Let $S$ be a semigroup satisfying any of the d.c.c. $M_R$, $M_L$, $M_J$. Then $S$ is weakly periodic. If, additionally, all 0-simple principal factors of $S$ are completely 0-simple, then $S$ is strongly $\pi$-regular.

*Proof.* Let $s \in S$. Consider the chain of principal ideals $S^1 s S^1 \supseteq S^1 s^2 S^1 \supseteq S^1 s^3 S^1 \supseteq \cdots$. The hypothesis on $S$ implies that $S^1 s^n S^1 = S^1 s^{n+1} S^1 = \cdots = S^1 s^{2n} S^1$ for some $n \geq 1$. Since we always have $S^1 s^{2n} S^1 \subseteq (S^1 s^n S^1)^2 \subseteq S^1 s^n S^1$, then $S$ is weakly periodic. If all 0-simple principal factors of $S$ are completely 0-simple, then the principal factor $S_{s^n} = (S^1 s^n S^1)/I_{s^n}$ is completely 0-simple. Since $(s^n)^2 \notin I_{s^n}$, then $s^n$ lies in a subgroup of $S_{s^n}$ (see Lemma 4 in Chapter 1), and so it lies in a subgroup of $S$. Thus, $S$ is strongly $\pi$-regular.

Now, it is easy to see that we have the following "inclusion graph" on the classes of semigroups with the finiteness conditions considered before and those defined above. (Here $\rightarrow$ stands for $\supset$).

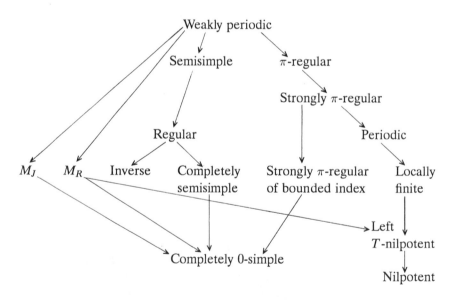

Let us also note that the following characterization of 0-simple semigroups that are strongly $\pi$-regular is well known; see [26], Theorem 2.55.

**Lemma 2** Let $S$ be a 0-simple semigroup. Then $S$ is strongly $\pi$-regular if and only if $S$ is completely 0-simple.

Weakly periodic semigroups of a special type will often be encountered in the sequel. This class is described below.

**Theorem 3** Let $S$ be a weakly periodic semigroup. Then $S$ has finitely many $\mathcal{J}$-classes determining 0-simple principal factors if and only if there exists a chain of ideals $J_1 \subseteq J_2 \subseteq \cdots \subseteq J_n = S$ of $S$ such that $J_1$ and all $J_i/J_{i-1}$, $i > 1$, are 0-simple or nil. Moreover, if $k$ denotes the number of $\mathcal{J}$-classes determining 0-simple principal factors of $S$, then the ideals $J_i$ can be chosen so that $n \le 2k$. If $S$ is a strongly $\pi$-regular semigroup of this type, then the non-nil semigroups $J_1$, $J_i/J_{i-1}$ are completely 0-simple.

*Proof.* It is clear that, if a desired chain of ideals of $S$ of length $n$ exists, then $S$ has, at most, $n$ $\mathcal{J}$-classes determining 0-simple principal factors of $S$. Thus, assume that $S$ has finitely many such $\mathcal{J}$-classes. We proceed by induction on the number $m(S)$ of nonzero $\mathcal{J}$-classes of this type. If $m(S) = 0$, then the fact that $S$ is weakly periodic means that $\theta \in S$, and it is the only 0-simple principal factor of $S$, so that $S$ is a nil semigroup. Thus, in this case, we may put $n = 2$ and $J_1 = \theta$, $J_2 = S$, ($n = 1$, $J_1 = S$ also might be taken). Assume that $m(S) > 0$. Then there exists a nonzero principal ideal $S^1 s S^1$ of $S$, $s \in S$, which is minimal among all nonzero ideals $S^1 x S^1$, $x \in S$, such that the corresponding principal factor $S_x$ is 0-simple. Then $S^1 s S^1 = SsS$. Since $S$ is weakly periodic, then either $I_s = \varnothing$, that is, $SsS$ is a simple ideal of $S$, or $I_s$ is a nil ideal of $S$. We distinguish these two cases:

(i)  $I_s = \varnothing$. Then we put $J_2 = SsS$.
(ii) $I_s \ne \varnothing$. Then we define $J_1 = I_s$, $J_2 = SsS$.

Since $S/(SsS)$ is weakly periodic and $m(S) > m(S/(SsS))$, then the induction hypothesis applies to this semigroup, providing a chain $J_3' \subseteq \cdots \subseteq J_t'$, $t \ge 3$, of ideals of $S/(SsS)$ with the desired property. Let $J_3, \ldots, J_t = S$ be the inverse images of $J_3', \ldots, J_t'$, respectively, in $S$. Then $J_i'/J_{i-1}' \cong J_i/J_{i-1}$ for $i = 4, \ldots, t$. Moreover, $J_3/(SsS) \cong J_3'$. The chains $J_2 \subseteq J_3 \subseteq \cdots \subseteq J_t$, $J_1 \subseteq J_2 \subseteq J_3 \subseteq \cdots \subseteq J_t$ satisfy the desired conditions in cases (i) and (ii), respectively.

It is clear from the above proof that we must have $n \le 2m(S) + 2$, so that $n \le 2k$, where $n$ denotes the length of the constructed chain.

Now, assume that $S$ also is strongly $\pi$-regular. Let $T$ be a 0-simple semigroup of the form $J_1$ or $J_i/J_{i-1}$, $i \in \{2, \ldots, n\}$. Then $T$ is strongly $\pi$-regular, too. From Lemma 2 it follows that $T$ is completely 0-simple, which completes the proof of the theorem.

**Corollary 4** Let $S$ be a $\pi$-regular semigroup. Then every $\mathcal{J}$-class of $S$ determining a 0-simple principal factor contains an idempotent. Moreover, if $S$ has $k$ $\mathcal{J}$-classes containing idempotents, $k < \infty$, then there exists a chain $J_1 \subseteq J_2 \subseteq \cdots \subseteq J_n = S$, $n \le 2k$, of ideals of $S$ such that $J_1$, and all

$J_i/J_{i-1}$, $i > 1$, are 0-simple or nil. If $S$ is a periodic semigroup, then the non-nil factors are completely 0-simple.

*Proof.* Let $T$ be a nonzero 0-simple principal factor of $S$. Then $T$ is $\pi$-regular, and so it contains a nonzero idempotent because, otherwise, $T$ is a nil semigroup, which is impossible by Lemma 1, in Chapter 1. The same lemma implies that $T$ is completely 0-simple whenever $S$ is periodic. Therefore, the result follows from Theorem 3.

The above result will now be applied to the class of linear semigroups. Let $K$ be a field, and let $n \geq 1$ be an integer. If $a \in M_n(K)$, then treating $a$ as a linear transformation of an $n$-dimensional vector space over $K$ with a fixed basis, we denote by $\wedge^j(a)$, $1 \leq j \leq n$, the $j$th exterior power of $a$, and treat it (in a usual way) as an element of $M_{\binom{n}{j}}(K)$. The following auxiliary result is well known; see [15], §.5, Exercise 11.

**Lemma 5** Let $1 \leq j \leq n$. Then $\wedge^j : M_n(K) \to M_{\binom{n}{j}}(K)$ is a homomorphism of multiplicative semigroups. Moreover, $\rho(\wedge^j(a)) = 0$ if $\rho(a) < j$, and $\rho(\wedge^j(a)) = \binom{\rho(a)}{j}$ if $\rho(a) \geq j$.

We start with a result on matrices of rank one. Here we use the well-known fact that if $a_1 \ldots a_m = 0$ for some $m > 1$ and some matrices $a_1, \ldots, a_m \in M_n(K)$ of rank one, then there exists $j \in \{1, \ldots, m-1\}$ such that $a_j a_{j+1} = 0$. This fact is, for example, a consequence of the structure theorem for the multiplicative semigroup of the algebra $M_n(K)$; see Chapter 1.

**Lemma 6** Let $T \subseteq M_n(K)$ be a set of idempotents of rank one. Assume that $(exf)^2 = 0$ for any $x \in \langle T \rangle$, and any $e, f \in T$ with $e \neq f$. Then $|T| \leq n$.

*Proof.* Observe first that, in view of the preceding remark, the condition $(exf)^2 = 0$ implies that $ex = 0$ or $xf = 0$ or $fe = 0$, and, consequently,

(a)             $exfe = 0$ for any $x \in \langle T \rangle$ and any $e, f \in T, e \neq f$

In particular, putting $x = f$, we also get

(b)                  $ef = 0$ or $fe = 0$ for any $e, f \in T, e \neq f$

We will proceed by induction on $n$. The result is obvious for $n = 1$. Let $n > 1$. Fix an element $e \in T$. From (b) it follows that $(1 - e)f(1 - e)$ is an idempotent for any $f \in T \setminus \{e\}$. If $(1-e)f(1-e) = 0$, then $f = ef + fe$ because $efe = 0$. Hence, $f = f^2 = (ef + fe)f = ef$ and, similarly, $f = f(ef + fe) = fe$ because $fef = 0$. Thus, $f = (ef)(fe) = efe = 0$, a contradiction. Define $f' = (1-e)f(1-e)$ for $f \in T \setminus \{e\}$, $T' = \{f' \mid f \in T \setminus \{e\}\}$. Since $\rho(1-e) = n-1$, $T'$

embeds into $M_{n-1}(K) \cong (1-e)M_n(K)(1-e)$ and, by the above, $T'$ consists of idempotents of rank one. Let $f, g \in T \setminus \{e\}, f \neq g$, and let $x \in \langle T' \rangle$. While the element $z = f(1-e)x(1-e)g(1-e)f$ is a sum of elements of the form $\pm fygf, \pm fyef$ for some $y \in \langle T \rangle$, then (a) implies that $z = 0$. Consequently, $(f'xg')^2 = (1-e)z(1-e)x(1-e)g(1-e) = 0$, which shows that $T'$ satisfies the induction hypothesis as a subset of $(1-e)M_n(K)(1-e) \cong M_{n-1}(K)$. Moreover, with $x = g'$, we have $(f'g')^2 = 0$, proving that $f' \neq g'$. In particular, $|T'| = |T \setminus \{e\}| = |T| - 1$. Thus, by the induction hypothesis, we get $|T| = |T'| + 1 \leq (n-1) + 1 = n$.

Observe that the above result generalizes the well-known fact that $M_n(K)$ has no set of orthogonal idempotents of cardinality exceeding $n$.

**Corollary 7** Let $T \subseteq M_n(K)$ be a set of idempotents of rank $j$, $j \leq n$. Assume that $\rho((exf)^2) < j$ for any $e, f \in T$ with $e \neq f$, and any $x \in \langle T \rangle$. Then $|T| \leq \binom{n}{j}$.

*Proof.* Consider the set $\wedge^j(T) \subseteq M_{\binom{n}{j}}(K)$. By Lemma 5 for any $e, f \in T$, $e \neq f$, and any $x \in \langle T \rangle$, we have $(\wedge^j(e)\wedge^j(x)\wedge^j(f))^2 = \wedge^j((exf)^2) = 0$ and $\wedge^j(\langle T \rangle) = \langle \wedge^j(T) \rangle$. In particular, $\wedge^j(e)\wedge^j(f) = 0$, and so $\wedge^j(e) \neq \wedge^j(f)$, implying that $|\wedge^j(T)| = |T|$. Thus, the inequality $|T| \leq \binom{n}{j}$ follows from Lemmas 6 and 5.

The hypothesis of Corollary 7 appears to be connected with the $\mathcal{J}$-classes of subsemigroups of $M_n(K)$ of certain types. This is, in fact, the essence of the following result.

**Proposition 8** Let $S \subseteq M_n(K)$ be a $\pi$-regular semigroup. Then $S$ has at most $2^n$ $\mathcal{J}$-classes containing idempotents.

*Proof.* Let $1 \leq j \leq n$. Assume that some idempotents $e, f \in S$ of rank $j$ are given with $e, f$ lying in distinct $\mathcal{J}$-classes of $S$. Suppose that $\rho((exf)^2) = j$ for some $x \in S$. By the assumption on $S$, there exists $k \geq 1$ such that $(exf)^k$ is a regular element in $S$. Therefore, $(exf)^k S^1 = tS^1$ for some $t = t^2 \in S$. Then $\rho(t) = \rho((exf)^k)$. Thus, $\rho(t) = j$ because $\rho((exf)^2) = j = \rho(exf)$ implies that $\rho((exf)^k) = j$. Now, from $t = t^2$ and $t = et$, it follows that $e = te$; see Corollary 7 of Chapter 1. Hence, $e = ete \in e(exf)^k S^1 e \subseteq SfS$. A symmetric argument yields $f \in SeS$, contradicting the choice of $e, f$. Therefore, $\rho((exf)^2) < j$ for all $x \in S$. From Corollary 7 it then follows that there are, at most, $\binom{n}{j}$ $\mathcal{J}$-classes of $S$ containing idempotents of rank $j$. Hence, there are, at most, $2^n$ $\mathcal{J}$-classes of $S$ containing idempotents.

Let $S \subset M_n(K)$ be the semigroup of all diagonal idempotents. Then $S$ is a commutative regular and strongly $\pi$-regular semigroup with $|S| = 2^n$. This shows that the bound obtained in Proposition 8 cannot be improved because the $\mathcal{J}$-classes of $S$ are singletons.

We now show that, in the case of linear semigroups, a quantitative information on the nil factors arising from Corollary 4 may also be given.

**Proposition 9** Let $S \subseteq M_n(K)$ be a $\pi$-regular semigroup. If $I$ is an ideal of $S$ such that $S/I$ is a nil semigroup, then $(S/I)^m = \theta$, where $m = \prod_{j=1}^{n} \binom{n}{j}$.

*Proof.* Let $I_k = \{a \in M_n(K) \mid \rho(a) \leq k\}$ for $k = 0, 1, \ldots, n$. Define $H$ as the subsemigroup of $S$ generated by $S \setminus I$. Let $j$ be the least integer such that $S \subseteq I_j$. If $h = s_1 \ldots s_p$, $p \geq 1$, $s_i \in S \setminus I$, then $h^t \in I$ for some $t \geq 1$ by the hypothesis. Moreover, the assumption on $S$ implies that there exists $r \geq 1$ such that $(h^t)^r y = e$ for some $y \in S$, $e = e^2 \in S$, with $\rho(e) = \rho(h^{tr})$. Suppose that $\rho(e) = j$. Then $\rho(h) = j$ by the choice of $j$. From Corollary 7 in Chapter 1 it follows that $s_1 = es_1$. Since $I$ is an ideal of $S$ and $s_1 \notin I$, then $e \notin I$. This contradicts the fact that $h^t \in I$ and shows that $\rho(h^{tr}) = \rho(e) < j$. While $h \in H$ is an arbitrary element, then $H/(H \cap I_{j-1})$ is a nil semigroup.

Consider the semigroup $\mathcal{N}^j(H) \subseteq M_{\binom{n}{j}}(K)$. Since, by Lemma 5, $\mathcal{N}^j(I_{j-1}) = 0$, the first part of the proof implies that $\mathcal{N}^j(H)$ is a nil semigroup. It is well known that linear nil semigroups are nilpotent, more specifically, that $\mathcal{N}^j(H^{\binom{n}{j}}) = (\mathcal{N}^j(H))^{\binom{n}{j}} = 0$; see Proposition 13 in Chapter 2. This means that $H^{\binom{n}{j}} \subseteq I_{j-1}$. It follows that $S/(I \cup (I_{j-1} \cap S))$ is a nilpotent semigroup of index not exceeding $\binom{n}{j}$. On the other hand, we have

$$S/I \supseteq (I \cup (I_{j-1} \cap S))/I \cong (I_{j-1} \cap S)/(I \cap I_{j-1} \cap S) = (I_{j-1} \cap S)/(I_{j-1} \cap I)$$

Put $S' = I_{j-1} \cap S$, $I' = I_{j-1} \cap I$. Then $S'$ is an ideal of $S$, and so $S'$ inherits the hypothesis on $S$. Hence, we may repeat the above procedure with respect to the nil semigroup $S'/I'$. Since $S' \subseteq I_{j-1}$, this yields $S'^{\binom{n}{j-1}} \subseteq I' \cup (I_{j-2} \cap S') \subseteq I \cup (I_{j-2} \cap S)$. Thus, after, at most, $j$ such steps, we come to

$$S^{\binom{n}{j}\binom{n}{j-1}\cdots\binom{n}{1}} \subseteq I$$

Hence, $(S/I)^m = \theta$.

We now state the main result of this section, which was obtained for strongly $\pi$-regular semigroups by Okniński in [182].

**Theorem 10** Let $S \subseteq M_n(K)$ be a $\pi$-regular semigroup. Then there exists a chain of ideals $J_1 \subset J_2 \subset \cdots \subset J_t = S$ of $S$, $t \leq 2^{n+1}$, such that $J_1$ and all factors $J_i/J_{i-1}$, $i = 2, \ldots, t$, are completely 0-simple or nilpotent of index not exceeding $m = \prod_{j=1}^{n} \binom{n}{j}$.

*Proof.* Clearly, we may assume that $S$ is a nonzero semigroup. Let $T$ be a nonzero 0-simple principal factor of $S$. Since $M_n(K)$ has no infinite chain of idempotents but $T$ has nonzero idempotents by Corollary 4, then $T$ has a primitive idempotent. Hence, $T$ must be completely 0-simple. While $S$ has, at most, $2^n$ $\mathcal{J}$-classes containing idempotents by Proposition 8, this implies that $S$ has, at most, $2^n$ $\mathcal{J}$-classes determining completely 0-simple principal factors. Thus, Theorem 3 implies that a chain of ideals $J_1 \subset J_2 \subset \cdots \subset J_t = S$ of $S$, $t \leq 2^{n+1}$, exists, with $J_1$ and all factors $J_i/J_{i-1}$ being completely 0-simple or nil.

The remaining assertion on those that are nil follows from Proposition 9.

Since an element of a completely 0-simple semigroup $T$ either lies in a subgroup of $T$ or its square is equal to zero, then the following is an immediate consequence of Theorem 10.

**Corollary 11**   Let $S \subseteq M_n(K)$ be a $\pi$-regular semigroup. Then $S$ is strongly $\pi$-regular of bounded index, which is dependent on $n$ only.

We derive an application of Theorem 10 to locally finite semigroups.

**Corollary 12**   Let $S \subseteq M_n(K)$ be a $\pi$-regular semigroup. If any subgroup of $S$ is torsion, then $S$ is locally finite.

*Proof.* With the notation of Theorem 10, and in view of Proposition 2 in Chapter 2, it is enough to show that $J_1$ and all $J_i/J_{i-1}$ are locally finite. While the nilpotent case is obvious, the completely 0-simple case follows from Lemma 3 in Chapter 2 and the Burnside theorem on local finiteness of torsion linear groups; see [86], Theorem 2.3.5.

We now show that $\pi$-regular linear semigroups satisfy the minimal conditions $M_J$, $M_R$, $M_L$. The following result, due to Putcha [226], shows even more.

**Theorem 13**   Let $S \subseteq M_n(K)$ be a $\pi$-regular semigroup, and let $t$, $m$ be the integers determined for $S$ in Theorem 10. Then,

(i)   $S$ has no chain of more than $m^t$ principal right (left) ideals,
(ii)  $S$ has no chain of more than $m^{2t}$ principal two-sided ideals.

*Proof.* Let $J_1 \subset J_2 \subset \cdots J_t = S$ be the chain of ideals of $S$ resulting from Theorem 10, and let $J_0 = \varnothing$. Put $q = m^t$. Suppose, first, that there exist

$s_1, \ldots, s_{q+1} \in S$ such that $s_1 S^1 \supset s_2 S^1 \supset \cdots \supset s_{q+1} S^1$. Then $s_j w_j = s_{j+1}$ for some $w_j \in S$, $j = 1, \ldots, q$.

We first claim that there exist $1 \leq \alpha < \beta < \gamma \leq q$ such that $w_\alpha \ldots w_\beta S^1 = w_\alpha \ldots w_\gamma S^1$. We prove this by induction on $t$. Let $u_1 = w_1 \cdots w_m$, $u_2 = w_{m+1} \cdots w_{2m}$, .... If each $u_j \in J_{t-1}$, $j = 1, \ldots, m^{t-1}$, then the chain $s_1 S^1 \supset s_{m+1} S^1 \supset s_{2m+1} S^1 \supset \cdots \supset s_{q+1} S^1$, together with the $u_j$, satisfies the induction hypothesis. Suppose that some $u_j \notin J_{t-1}$ (this handles the case $t = 1$ in particular). If $S/J_{t-1}$ is nilpotent, then, by Theorem 10, $u_j \in S^m \subseteq J_{t-1}$ if $t > 1$, or $S^m = \theta$ and $s_m = s_1 w_1 \ldots w_{m-1} = \theta = s_1 w_1 \ldots w_m = s_{m+1}$ if $t = 1$, a contradiction. Hence, $S/J_{t-1}$ is completely 0-simple. Clearly, $w_{jm}, w_{jm-1}, w_{jm-1} w_{jm} \in S \setminus J_{t-1}$ because $u_j \notin J_{t-1}$. Thus, by Lemma 4 in Chapter 1, $w_{jm-1} w_{jm} S^1 = w_{jm-1} S^1$, proving our claim.

If follows that $s_{\beta+1} S^1 = s_\alpha w_\alpha \ldots w_\beta S^1 = s_\alpha w_\alpha \ldots w_\gamma S^1 = s_{\gamma+1} S^1$. This contradicts our supposition and proves (i) for principal right ideals. The left-right symmetric case goes similarly.

Let $r = q^2 + 1$. Suppose that there exist $s_1, \ldots, s_r \in S$ such that $S^1 s_1 S^1 \supset \cdots \supset S^1 s_r S^1$. Then $x_j s_j y_j = s_{j+1}$ for some $x_j, y_j \in S^1$, $j = 1, \ldots, r-1$. Put $x_0 = y_0 = 1$, and define $t_{j+1} = s_1 y_0 \ldots y_j$, $z_{j+1} = x_j \ldots x_0 s_1$ for $j = 0, \ldots, r-1$. Then $t_1 S^1 \supseteq t_2 S^1 \supseteq \cdots \supseteq t_r S^1$ and $S^1 z_1 \supseteq S^1 z_2 \supseteq \cdots \supseteq S^1 z_r$. Let $X = \{t_1 S^1, \ldots, t_r S^1\}$, $Y = \{S^1 z_1, \ldots, S^1 z_r\}$. From (i) we know that $|X| \leq q$ and $|Y| \leq q$. Let $W = \{1, \ldots, r\}$, and let us define a function $\phi : W \to X \times Y$ by $\phi(j) = (t_j S^1, S^1 z_j)$. Since $r = q^2 + 1$, there exist $\alpha, \beta \in W$, $\alpha \neq \beta$ such that $\phi(\alpha) = \phi(\beta)$. Thus, $t_\alpha S^1 = t_\beta S^1$, $S^1 z_\alpha = S^1 z_\beta$, and so

$$
\begin{aligned}
S^1 s_\alpha S^1 &= S^1 x_{\alpha-1} \ldots x_0 s_1 y_0 \ldots y_{\alpha-1} S^1 = S^1 z_\alpha y_0 \ldots y_{\alpha-1} S^1 \\
&= S^1 z_\beta y_0 \ldots y_{\alpha-1} S^1 = S^1 x_{\beta-1} \ldots x_0 s_1 y_0 \ldots y_{\alpha-1} S^1 = S^1 x_{\beta-1} \ldots x_0 t_\alpha S^1 \\
&= S^1 x_{\beta-1} \ldots x_0 t_\beta S^1 = S^1 x_{\beta-1} \ldots x_0 s_1 y_0 \ldots y_{\beta-1} S^1 = S^1 s_\beta S^1
\end{aligned}
$$

This contradicts our supposition and proves (ii).

The above result, together with Lemma 1 and Corollary 11, shows that the classes of weakly periodic semigroups, strongly $\pi$-regular semigroups, and semigroups satisfying $M_R$ (or $M_J$ or $M_L$) coincide when restricted to linear semigroups, all of which 0-simple principal factors are completely 0-simple.

We continue with an example of a finite linear semigroup, which will turn out to be useful in the sequel.

**Lemma 14** Let $S$ be the subset of the matrix algebra $M_n(K)$, $n \geq 1$, consisting of all matrices $s = (s_{ij})$ such that $s_{ij} \in \{0, 1\}$, and any row and any column of $s$ contains at most one nonzero entry. Then $S$ is a finite semigroup such that $s^{n!}$ is an idempotent for any $s \in S$.

*Proof.* It is clear that $S$ is finite and closed under multiplication. Let $s \in S$, and let $k$ be the least integer such that $\rho(s^k) = \rho(s^{k+1})$. We may assume that $\rho(s) < n$. Put $m = \rho(s^k)$. Then, from Corollary 7 in Chapter 1, we know that $\rho(s^j) = \rho(s^k)$ for every $j \geq k$. It is easy to see that this yields $k - 1 + m < n$. Let $t = (t_{ij}) = s^k$. It is clear that $\{j \mid t_{ij} \neq 0 \text{ for some } i\} = \{i \mid t_{ij} \neq 0 \text{ for some } j\}$. Therefore, $t$ may be treated as an $m \times m$ permutation matrix, and so $t^{m!}$ is an idempotent. Since $k \leq n - m$, then $k$ divides the product $(m + 1)(m + 2)\ldots n$. Hence, $s^{n!} = t^{m!}$ is an idempotent, too.

We close this chapter with an example showing that, in general, weakly periodic linear semigroups are not of the form described in Theorem 10. Much more can be said: there exist simple cancellative linear semigroups that are irreducible but are not groups. The first example of this type was constructed by Kelarev in [109] answering a question asked by Ponizovskii in his survey paper [215a].

**Example 15**   The construction is based on two independent observations. First, for every field $K$ of characteristic zero, the full linear group $GL_n(K)$, $n \geq 2$, contains a free noncommutative subgroup $H$. Therefore, $GL_n(K)$ has an infinitely generated free subgroup $G$. Specifically, it can be verified that the subgroups generated by the matrices

$$x = \begin{pmatrix} 1 & 2 \\ 0 & 1 \end{pmatrix} \qquad y = \begin{pmatrix} 1 & 0 \\ 2 & 1 \end{pmatrix}$$

and

$$x_n = x^n y x^n = \begin{pmatrix} 1 & 2 \\ 0 & 1 \end{pmatrix}^n \begin{pmatrix} 1 & 0 \\ 2 & 1 \end{pmatrix} \begin{pmatrix} 1 & 2 \\ 0 & 1 \end{pmatrix}^n \qquad n = 1, 2, \ldots$$

in $GL_2(\mathbb{Q})$ are of this type. The second general fact establishes the existence of a proper simple subsemigroup $S$ of $G$ containing $X = \langle x_1, x_2, \ldots \rangle$. We define a chain of subsemigroups of $G$ by $S_1 = \langle x_1 \rangle$, and $S_{n+1} = \langle x_{n+1}, S_n x_{n+1}^{-1} G_n \rangle$ for $n \geq 2$, where $G_n$ is the subgroup generated by $S_n$. Clearly, $S_n = (S_n x_{n+1}^{-1}) x_{n+1} \subseteq S_{n+1}$, and $G_n$ is the free group generated by $x_1, \ldots, x_n$. Put $S = \bigcup_{n \geq 1} S_n$. If $s, t \in S$, then $sx_{n+1}^{-1} t^{-1} \in S$, where $n \geq 1$ is such that $s, t \in S_n$. Therefore $s = (sx_{n+1}^{-1} t^{-1}) t x_{n+1} \in StS$, which shows that $S$ is a simple semigroup. By induction on $n$, we show that $S_n$ does not contain the identity $e$ of $G$. This is clear for $S_1 = \langle x_1 \rangle$. Assume that $e \notin S_n$, $n \geq 1$. Every element $t \in S_{n+1}$ has a unique presentation of shortest length $t = t_1 \ldots t_m$ with $t_i \in G_n \cup \langle x_{n+1}, x_{n+1}^{-1} \rangle$. Let $t = y_1 \ldots y_r$, where $r \geq 1$ and all $y_i$ are in $\{x_{n+1}\} \cup S_n x_{n+1}^{-1} G_n$. By induction on $r$, we first show that

(i)   $t_1 \in \langle x_{n+1} \rangle \cup S_n$.

(ii)   If some $t_i \in \langle x_{n+1} \rangle$, then $i = m$ or $t_{i+1} \in S_n$.

If $r = 1$, then $t = x_{n+1}$ or $t \in S_n x_{n+1}^{-1} G_n$. Since $e \notin S_n$, then, in the latter case, $t_1 \in S_n$, $t_2 = x_{n+1}^{-1}$, and $m = 2$ or $m = 3$. Hence, (i) and (ii) are satisfied. Assume that $r > 1$. If $y_1 = x_{n+1}$, then the induction hypothesis used for $y_2 \ldots y_r \in S_{n+1}$ easily implies that (i) and (ii) hold.

Assume that $y_1 = s x_{n+1}^{-1} g$ for some $s \in S_n$, $g \in G_n$. It is easy to see that (ii) holds in this case. Suppose that $t_1 \notin \langle x_{n+1} \rangle \cup S_n$. If $y_2 \ldots y_r = t_1' \ldots t_p'$ is the shortest presentation with $t_i' \in G_n \cup \langle x_{n+1}, x_{n+1}^{-1} \rangle$, then we must have $t_1' = g^{-1}$ and $t_2' = x_{n+1}$ or $g = 1$ and $t_1' = x_{n+1}$. The induction hypothesis (condition (ii) applied to $y_2 \ldots y_r$) implies then that $t_3' \in S_n$ and $t = s t_3' \ldots t_p'$ or that $t_2' \in S_n$ and $t = s t_2' \ldots t_p'$, respectively. Hence, $t_1 = s t_3'$ or $t_1 = s t_2'$ is in $S_n$. This contradicts our supposition and proves the claim.

From (i) it now follows that $S_{n+1}$ does not contain the identity of $G$, because $e \notin S_n$ by our hypothesis. Therefore, $S = \bigcup_{n \geq 1} S_n$ is not a group.

It is easy to check that $S \subseteq M_2(\mathbb{Q})$ is an irreducible semigroup. For example: if this is not the case, then there exists a one-dimensional $S$-invariant submodule in $\mathbb{Q}^2$. Thus, in the corresponding basis, $S$ can be presented as a semigroup consisting of triangular matrices. This implies that the subgroup of $GL_2(\mathbb{Q})$ generated by $S$ is solvable, which contradicts the fact that $S \supseteq \langle x_1, x_2, \ldots \rangle$.

Finally, we note that examples of this type cannot be constructed with $S$ finitely generated. Namely, if $S = \langle t_1, \ldots, t_n \rangle$ is cancellative and simple, then, for $s \in S$, we choose $s_1^{(1)}, s_1^{(2)} \in \{t_1, \ldots, t_n\}$ such that $s = s_1^{(1)} s_1^{(2)}$. Then, similarly, $s_1^{(1)} = s_2^{(1)} s_2^{(2)}$, and $s_l^{(1)} = s_{l+1}^{(1)} s_{l+1}^{(2)}$ for $l \geq 1$, where $s_i^{(j)} \in \{t_1, \ldots, t_n\}$. There exist $m > k$ such that $s_m^{(1)} = s_k^{(1)}$. The cancellativity in $S$ implies that $s_m^{(2)} s_{m-1}^{(2)} \ldots s_{k+1}^{(2)}$ is the identity $e$ of $S$. If $st = e$, then $ts$ is an idempotent, so that $ts = e$. If follows that one-sided units in $S$ are two-sided. Therefore, $S \setminus U(S)$ is an ideal of $S$ whenever it is nonempty. Hence, $S = U(S)$ is a group.

## Comments on Chapter 3

The strong $\pi$-regularity condition first appeared in [6], [50], and [156] in the context of associative rings and semigroups as a generalization of von Neumann regularity. Semigroups of this type, also called "quasiperiodic" or "group-bound," have been studied by many authors. A new impulse for the study was the fact that any linear algebraic semigroup is strongly $\pi$-regular; see [227], Theorem 3.18. The weak periodicity was introduced by Hotzel in

[88] for an investigation of the a.c.c. conditions on one-sided congruences of semigroups and of their connections with some other finiteness conditions.

We note that Theorem 3 extends [225], Theorem 1.7, proved for strongly $\pi$-regular semigroups. Moreover, the assertion of Theorem 10 was established for strongly $\pi$-regular semigroups in another paper of Putcha [226], with no quantitative bounds on the length of the chain of the ideals and on the nilpotency indices on the nil factors. The Burnside theorem for periodic linear semigroups (see Corollary 12) was first established by McNaughton and Zalcstein [150]; see also [132], Section 10.2. The assertion of Proposition 8 was proved for regular semigroups by Hofmann and Skryago in [87] and also through an exterior power argument. Finally, we refer to the monograph of Putcha [227] for the recent trends and bibliography in the study of algebraic monoids.

# 4

# Semigroup Algebras: General Results and Techniques

We will deal with semigroup algebras with coefficients being fields only. Thus, throughout, $K$ will stand for a field. By the semigroup algebra $K[S]$ of a semigroup $S$ over $K$, we mean the set of all functions $f : S \to K$ such that $f(s) = 0$ for all but finitely many $s \in S$, with operations defined for every $f, g \in K[S]$, $s \in S$, $\lambda \in K$ as follows:

$$(f + g)(s) = f(s) + g(s)$$
$$(\lambda f)(s) = \lambda f(s)$$
$$(fg)(s) = \begin{cases} \displaystyle\sum_{(t,u)\in A(s)} f(t)g(u) & \text{if } A(s) \neq \varnothing \\ 0 & \text{if } A(s) = \varnothing \end{cases}$$

where $A(s) = \{(t,u) \in S \times S \mid tu = s\}$. $K[S]$ is an associative $K$-algebra subject to these operations. For any $s \in S$, let $f_s : S \to K$ be the function such that $f_s(s) = 1$, $f_s(t) = 0$ if $t \neq s$. Then $\{f_s \mid s \in S\}$ is a subsemigroup of the multiplicative semigroup of $K[S]$, which is a $K$-basis of $K[S]$. Moreover

33

$s \to f_s$ is a semigroup isomorphism. Thus, as usual, $K[S]$ will be identified with the set of all finite sums $\sum \alpha_s s$, $\alpha_s \in K$, $s \in S$, so that it is a $K$-space with a basis $S$ and the multiplication induced by the multiplication in $S$.

If $a = \sum \alpha_s s \in K[S]$, then, by the support of the element $a$, we mean the set $\{s \in S \mid \alpha_s \neq 0\}$ and denote it by $\mathrm{supp}(a)$.

We list a few important obvious examples of semigroup algebras. If $S$ is a free commutative monoid of rank $n$, then $K[S]$ is isomorphic to the polynomial algebra $K[x_1, \ldots, x_n]$. If $S$ is a free monoid of rank $n$, then $K[S]$ is isomorphic to the free algebra $K\{x_1, \ldots, x_n\}$, that is, the algebra of polynomials in $n$ noncommuting indeterminates. The algebras $K[S]$ for $S$ being groups, called group algebras, form another extensively studied class of basic importance for our considerations.

As in the commutative case (see [64], Theorem 7.11), it may be easily shown that the class of semigroup algebras over $K$ coincides, to within isomorphism, with the class of all factor algebras $K[X]/I$, where $X$ is a free monoid and $I$ an ideal generated by a set of elements of the form $x - y$, with $x, y \in X$.

The lattices of the right, left, and two-sided ideals of $K[S]$ will be denoted by $\mathcal{R}(K[S])$, $\mathcal{L}(K[S])$, and $\mathcal{T}(K[S])$, respectively.

Let $Z$ be a nonempty set. As a notational convenience, $K[Z]$ will stand for the set of all finite sums $\sum \alpha_z z$, $z \in Z$, $\alpha_z \in K$. Let $\phi : S \to Z$ be any mapping. By $\bar{\phi}$ we mean the extension of $\phi$ to the mapping of $K[S]$ into $K[Z]$ given by the formula $\bar{\phi}(\sum \alpha_s s) = \sum \alpha_s \phi(s)$. Clearly, if $Z$ is a semigroup and $\phi$ a semigroup homomorphism, then $\bar{\phi}$ is a homomorphism of the corresponding semigroup algebras.

Let $\rho$ be a right congruence on $S$, that is, $\rho$ is an equivalence relation such that, for any $s$, $t$, $x \in S$, we have $(sx, tx) \in \rho$ whenever $(s, t) \in \rho$. If $\phi_\rho : S \to S/\rho$ is the natural mapping onto the set $S/\rho$ of $\rho$-classes in $S$, then we denote by $I(\rho)$ the right ideal of $K[S]$ generated by the set $\{s - t \mid s, t \in S, (s, t) \in \rho\}$. Since $\rho$ is a right congruence on $S$, then $I(\rho)$ coincides with the $K$-subspace generated by the set $\{s - t \mid s, t \in S, (s, t) \in \rho\}$. Moreover, $K[S/\rho]$ is a right $K[S]$-module under the natural action defined by $\phi_\rho(s) \circ t = \phi_\rho(st)$ for $s$, $t \in S$. With this notation, we have the following result.

**Lemma 1** For any right congruence $\rho$ on $S$, $\bar{\phi}_\rho : K[S] \to K[S/\rho]$ is a homomorphism of right $K[S]$-modules such that

$$\ker(\bar{\phi}_\rho) = I(\rho) = \sum_{s \in S} \omega_s(\rho)$$

where $\omega_s(\rho) = \{\sum_{i=1}^{m} \alpha_i s_i \in K[S] \mid m \geq 1, \sum_{i=1}^{m} \alpha_i = 0, (s, s_i) \in \rho$ for all $i = 1, 2, \ldots, m\}$, and $K[S/\rho] \cong K[S]/I(\rho)$ as right $K[S]$-modules. Moreover, the correspondence $\rho \to I(\rho)$ establishes a one-to-one order-preserving mapping of the lattice $\mathcal{R}(S)$ of right congruences on $S$ into the lattice $\mathcal{R}(K[S])$ of right ideals of $K[S]$.

*Proof.* Let $s, t \in S$. Then $\bar{\phi}_\rho(st) = \phi_\rho(st) = \phi_\rho(s) \circ t = \bar{\phi}_\rho(s) \circ t$, which implies that $\bar{\phi}_\rho$ is a homomorphism of right $K[S]$-modules.

Assume that $(s, t) \in \rho$. Then $\bar{\phi}_\rho(s) = \phi_\rho(s) = \phi_\rho(t) = \bar{\phi}_\rho(t)$, and so $s - t \in \ker(\bar{\phi}_\rho)$. From the definition of $I(\rho)$, it then follows that $I(\rho) \subseteq \ker(\bar{\phi}_\rho)$.

Assume that $a \in \ker(\bar{\phi}_\rho)$. Write $a = a_1 + \cdots + a_m$, where $m \geq 1$ and, for all $i = 1, \ldots, m$, supp($a_i$) lies in a $\rho$-class of $S$. Then $0 = \bar{\phi}_\rho(a) = \bar{\phi}_\rho(a_1) + \cdots + \bar{\phi}_\rho(a_m)$ implies that $\bar{\phi}_\rho(a_i) = 0$ for all $i$. Fix some $i \in \{1, \ldots, m\}$. If $a_i = \sum \alpha_s^{(i)} s$ for some $\alpha_s^{(i)} \in K$, then $\bar{\phi}_\rho(a_i) = \sum \alpha_s^{(i)} \phi_\rho(s)$ and $\phi_\rho(s) = \phi_\rho(t)$ for any $s, t \in S$ with $\alpha_s^{(i)}, \alpha_t^{(i)} \neq 0$. Therefore, $\sum \alpha_s^{(i)} = 0$, which shows that $a_i \in \omega_t(\rho)$ for every $t \in$ supp($a_i$). Hence, $\ker(\bar{\phi}_\rho) \subseteq \sum_{s \in S} \omega_s(\rho)$.

It is easy to see that $\omega_s(\rho) \subseteq I(\rho)$ for every $s \in S$. This proves that $\ker(\bar{\phi}_\rho) = I(\rho) = \sum_{s \in S} \omega_s(\rho)$. It is now clear that

$$K[S/\rho] \cong K[S]/\ker(\bar{\phi}_\rho) \cong K[S]/I(\rho)$$

Finally, we claim that

(a) $$\rho = \{(s, t) \in S \times S \mid s - t \in I(\rho)\}$$

In fact, $s - t \in I(\rho)$ implies that $\bar{\phi}_\rho(s - t) = 0$ and, hence, $\phi_\rho(s) = \phi_\rho(t)$. This means that $(s, t) \in \rho$, yielding (a).

From (a) it is clear that the right ideal $I(\rho)$ determines the right congruence $\rho$, from which the remaining assertion follows. $\blacksquare$

Combining Lemma 1 with its left-right symmetric analog, we derive the following consequence.

**Corollary 2** For any congruence $\rho$ on $S$, $\bar{\phi}_\rho : K[S] \to K[S/\rho]$ is a homomorphism of algebras such that $\ker(\bar{\phi}_\rho) = I(\rho)$ and $K[S/\rho] \cong K[S]/I(\rho)$ as $K$-algebras. Consequently, $\rho \to I(\rho)$ is an order-preserving mapping of $\mathcal{T}(S)$ into $\mathcal{T}(K[S])$.

It is clear that the trivial congruence $\vartheta$ on $S$ determines the zero ideal $I(\vartheta) = 0$ of $K[S]$. On the other hand, the universal congruence $\iota = S \times S$ on $S$ leads to the ideal $I(\iota) = \{s - t \mid s, t \in S\}K = \{\sum \alpha_s s \in K[S] \mid \sum \alpha_s = 0\}$. This ideal is usually denoted by $\omega(K[S])$ and is called the augmentation ideal of $K[S]$, while the corresponding homomorphism $K[S] \to K$ is called the augmentation map.

**Lemma 3** Let $\{\rho_\beta \mid \beta \in B\}$ be a nonempty family of congruences on $S$. Assume that the following condition is satisfied:

(b)       For any $\beta, \gamma \in B$, there exists $\delta \in B$ such that $\rho_\delta \subseteq \rho_\beta \cap \rho_\gamma$

Then $I(\bigcap_{\beta \in B} \rho_\beta) = \bigcap_{\beta \in B} I(\rho_\beta)$ and $K[S/\bigcap_{\beta \in B} \rho_\beta]$ is a subdirect product of the algebras $K[S/\rho_\beta]$, $\beta \in B$.

*Proof.* Let $\rho = \bigcap_{\beta \in B} \rho_\beta$. If $s, t \in S$ are such that $(s,t) \in \rho$, then $(s,t) \in \rho_\beta$ for any $\beta \in B$, which implies that $s - t \in \bigcap_{\beta \in B} I(\rho_\beta)$. Thus, $I(\rho) \subseteq \bigcap_{\beta \in B} I(\rho_\beta)$. Therefore, passing to the homomorphic image $K[S/\rho]$ of $K[S]$, we may assume, in view of Corollary 2, that $\rho$ is the trivial congruence on $S$ and $I(\rho) = 0$. Let $0 \neq a \in K[S]$. Write $a = \sum_{i=1}^m \lambda_i s_i$ for some $m \geq 1$, $0 \neq \lambda_i \in K$, $s_i \in S$. Let $\phi_\beta : S \to S/\rho_\beta$, $\beta \in B$, denote the natural homomorphism, and let $\bar{\phi}_\beta$ be the induced homomorphism of semigroup algebras. If $m = 1$, then $\bar{\phi}_\beta(a) = \lambda_1 \phi_\beta(s_1) \neq 0$ for any $\beta \in B$ because $\lambda_1 \neq 0$ and $\phi_\beta(s_1) \in S/\rho_\beta$. Assume that $m \geq 2$. Since $\rho = \bigcap_{\beta \in B} \rho_\beta$ is the trivial congruence on $S$, then, for any $i, j \in \{1, \dots, m\}$, $i \neq j$, there exists $\beta_{ij} \in B$ such that $(s_i, s_j) \notin \rho_{\beta_{ij}}$. Now, (b) implies that there exists $\delta \in B$ with $\rho_\delta \subseteq \bigcap \rho_{\beta_{ij}}$, where the intersection runs over all pairs $(i,j)$ such that $i, j \in \{1, \dots, m\}$ and $i \neq j$. It is clear that $|\mathrm{supp}(\bar{\phi}_\delta(a))| = m$. In particular, $\bar{\phi}_\delta(a) \neq 0$. Therefore, in any case, $\bar{\phi}_\gamma(a) \neq 0$ for some $\gamma \in B$. Since $\ker(\bar{\phi}_\gamma) = I(\rho_\gamma)$ by Corollary 2, then it follows that $\bigcap_{\beta \in B} I(\rho_\beta) = \bigcap_{\beta \in B} \ker(\bar{\phi}_\beta) = 0 = I(\rho)$, which proves the result.

We derive a useful consequence concerning the Jacobson radical $\mathcal{J}(K[S])$ of the algebra $K[S]$.

**Corollary 4** Assume that $S$ is a semigroup with a family of congruences $\rho_\beta$, $\beta \in B$, satisfying condition (b) of Lemma 3 and such that $\bigcap \rho_\beta$ is the trivial congruence on $S$. If for any $\beta \in B$, $\mathcal{J}(K[S/\rho_\beta])$ is a nil ideal of index bounded by an integer $n \geq 1$ (not dependent on $\beta$), then $\mathcal{J}(K[S])$ is a nil ideal of index bounded by $n$. In particular, $\mathcal{J}(K[S]) = 0$ whenever $\mathcal{J}(K[S/\rho_\beta]) = 0$ for any $\beta \in B$.

*Proof.* Since $\mathcal{J}(K[S])$ is mapped onto $\mathcal{J}(K[S/\rho_\beta])$ under the natural homomorphism $\bar{\phi}_\beta$, $\beta \in B$, used in the proof of Lemma 3, then the assertion is a direct consequence of this lemma.

It is easy to see that, in the above result, "nil of bounded index" may be replaced by any identity holding in all algebras $\mathcal{J}(K[S/\rho_\beta])$, $\beta \in B$.

Now, we show that any right ideal of $K[S]$ determines a right congruence on $S$. Let $J \in \mathcal{R}(K[S]) \cup \mathcal{L}(K[S])$. Define a relation $\rho_J$ on $S$ by $\rho_J = \{(s,t) \in S \times S \mid s - t \in J\}$.

**Lemma 5** Let $J$ be a right ideal of $K[S]$. Then,

(i)    $\rho_J$ is a right congruence on $S$ such that $I(\rho_J) \subseteq J$.

(ii)   There exist natural homomorphisms of right $K[S]$-modules, $K[S] \to K[S/\rho_J] \to K[S]/J$.

(iii)  If $J$ is a two-sided ideal of $K[S]$, then $\rho_J$ is a congruence on $S$, the mappings in (ii) are homomorphisms of $K$-algebras, and the semigroup $S/\rho_J$ embeds into the multiplicative semigroup of the algebra $K[S]/J$.

Moreover, $J \to \rho_J$ is an order-preserving $\wedge$-complete semilattice homomorphism of $\mathcal{R}(K[S])$ onto $\mathcal{R}(S)$ and $\rho_{I(\rho)} = \rho$ for any $\rho \in \mathcal{R}(S)$.

*Proof.* It is clear that $\rho_J$ is an equivalence relation. Since $sx - tx = (s-t)x \in J$ for any $s, t, x \in S$ with $(s,t) \in \rho_J$, then $\rho_J$ is a congruence. The inclusion $I(\rho_J) \subseteq J$ is obvious. Thus, (i) holds. Now, (ii) is a direct consequence of (i) and Lemma 1. Finally, (iii) is derived through a left-right symmetric argument and Corollary 2.

Let $J_\beta$, $\beta \in B$, be a nonempty family of right ideals of $K[S]$. Put $J = \bigcap_{\beta \in B} J_\beta$. Then $(s,t) \in \rho_J$ if and only if $s - t \in J_\beta$ for all $\beta \in B$, the latter being equivalent to the fact that $(s,t) \in \bigcap_{\beta \in B} \rho_{J_\beta}$. Thus, $\rho_J = \bigcap_{\beta \in B} \rho_{J_\beta}$, showing that the considered mapping $\mathcal{R}(K[S]) \to \mathcal{R}(S)$ is a complete $\wedge$-semilattice homomorphism. The fact that $\rho_{I(\rho)} = \rho$ for any $\rho \in \mathcal{R}(S)$ is a reformulation of the claim (a) used in the proof of Lemma 1. Consequently, any right congruence $\rho$ on $S$ is of the form $\rho_J$, $J \in \mathcal{R}(K[S])$, and so the mapping is "onto."

We will often use the congruence $\rho_{\mathcal{J}(K[S])}$ coming from the Jacobson radical $\mathcal{J}(K[S])$ of $K[S]$. In the following, more general observation, we deal with the radical properties in the sense of Kurosh and Amitsur; see [45].

**Corollary 6** Let $\mathcal{P}$ be a Kurosh–Amitsur radical. Then $\rho_{\mathcal{P}(K[S/\rho_{\mathcal{P}(K[S])}])}$ is the trivial congruence on $S/\rho_{\mathcal{P}(K[S])}$.

*Proof.* Let $s, t \in S$ be such that $s' - t' \in \mathcal{P}(K[S/\rho_{\mathcal{P}(K[S])}])$, where $s'$, $t'$ denote the images of $s$, $t$ in $S/\rho_{\mathcal{P}(K[S])}$. From Lemma 5 it follows that $I(\rho_{\mathcal{P}(K[S])}) \subseteq \mathcal{P}(K[S])$, and so $s - t \in \mathcal{P}(K[S])$ by Corollary 2. Hence, $s' = t'$, and the result follows.

Let $S$ be a semigroup with zero $\theta$. By the contracted semigroup algebra of $S$ over $K$, denoted by $K_0[S]$, we mean the factor algebra $K[S]/K\theta$. Thus, $K_0[S]$ may be identified with the set of finite sums $\sum \alpha_s s$ with $\alpha_s \in K$, $s \in S \setminus \{\theta\}$, subject to the componentwise addition and multiplication given by the rule

$$s \circ t = \begin{cases} st & \text{if } st \neq \theta \\ 0 & \text{if } st = \theta \end{cases}$$

defined on the basis $S \setminus \{\theta\}$. If $S$ has no zero element, then we put $K_0[S] = K[S]$. For any $a = \sum \alpha_s s$, $\alpha_s \in K$, by $\text{supp}_0(a)$ we mean the set $\{s \in S \setminus \{\theta\} \mid \alpha_s \neq 0\}$. Thus, $\text{supp}_0(a) = \text{supp}(a) \setminus \{\theta\}$.

From the definition it follows directly that, for any semigroup $S$, we have $K_0[S^0] \cong K[S^0]/K\theta \cong K[S]$. We will often use the following extension of this fact.

**Lemma 7** Let $I$ be an ideal of a semigroup $S$. Then $K_0[S/I] \cong K[S]/K[I]$.

*Proof.* Let $\phi : K[S] \to K_0[S/I]$ be the mapping defined by $\phi(\sum \alpha_s s) = \sum_{s \notin I} \alpha_s s$. It is easy to see that $\phi$ is the composition of the natural homomorphisms $K[S] \to K[S/I] \to K_0[S/I]$. Clearly, $\ker(\phi) = K[I]$ and $\phi$ is an "onto" homomorphism, so the assertion follows.

If $J$ is an ideal of an algebra $R$, and $J$ has an identity $e$, then, for every $x \in R$, we have $ex = (ex)e = e(xe) = xe$ because $ex, xe \in J$. Consequently, $e$ is a central idempotent in $R$, and $R \cong eR \oplus (1-e)R$ as algebras. This fact is exploited in the following useful observation.

**Corollary 8** Let $I$ be an ideal of $S$ such that $K[I]$ is an algebra with an identity $e$. Then $\phi : K[S] \to K[I] \oplus K_0[S/I]$ defined by $\phi(x) = ex + \varphi(x)$, where $\varphi : K[S] \to K_0[S/I]$ is the natural homomorphism, is an isomorphism of algebras.

*Proof.* We know that $e$ is a central idempotent of $K[S]$, $K[I] = eK[S]$ and $K[S] = K[I] \oplus (1-e)K[S]$ as algebras. From Lemma 7 it follows that $(1-e)K[S] \cong K[S]/K[I] \cong K_0[S/I]$. Specifically, since, for any $x \in K[S]$, $\varphi((1-e)x) = \varphi(x) - \varphi(ex) = \varphi(x)$ and $\ker(\varphi) \cap (1-e)K[S] = K[I] \cap (1-e)K[S] = 0$, then the restriction of $\varphi$ to $(1-e)K[S]$ is an isomorphism onto $K_0[S/I]$. The result follows.

If $S$ has a zero element $\theta$, then $K\theta$ is an ideal of $K[S]$ with an identity. Hence, the following is immediate.

**Corollary 9** Assume that $S$ has a zero element. Then $K[S] \cong K \oplus K_0[S]$.

This and the fact that, for any $S$, $K[S^0] \cong K \oplus K_0[S^0] \cong K \oplus K[S]$ will allow us to switch from one of the algebras $K[S]$, $K_0[S]$ to the other whenever convenient.

Some important natural classes of algebras may be treated as contracted semigroup algebras but not ordinary semigroup algebras. For example, let

$n > 1$ be an integer, and let $S$ be the semigroup of $n \times n$ matrix units, that is, $S = \{e_{ij} \mid i,j = 1,\ldots,n\} \cup \{\theta\}$ subject to the multiplication

$$e_{ij}e_{kl} = \begin{cases} e_{il} & \text{if } j = k \\ \theta & \text{if } j \neq k \end{cases}$$

Then $K_0[S]$ is isomorphic, in a natural way, to the matrix algebra $M_n(K)$. However, $M_n(K)$ is not an ordinary semigroup algebra because it is simple, whereas the augmentation ideal of any semigroup algebra of a nontrivial semigroup always is a proper ideal. An extension of this example leads to an important class of algebras considered in Chapter 5.

Another useful class of contracted semigroup algebras arises as $K_0[X/I] \cong K[X]/K[I]$, where $I$ is an ideal of a free semigroup $X$. Algebras of this type, called monomial algebras, are discussed in Chapter 24. We note here that, while an arbitrary semigroup algebra arises as $K[X]/J$ for an ideal $J$ of a free algebra $K[X]$ generated by some elements of the form $x - y$, where $x, y \in X$, contracted semigroup algebras are of the form $K[X]/J$ for some $J$ generated by elements of the form $x - y$, $w$, where $x, y, w \in X$.

We continue with an easy observation on the behavior of semigroup algebras under the tensor multiplication.

**Lemma 10**   Let $L$ be a field extension of $K$, and let $S$, $T$ be semigroups. Then $L[S] \cong L \otimes_K K[S]$ and $K[S] \otimes_K K[T] \cong K[S \times T]$. Further, if $S$, $T$ have zero elements $\theta_S$, $\theta_T$, respectively, then $L_0[S] \cong L \otimes_K K_0[S]$ and $K_0[S] \otimes_K K_0[T] \cong K_0[(S \times T)/I]$, where $I = \{(s,t) \in S \times T \mid s = \theta_S \text{ or } t = \theta_T\}$.

*Proof.*   The former is standard, and similar to the group algebra case; see [203], Lemmas 1.3.4 and 1.3.6. If $S$, $T$ have zero elements $\theta_S$, $\theta_T$, then $L\theta_S \cong L \otimes_K K\theta_S$ and $L_0[S] \cong (L \otimes_K K[S])/(L \otimes_K K\theta_S) \cong L \otimes_K K_0[S]$. Further, using Lemma 7, we come to

$$\begin{aligned} K_0[(S \times T)/I] &\cong K[S \times T]/K[I] \\ &\cong (K[S] \otimes_K K[T])/(K\theta_S \otimes_K K[T] + K[S] \otimes_K K\theta_T) \\ &\cong (K[S]/K\theta_S) \otimes_K (K[T]/K\theta_T) = K_0[S] \otimes_K K_0[T] \end{aligned}$$

It is often convenient for technical reasons to deal with unitary algebras. There is a standard way of extending an arbitrary $K$-algebra $R$ to a $K$-algebra $R^1$ with unity, so that $R$ is an ideal of $R^1$ and $R^1/R \cong K$, see [29], Theorem 10.4.1. From the construction, it is clear that $K[S]^1 \cong K[S^1]$. We state an easy consequence of this fact.

**Lemma 11**   If $\mathcal{P}$ is a Kurosh–Amitsur radical such that $\mathcal{P}(K) = 0$, then $\mathcal{P}(K[S^1]) = \mathcal{P}(K[S])$.

*Proof.* Since $K[S]$ is an ideal of $K[S^1]$ with $K[S^1]/K[S]$ being $\mathcal{P}$-semisimple, then $\mathcal{P}(K[S^1]) \subseteq K[S]$, so that $\mathcal{P}(K[S^1]) \subseteq \mathcal{P}(K[S])$. The converse inclusion is a consequence of the fact that, by the Anderson–Divinsky–Sulinski theorem [2], $\mathcal{P}(K[S])$ is an ideal of $K[S^1]$.

Clearly, the above result applies to the Jacobson radical, and the prime radical $\mathcal{B}(K[S])$ of $K[S]$.

If $S$ is a monoid, then let $U(S)$ denote the group of units of $S$.

**Lemma 12** Assume that $S$ is a monoid that is not a group. Let $I = S \setminus U(S)$. If any right unit of $S$ is a unit of $S$, then $I$ is an ideal of $S$ and $K_0[S/I] \cong K[U(S)]$. In particular, this is the case if $S$ is a periodic or cancellative semigroup, $K[S]$ is right noetherian, or $K[S]$ satisfies a polynomial identity.

*Proof.* Assume that $st \in U(S)$ for some $s, t \in S$. Then $s$ is right invertible in $S$, and so, by the hypothesis, $s \in U(S)$. Hence, $t \in U(S)$, showing that $I$ is an ideal of $S$. From Lemma 7, it follows that $K_0[S/I] \cong K[S]/K[I] \cong K[U(S)]$.

Assume that $st = 1$. If $S$ is periodic, then $1 = s^n t^n = e t^n$ for an integer $n \geq 1$ such that $e = s^n$ is an idempotent. Then $1 = e(e t^n) = e$, and so $s \in U(S)$ because $s^{-1} = s^{n-1}$. If $S$ is cancellative, then $ts$ is an idempotent, so that $ts = 1$ and, again, $s \in U(S)$.

It is well known that a right unit of an algebra $R$ that is not a unit leads to an infinite set of matrix units, that is, a set $\{e_{ij} \mid i,j = 1,2,\ldots\}$ such that $e_{ij}e_{jl} = e_{il}$, $e_{ij}e_{kl} = 0$ for $j \neq k$; see [93], Proposition III.7.6. Thus, $R$ cannot be right noetherian because it contains an infinite set of orthogonal idempotents, and by the Amitsur–Levitzki theorem (see [203], Lemma 5.1.4), $R$ cannot satisfy polynomial identities because it contains matrix algebras of arbitrary size.

While, in general, the assertion of Lemma 12 does not hold, the following result due to Kozhukhov [117] allows us, in view of Lemmas 1 and 5 herein, to connect the structures of $K[S]$ and $K[G]$ for every subgroup $G$ of $S$. Recall that the lattice of right congruences of a group $G$ may be identified with the lattice of subgroups of $G$.

**Proposition 13** Let $G$ be a subgroup of a semigroup $S$. Then there exists a lattice embedding $\tau$ of the lattice $\mathcal{R}(G)$ of right congruences on $G$ into the lattice $\mathcal{R}(S)$ of right congruences on $S$ such that $\tau(\rho)_{|G} = \rho$ for $\rho \in \mathcal{R}(G)$.

*Proof.* Let $\rho$ be a right congruence on $G$, and let $H$ be the subgroup of $G$ determined by $\rho$. We define a binary relation $\tau_H$ on $S$ by: $(x,y) \in \tau_H$ if and only if any of the following conditions holds for $x, y$.

(1)  $x = y$.
(2)  $x = hy$ and $y = h^{-1}x$ for some $h \in H$.
(3)  $x = gy, y = g^{-1}x$ for some $g \in G$, and $g_1x = g_2x$ for some $g_1, g_2 \in G$, $g_1 \neq g_2$.

(In the above, $h^{-1}, g^{-1}$ stand for the inverses of $h, g$ in $G$.)

If (3) holds, then $(g_1g)y = (g_2g)y$ and $g_1g \neq g_2g$, which shows that $\tau_H$ is symmetric. Assume that $(x,y), (y,z) \in \tau_H$ for some $x, y, z \in S$. Let $x = hy$, $y = h^{-1}x, y = fz, z = f^{-1}y$ for some $h, f \in G$. Then $x = (hf)z, z = (hf)^{-1}x$, and $hf \in G$. Thus, it is easy to see that if (3) holds for $x, y$ or $y, z$, then (3) holds for $x, z$. If (2) holds for $x, y$ and for $y, z$, then we may assume that $hf \in H$, and so (2) holds for $x, z$. It follows that $\tau_H$ is a transitive relation. It is straightforward to check that any one of the conditions (1), (2), or (3) is inherited by the pair $(xs,ys), s \in S$, and so $\tau_H$ is a right congruence on $S$.

If $x, y \in G$ are such that $(x,y) \in \tau_H$, then (3) cannot hold. This shows that $\tau_H$ restricted to $G$ coincides with the right congruence $\rho$ determined by $H$ on $G$. Thus, $\tau : \mathcal{R}(G) \to \mathcal{R}(S)$ given by $\tau(\rho) = \tau_H$ satisfies the condition $\tau(\rho)_{|G} = \rho$.

Let $F$ be a subgroup of $G$. From the above it follows that $\tau$ is an order-preserving mapping. Consequently, $\tau_{F \cap H} \subseteq \tau_F \cap \tau_H$ and $\tau_F \cup \tau_H \subseteq \tau_{(FH)}$, where $(FH)$ is the subgroup of $G$ generated by $FH$. Assume that $(x,y) \in \tau_F \cap \tau_H, x, y \in S$. Since conditions (1) and (3) are independent of $H$, then it is enough to consider the case in which (2) holds for $x$ and $y$ with respect to $F$ and $H$, and (3) does not hold for $x, y$. Then $x = fy, x = hy$, $y = f^{-1}x$ for some $f \in F, h \in H$, so $fy = hy$. Therefore, $f = h \in F \cap H$, which shows that $(x,y) \in \tau_{F \cap H}$. Hence, $\tau_{F \cap H} = \tau_F \cap \tau_H$ as desired. Finally, if $\sigma$ is a right congruence on $S$ such that $\tau_F \cup \tau_H \subseteq \sigma$, then $(x,y) \in \sigma$ whenever $x = gy, y = g^{-1}x$ for some $g \in (FH)$. Therefore, $\tau_{(FH)} \subseteq \sigma$, and taking $\sigma = \tau_F \cup \tau_H$, we get $\tau_F \cup \tau_H = \tau_{(FH)}$. This completes the proof.

We now introduce a type of subsemigroups of a given semigroup $S$ that is very useful when dealing with the Jacobson radical and some other important objects in $K[S]$. Let $Z$ be a nonempty subset of $S$. We say that $Z$ is a left group-like subset of $S$ if for any $z \in Z, s \in S$, we have $s \in Z$ whenever $zs \in Z$. We say that $Z$ is a left group-like subsemigroup of $S$ if $Z$ is a subsemigroup of $S$, which is a left group-like subset. Right group-like subsemigroups are defined symmetrically. Note that the group of units $U(S)$ of a monoid $S$ always is left and right group-like in $S$.

Clearly, this notion plays a role similar to that of a subgroup of a given group in the case of group algebras. Thus, defining $\pi_T : K[S] \to K[T]$ by

$\pi_T(\sum \alpha_s s) = \sum_{s \in T} \alpha_s s$ for any subset $T$ of $S$, we get the following extension of [203], Lemma 1.1.2.

**Lemma 14**   Let $Z$ be a left group-like subsemigroup of $S$. Then $\pi_Z$ is a $K$-linear map such that $\pi_Z(ab) = a\pi_Z(b)$ for any $a \in K[Z]$, $b \in K[S]$.

*Proof.*   Let $b = b_1 + b_2$, where $\text{supp}(b_1) \subseteq Z$ and $\text{supp}(b_2) \cap Z = \varnothing$. If $s \in \text{supp}(b_2)$, $z \in \text{supp}(a)$, then $zs \notin Z$, and so $\pi_Z(zs) = 0$. Since $\pi_Z$ is clearly linear, then $\pi_Z(ab_2) = 0$, and the result follows because $\pi_Z(ab_1) = ab_1 = a\pi_Z(b_1) = a\pi_Z(b)$.

The main advantage of dealing with subsemigroups of the above type comes from the following observation.

**Lemma 15**   Let $Z$ be a subsemigroup of a semigroup $S$. Then,

(i)   If $Z$ is left group-like in $S$, then $K[Z]$ is a direct summand of the left $K[S]$-module $K[S]$.

(ii)   If the elements of $S$ are not zero divisors in $K[S]$, then the converse holds.

*Proof.*   (i) Define $V$ as the $K$-subspace of $K[S]$ spanned by $S \setminus Z$. Since for any $v \in V$, $z \in Z$, we have $\pi_Z(zv) = z\pi_Z(v) = 0$, then $zv \in V$. Thus, $V$ is a left $K[Z]$-submodule of $K[S]$, and so $K[S] = V \oplus K[Z]$ as $K[S]$-modules.

(ii) Let $K[S] = V \oplus K[Z]$ for a left $K[Z]$-module $V$. Assume that $s \in S$ and that $xs = y$ for some $x$, $y \in Z$. By the hypothesis, $s$ may be written as $s = a + b$ for some $a \in V$, $b \in K[Z]$. Then $xa + xb = xs = y \in K[Z]$ and, consequently, $xa = y - xb \in K[Z] \cap V = 0$. Thus, $xa = 0$ and, by the hypothesis on $S$, $a = 0$. Hence, $s = b \in K[Z]$, and so $s \in Z$, which proves that $Z$ is a left group-like subsemigroup in $S$.

**Corollary 16**   Let $Z$ be a left group-like subsemigroup of $S$. Then, for every subalgebra $R$ of $K[S]$, we have $\mathcal{J}(R) \cap K[Z] \subseteq \mathcal{J}(R \cap K[Z])$. In particular, $\mathcal{J}(K[S]) \cap K[Z] \subseteq \mathcal{J}(K[Z])$.

*Proof.*   Let $a \in \mathcal{J}(R) \cap K[Z]$. Then $ab = ba = a + b$ for some $b \in \mathcal{J}(R)$, and $1 - a$ is the inverse of $1 - b$ in $K[S^1]$. Moreover, by Lemma 14, $a + \pi_Z(b) = \pi_Z(a + b) = \pi_Z(ab) = a\pi_Z(b)$. Therefore, $(1 - a)(1 - \pi_Z(b)) = 1$, which implies that $b = \pi_Z(b) \in \mathcal{J}(R) \cap K[Z]$. It follows that $\mathcal{J}(\mathcal{J}(R) \cap K[Z]) = \mathcal{J}(R) \cap K[Z]$. Since the latter is an ideal in $R \cap K[Z]$, this establishes the assertion.

It is a "folklore" result that if $A$ is a subalgebra of a $K$-algebra $B$, and $A$ is a direct summand of $B$ as a left $A$-module, then, for any right ideal $I$ of $A$ that is a subspace, we have $IB^1 \cap A = I$. In fact, the inclusion $I \subseteq IB^1 \cap A$ is clear. Assume that $a = \sum_{j=1}^{n} i_j b_j \in A$ for some $n \geq 1$, $i_j \in I$, $b_j \in B^1$. If $B = A \oplus C$ for an $A$-submodule $C$ of $B$, then every $b_j$ may be written as $a_j + c_j$, where $a_j \in A^1$, $c_j \in C$. Now $\sum_j i_j a_j + \sum_j i_j c_j = a \in A$, $\sum_j i_j c_j \in C$, and $\sum_j i_j a_j \in I$ because $I$ is a right algebra ideal of $A$. Therefore, $\sum_j i_j c_j \in A \cap C = 0$, and so $a = \sum_j i_j a_j \in I$, which proves the converse inclusion.

This allows us to derive the following important consequence of Lemma 15.

**Corollary 17**  Let $Z$ be a left group-like subsemigroup in $S$. Then, for any right ideal $I$ of $K[Z]$ that is a $K$-subspace, we have $IK[S^1] \cap K[Z] = I$. Moreover, the rule $\alpha(I) = IK[S^1]$ defines a one-to-one order-preserving additive mapping $\alpha : \mathcal{R}^K(K[Z]) \to \mathcal{R}^K(K[S])$ between the lattices of the $K$-algebra right ideals of $K[Z]$ and $K[S]$.

In some special cases with a given subsemigroup $T$ of $S$, one may associate a well-described group-like subsemigroup $Z$ of $S$ determined by $T$.

**Lemma 18**  Let $T$ be a subsemigroup of $S$ such that $tT \subseteq Tt$ for any $t \in T$. Then the set $T_l = \{x \in S \mid tx \in T$ for some $t \in T\}$ is a left group-like subsemigroup of $S$. Moreover, in this case, $T_l$ is the smallest left group-like subset of $S$ containing $T$.

*Proof.*  Let $zs \in T_l$ for some $z \in T_l$, $s \in S$. Then there exist $t_1, t_2 \in T$ such that $t_1 z s, t_2 z \in T$. Now, by the hypothesis on $T$, $t_2 t_1 = t_1' t_2$ for some $t_1' \in T$. Then $t_1' t_2 z s = t_2(t_1 z s) \in T$ and, since $t_1' t_2 z \in T$, this shows that $s \in T_l$. Consequently, $T_l$ is a left group-like subset in $S$.

If $y \in T_l$, then there exist $t_3, t_4 \in T$ such that $t_3 y \in T$ and $t_3(t_2 z) = t_4 t_3$. Therefore, $t_3(t_2 z) y = t_4 t_3 y \in T$ which shows, in view of $t_3 t_2 \in T$, that $zy \in T_l$. Thus, $T_l$ is a left group-like subsemigroup of $S$. From the definition, it is clear that any left group-like subset $Z$ of $S$ containing $T$ must satisfy $T_l \subseteq Z$.

It is straightforward that the intersection of a (nonempty) family of left group-like subsets of $S$ is a left group-like subset of $S$. Thus, for any subsemigroup $T$ of $S$, there exists the smallest group-like subsemigroup of $S$ containing $T$. However, in general, it is hard to find a useful description of this object.

We list the most interesting special cases of Lemma 18.

**Corollary 19**  Let $T$ be a subsemigroup of $S$. Then the set $T_l$, defined in Lemma 18, is a left group-like subsemigroup of $S$ in either of the following cases:

(i)   $T$ is a subgroup of $S$.

(ii)  $T$ is a commutative semigroup.

Similarly, one shows that the set $T_r = \{x \in S \,|\, xt \in T \text{ for some } t \in T\}$ is the smallest right group-like subsemigroup of $S$ containing $T$, provided that $tT \supseteq Tt$ for any $t \in T$. Since, in general, $T_r \neq T_l$ (for example, consider a nonzero semigroup $S$ satisfying the identity $xy = y$, and $T = \{s\}$ for some $s \in S$), one cannot expect to construct (left and right) group-like subsemigroups this way. However, the following symmetric construction turns out to be useful because it allows us to prove an analog of Lemma 14.

**Lemma 20**   Let $T$ be a subsemigroup of $S$ such that $tT = Tt$ for any $t \in T$. Define $T_d = \{x \in S \,|\, sxt \in T \text{ for some } s,t \in T\}$. Then $T_r \subseteq T_d$, $T_l \subseteq T_d$ and $\pi_{T_d}(ab) = \pi_{T_d}(a)b$, $\pi_{T_d}(ba) = b\pi_{T_d}(a)$ for any $a \in K[S]$, $b \in K[T]$.

*Proof.*   Let $y \in T$. Assume that $xy \in T_d$ for some $x \in S$. Then there exist $s$, $t \in T$ such that $sxyt \in T$. Since $yt \in T$, this shows that $x \in T_d$. Conversely, let $x \in T_d$. Then $sxt \in T$ for some $s$, $t \in T$. Since, by the hypothesis, there exists $t' \in T$ such that $ty = yt'$, then $sxyt' = (sxt)y \in T$, and so $xy \in T_d$.

It follows that $\pi_{T_d}(ab) = \pi_{T_d}(a)b$. The second equality may be derived through a symmetric argument. It is clear that $T_r$, $T_l \subseteq T_d$.

It is sometimes convenient to use an analog of group-like subsemigroups, which can work for the contracted semigroup algebras as well. For this we say that a subsemigroup $T \subseteq S$ is a left 0-group-like subsemigroup in $S$ if, for any $z \in T$, $s \in S$, such that $\theta \neq zs \in T$ we have $s \in T$. Then, repeating the above reasoning, we derive the following result.

**Lemma 21**   Assume that $T \subseteq S$ is a left 0-group-like subsemigroup in $S$. Then $K_0[T]$ is a direct summand of the left $K_0[T]$-module $K_0[S]$, and $IK_0[S^1] \cap K_0[T] = I$ for any right ideal $I$ of $K_0[T]$ that is a $K$-subspace.

The above observation is motivated by the following important example.

**Example 22**   Let $S = \mathfrak{M}^0(G,I,M,P)$ be a semigroup of matrix type. If $z = (g,i,m)$, $s = (h,j,n) \in S$ are such that $zs \neq \theta$, then $zs \in S_{(i)}^{(n)} \setminus \{\theta\}$, and $zs \notin S_{(i')}^{(n')}$ for any $i' \in I$, $n' \in M$, such that $(i',n') \neq (i,n)$; see Chapter 1. Therefore, for any subsets $J \subseteq I$, $N \subseteq M$, the semigroup $S_{(J)}^{(N)} = \{(g,i,m) \in S \,|\, i \in J, m \in N\}$ is left 0-group-like in $S_{(J)} = \{(h,j,n) \in S \,|\, j \in J\}$.

## Comments on Chapter 4

An extensive study of group algebras of not necessarily finite groups started after the first book of Passman [201] was published. His later monograph [203] covered most of the material available at that time. Somewhat earlier, two more textbooks on group rings by Mikhalev and Zalesskii [153a] and Bovdi [16a] appeared in Russian. Some special aspects of the commutative group rings and certain specific topics in group rings were presented in the books of Karpilovsky [107] and Sehgal [243]. Then, a theory of the commutative semigroup rings was developed in Gilmer's monograph [64]. We note that some basic facts on the algebras $K[S]$, $K_0[S]$ for finite semigroups $S$ are presented in [26], §.5.2. An extensive bibliography on various aspects of semigroup rings was given in Ponizovskii's survey paper [221]. We also mention the survey of Munn on semigroup algebras of inverse semigroups [162], and Okniński's survey on radicals of group and semigroup rings [185]. An account of the interplay of the ring and semigroup techniques in ring theory was given by Petrich in [208].

Most of the material in this chapter generalizes some useful facts and methods in group algebras and semigroup algebras of commutative semigroups. Here, and throughout the rest of the text, our main general references are [203] and [64].

We note that the notion of a left group-like subset coincides with that of "a left unitary subset," introduced by Dubreil in [53] and adopted by semigroup theorists. We use the former name because it was originally used in the context of group algebras by Schneider and Weissglass [242]. The essence of this notion (our Lemma 15) was pointed out by Krempa and Sierpińska [129].

# 5

# Munn Algebras

In this chapter, we describe an important class of semigroup algebras arising from completely 0-simple semigroups. They are crucial for investigating "local" properties of arbitrary semigroup algebras. In particular, this is one of the main tools in Part III.

Let $R$ be an associative $K$-algebra. Let $I$, $M$ be nonempty sets and $P = (p_{mi})_{m \in M, i \in I}$ a generalized $M \times I$ matrix with $p_{mi} \in R$. Consider the set $\mathfrak{M}(R, I, M, P)$ of all generalized $I \times M$ matrices over $R$ with finitely many nonzero entries. For any $A = (a_{im})$, $B = (b_{im}) \in \mathfrak{M}(R, I, M, P)$, addition and multiplication are defined as follows:

$$A + B = (c_{im}) \quad \text{where } c_{im} = a_{im} + b_{im} \text{ for } i \in I, m \in M$$
$$AB = A \circ P \circ B \quad \text{where } \circ \text{ stands for the usual product of matrices}$$
$$\lambda A = (\lambda a_{im}) \quad \text{for } \lambda \in K$$

$\mathfrak{M}(R,I,M,P)$, subject to these operations, becomes an associative $K$-algebra, called an algebra of matrix type over $R$. The crucial example and motivation comes from the following observation.

**Lemma 1** Let $G$ be a group, with $I$, $M$ nonempty sets and $P$ an $M \times I$ matrix over $G^0$. Then the contracted semigroup algebra $K_0[\mathfrak{M}^0(G,I,M,P)]$ of the semigroup of matrix type $\mathfrak{M}^0(G,I,M,P)$ is isomorphic to the algebra of matrix type $\mathfrak{M}(K[G],I,M,P)$ over the group algebra $K[G]$.

*Proof.* Define a mapping $\varphi : \mathfrak{M}^0(G,I,M,P) \to \mathfrak{M}(K[G],I,M,P)$ by $\varphi((g,i,m)) = (a_{jn})$, where $a_{jn} = g$ if $j = i$, $n = m$ and, otherwise $a_{jn} = 0$. From the definitions, it follows that $\varphi$ is a homomorphism into the multiplicative semigroup of the algebra $\mathfrak{M}(K[G],I,M,P)$. Since the images of the nonzero elements of $\mathfrak{M}^0(G,I,M,P)$ are linearly independent, then it is clear that the extension of $\varphi$ to a homomorphism of $K$-algebras $K_0[\mathfrak{M}^0(G,I,M,P)] \to \mathfrak{M}(K[G],I,M,P)$ is an isomorphism.

We will repeatedly use the above observation by identifying the contracted semigroup algebra of a semigroup of matrix type with an appropriate algebra of matrix type.

For brevity, an algebra of matrix type $\mathfrak{M}(R,I,M,P)$ will usually be denoted by $\hat{R}$. Further, we introduce the following notation, motivated by that of Chapter 1 and Lemma 1.

A matrix $(a_{im}) \in \hat{R}$ such that $a_{jn} = r$, $a_{im} = 0$ for $i \neq j$, $m \neq n$, is denoted by $(r,j,n)$. For any $i \in I$, $m \in M$, we put $\hat{R}_{(i)}^{(m)} = \{(r,i,m) \mid r \in R\}$. If $N \subseteq M$, $J \subseteq I$ are nonempty subsets, then $\hat{R}_{(J)}^{(N)} = \sum_{i \in J} \sum_{m \in N} \hat{R}_{(i)}^{(m)}$. Thus, $\hat{R}_{(J)}^{(N)} = \mathfrak{M}(R,J,N,P_{NJ})$, where $P_{NJ}$ is the $N \times J$ submatrix of $P$ defined in Chapter 1. If $|I| = k < \infty$, then we may assume that $I = \{1,2,\ldots,k\}$ and write $\mathfrak{M}(R,k,M,P)$ for $\hat{R}$. Similarly, $\mathfrak{M}(R,I,t,P)$ may be used if $|M| = t < \infty$. For any subset $Z$ of $\hat{R}$, and any $i \in I$, $m \in M$, we define $Z^{(m)} = Z \cap \hat{R}^{(m)}$, $Z_{(i)} = Z \cap \hat{R}_{(i)}$, $Z_{(i)}^{(m)} = Z_{(i)} \cap Z^{(m)}$, and $\tilde{Z}_{(i)}^{(m)} = \{r \in R \mid (r,i,m) \in Z_{(i)}^{(m)}\}$. Let $\varphi : R \to R'$ be a homomorphism of algebras. Then $\varphi(P)$ stands for the $M \times I$ matrix $(\varphi(p_{mi}))$. Further, $\hat{\varphi} : \hat{R} \to \hat{R}' = \mathfrak{M}(R',I,M,\varphi(P))$ denotes the induced homomorphism, that is, $\varphi(X) = (\varphi(x_{im}))$, where $X = (x_{im}) \in \hat{R}$. If $A$ is a subset of $R$, then, for any sets $J$, $N$, an $J \times N$ matrix $X$ over $R$ is said to lie over $A$ if all entries of $X$ are in $A$. We put $\mathfrak{M}(A,I,M,P) = \{X \in \hat{R} \mid X$ lies over $A\}$, and we write $\hat{A}$ for $\mathfrak{M}(A,I,M,P)$, if unambiguous. Thus, it is clear that $\ker(\hat{\varphi}) = \mathfrak{M}(\ker(\varphi),I,M,P)$ and $\hat{\varphi}(\hat{R}) = \mathfrak{M}(\varphi(R),I,M,\varphi(P))$.

Let $\hat{R} = \mathfrak{M}(R,I,M,P)$ be an algebra of matrix type. Define a semigroup of matrix type $S = \mathfrak{M}^0(\{1\},I,M,P')$, where $P' = (p'_{mi})$ is the $M \times I$ matrix given by

$$p'_{mi} = \begin{cases} 1 & \text{if } p_{mi} \neq 0 \\ \theta & \text{if } p_{mi} = 0 \end{cases}$$

For $s = (1,i,m) \in S$, put $\hat{R}_s = \hat{R}_{(i)}^{(m)}$, and let $\hat{R}_\theta = 0$ if $\theta \in S$ (that is, if $p_{mi} = 0$ for some $m \in M$, $i \in I$). Then $\hat{R}$ is a direct sum of its additive subgroups $\hat{R}_t$, $t \in S$. Moreover, for $t = (1,j,n) \in S$, we have

$$\hat{R}_s \hat{R}_t = \hat{R}_{(i)}^{(m)} \hat{R}_{(j)}^{(n)} \subseteq \hat{R}_{st} \text{ if } st \neq \theta \qquad \hat{R}_s \hat{R}_t = 0 \text{ if } st = \theta$$

Thus, while not any algebra of matrix type is a semigroup algebra, $\hat{R}$ may always be regarded as an $S$-graded algebra with homogeneous components $\hat{R}_s$, $s \in S$; see Chapter 6.

Recall that, if $Z \subseteq R$ is a nonempty subset of an algebra $R$, then by $l_R(Z), r_R(Z)$ we mean the left, respectively right, annihilator of $Z$ in $R$. We write $l(Z), r(Z)$ if unambiguous.

**Lemma 2** For any $m \in M$, we have $l(\hat{R}) = l_{\hat{R}}(\hat{R}^{(m)})$. Moreover, if $l(R) = 0$, then $l(\hat{R}) = \{A \in \hat{R} \mid A \circ P = 0\}$.

*Proof.* Assume that $A\hat{R}^{(m)} = 0$ for some $A \in \hat{R}$, $m \in M$. Then $A \circ P \circ \hat{R}^{(m)} = 0$, so that $A \circ P \circ (r,j,m) = 0$ for any $r \in R$, $j \in I$. Hence, $r$ annihilates, on the right, all columns of the matrix $A \circ P$. This implies that $A\hat{R} = A \circ P \circ \hat{R} = 0$. Hence, $l_{\hat{R}}(\hat{R}^{(m)}) \subseteq l(\hat{R})$, and the equality follows.

Assume that $l(R) = 0$, and let $A \in l(\hat{R})$. Since, as above, any element $r \in R$ annihilates on the right any column of the matrix $A \circ P$, then $A \circ P = 0$. Thus, $l(\hat{R}) \subseteq \{A \in R \mid A \circ P = 0\}$, the converse inclusion being obvious.

Let row$(P) \in R^M$ be the left $R$-submodule generated by the rows of the matrix $P$. In other words, row$(P) = \sum_{m \in M} R^I P_{(m)}$. Moreover, for any set $Z$, denote by $M_Z^{\text{row}}(R)$ the algebra of all $Z \times Z$ matrices over $R$ with finitely many nonzero rows subject to the natural addition and multiplication. $M_Z^{\text{col}}(R)$ is defined dually.

**Lemma 3** The rule $\varphi(A) = A \circ P$ defines a homomorphism of $K$-algebras $\varphi : \hat{R} \to M_I^{\text{row}}(R)$ such that

(1) $\varphi(\hat{R})$ is the subalgebra of $M_I^{\text{row}}(R)$ consisting of all matrices the rows of which lie in row$(P)$.

(2) If $l(R) = 0$, then $\ker(\varphi) = l(\hat{R})$.

*Proof.* For any $A, B \in \hat{R}$, $\lambda \in K$, we have

$$\varphi(AB) = \varphi(A \circ P \circ B) = A \circ P \circ B \circ P = \varphi(A) \circ \varphi(B)$$

$$\varphi(A + B) = (A + B) \circ P = (A \circ P) + (B \circ P) = \varphi(A) + \varphi(B)$$
$$\varphi(\lambda A) = (\lambda A) \circ P = \lambda(A \circ P) = \lambda\varphi(A)$$

Properties (1) and (2) follow directly from the definitions and Lemma 2.

If $A \circ P = 0$ for some $0 \neq A = (a_{im}) \in \hat{R}$, then, for any $j \in I$, $A_j \circ P = 0$, where $A_j = (b_{im})$, with $b_{im} = a_{im}$ if $i = j$ and $b_{im} = 0$ if $i \neq j$. It follows that a nontrivial left $R$-combination of rows of $P$ is zero. Thus, as a direct consequence of Lemma 2, we get the following corollary.

**Corollary 4**  Assume that the rows of $P$ are left $R$-independent as elements of the left $R$-module $R^M$. If $l(R) = 0$, then $l_{\hat{R}}(\hat{R}^{(m)}) = 0$ for any $m \in M$.

The $R$-submodule $\mathrm{col}(P)$ of the right $R$-module $R^I$ is defined dually to $\mathrm{row}(P)$; that is, $\mathrm{col}(P)$ is the submodule of $R^I$ generated by the columns of $P$. The following result shows that $l(\hat{R})$ is determined by some specific subsets of $\hat{R}$, which are also essential when representing the algebra $\hat{R}$ modulo $r(\hat{R})$.

**Lemma 5**  Let $\hat{R} = \mathfrak{M}(R,I,M,P)$ be an algebra of matrix type over an algebra $R$, and let $J \subseteq I$ be a subset such that the right $R$-submodules $\mathrm{col}(P) = \sum_{i \in I} P^{(i)} R^1$, $\sum_{i \in J} P^{(j)} R^1$ of $R^I$ coincide. Then, for $\hat{R}^{(M)}_{(J)} = \mathfrak{M}(R,J,M,P_{MJ})$, we have $\hat{R} = r(\hat{R}) + \hat{R}^{(M)}_{(J)}$ and $l(\hat{R}) = l_{\hat{R}}(\hat{R}^{(M)}_{(J)})$.

*Proof.*  Let $(r,i,m) \in \hat{R}$. From the hypothesis it follows that the column $P^{(i)}$ may be written as $\sum_{k=1}^{n} P^{(i_k)} r_k$ for some $n \geq 1, r_1, \ldots, r_n \in R^1, i_1, \ldots, i_n \in J$. Define an element $A \in \hat{R}^{(M)}_{(J)}$ by $A = \sum_{k=1}^{n}(r_k r, i_k, m) - (r,i,m)$. It is easy to see that $P \circ A = 0$, and so $A \in r(\hat{R})$. Since $\sum_{k=1}^{n}(r_k r, i_k, m) \in \hat{R}^{(J)}_{(M)}$, then $(r,i,m) \in r(\hat{R}) + \hat{R}^{(M)}_{(J)}$, proving the first equality. Now, $l(\hat{R}) = l_{\hat{R}}(r(\hat{R}) + \hat{R}^{(M)}_{(J)}) = l_{\hat{R}}(r(\hat{R})) \cap l_{\hat{R}}(\hat{R}^{(M)}_{(J)}) = \hat{R} \cap l_{\hat{R}}(\hat{R}^{(M)}_{(J)}) = l_{\hat{R}}(\hat{R}^{(M)}_{(J)})$.

**Remark 6**  Clearly, the right-left symmetric analog of Lemmas 2, 3, and 5 and Corollary 4 may be proved. For this, one has to interchange the roles of the left and right annihilators, the roles of the mappings $A \rightarrow A \circ P$, $A \rightarrow P \circ A$, and those of the $R$-modules $\mathrm{row}(P)$, $\mathrm{col}(P)$.

We will now be concerned with the relations between the lattices of ideals of $R$ and $\hat{R}$. To omit some pathologies, and having Lemma 1 in mind, we assume throughout this paragraph that $R$ has an identity and that $p_{mi}$ is a unit of $R$ for some $i \in I, m \in M$. Moreover, if every row and every column of $P$ contains a unit of $R$, then $\hat{R}$ will be referred to as a Munn algebra over $R$. Observe that every semigroup algebra $K_0[S] = \widehat{K[G]} = \mathfrak{M}(K[G],I,M,P)$

of a completely 0-simple semigroup $S = \mathfrak{M}^0(G,I,M,P)$, as well as any homomorphic image $\hat{\varphi}(\widehat{K[G]})$ determined by a homomorphism $\varphi$ of the group algebra $K[G]$, are Munn algebras, in view of Lemma 3 in Chapter 1.

Let $J$ be a right ideal of $R$. If $i \in I$, then it is straightforward that $\hat{J}_{(i)}$ and $\hat{J} = \sum_{i \in J} J_{(i)}$ are right ideals of $\hat{R}$. In fact, the mappings $J \to \hat{J}_{(i)}$, $J \to \hat{J}$ are complete lattice embeddings of the lattice $\mathcal{R}(R)$ of right ideals of $R$ into $\mathcal{R}(\hat{R})$. Similar embeddings are also given for the corresponding lattices of left and two-sided ideals. However, for investigating the structural connections between $R$ and $\hat{R}$, some other correspondence seems more useful. Recall that a matrix $Q$ lies over a subset $Z$ of $R$ if all entries of $Q$ are in $Z$.

**Lemma 7** Let $J$ be a right ideal of $R$. Then the set $\mathfrak{B}(J) = \{X \in \hat{R} \mid P \circ X \circ P$ lies over $J\}$ is a right ideal of $\hat{R}$. Moreover, the mapping $\mathfrak{B} : \mathcal{R}(R) \to \mathcal{R}(\hat{R})$ is a $\Lambda$-complete semilattice homomorphism. If, additionally, $J$ is a two-sided ideal of $R$, then $\mathfrak{B}(J) \supseteq \hat{J} = \mathfrak{M}(J,I,M,P)$ and $\hat{R}\mathfrak{B}(J)\hat{R} \subseteq \hat{J}$.

*Proof.* Let $X \in \mathfrak{B}(\hat{J})$, $Y \in \hat{R}$. Then $P \circ (XY) \circ P = P \circ X \circ P \circ Y \circ P$ and, since $P \circ X \circ P$ lies over $J$, $P \circ (XY) \circ P$ lies over $J$, too. It is clear that $\mathfrak{B}(J)$ is a subgroup of the additive group of $\hat{R}$, and so $\mathfrak{B}(J)$ is a right ideal of $\hat{R}$. If $J$ is a two-sided ideal of $R$, then, for any $Z \in \hat{R}$, $ZXY = Z \circ P \circ X \circ P \circ Y$ lies over $J$, and so $\mathfrak{B}(J)$ is a two-sided ideal of $\hat{R}$ such that $\hat{R}\mathfrak{B}(J)\hat{R} \subseteq \mathfrak{M}(J,I,M,P) = \hat{J}$. If $X \in \hat{J}$, then $X$ and also $P \circ X \circ P$ lie over $J$, which shows that $\hat{J} \subseteq \mathfrak{B}(J)$.

It remains to prove that $\mathfrak{B}$ is a $\wedge$-complete semilattice homomorphism. Let $J_\alpha$, $\alpha \in A$, be a nonempty set of right ideals of $R$. Then $X \in \mathfrak{B}(\bigcap_{\alpha \in A} J_\alpha)$ if and only if $P \circ X \circ P$ lies over $\bigcap_{\alpha \in A} J_\alpha$, which holds if and only if $P \circ X \circ P$ lies over all $J_\alpha$, $\alpha \in A$. Hence, $\mathfrak{B}(\bigcap_{\alpha \in A} J_\alpha) = \bigcap_{\alpha \in A} \mathfrak{B}(J_\alpha)$.

The (right) ideal defined for $J$ in Lemma 7, denoted by $\mathfrak{B}(J)$, or $\mathfrak{B}_{\hat{R}}(J)$ if ambiguous, is called the basic (right) ideal of $\hat{R}$ determined by $J$. The ideal $\mathfrak{B}(0)$ may be characterized through some annihilators arising from $\hat{R}$.

**Lemma 8** $\hat{R}/\mathfrak{B}(0) \cong (\hat{R}/l(\hat{R}))/r(\hat{R}/l(\hat{R})) \cong (\hat{R}/r(\hat{R}))/l(\hat{R}/r(\hat{R}))$.

*Proof.* Let $X \in \hat{R}$. Then $X$ lies in the kernel of the natural homomorphism $\hat{R} \to (\hat{R}/l(\hat{R}))/r(\hat{R}/l(\hat{R}))$ if and only if $\hat{R}X \subseteq l(\hat{R})$. The latter is equivalent to the fact that $\hat{R}X\hat{R} = 0$. Since $R$ has an identity, this happens if and only if $P \circ X \circ P = 0$, that is, $X \in \mathfrak{B}(0)$.

The second isomorphism is established similarly.

By Lemma 7, $\mathfrak{B}(T(R)) \subseteq T(\hat{R})$. To establish a corresponding mapping $T(\hat{R}) \to T(R)$ between the lattices of two-sided ideals of $R$, $\hat{R}$, we need the following observation.

**Lemma 9**  Let $(i,m) \in I \times M$ be such that $p_{mi}$ is a unit of $R$. Then,

(1) $E = (p_{mi}^{-1},i,m)$ is an idempotent of $\hat{R}$ such that $E\hat{R}E \cong R$.
(2) If $N$ is a right ideal of $\hat{R}$, then the set $\tilde{N}_{(i)}^{(m)} = \{r \in R \mid (r,i,m) \in N\}$ is a right ideal of $R$.
(3) If $N$ is a two-sided ideal of $\hat{R}$, then $\tilde{N}_{(i)}^{(m)}$ is a two-sided ideal of $R$, $N_{(j)}^{(n)} \subseteq \hat{R}N_{(i)}^{(m)}\hat{R}$ for any $j \in I$, $n \in M$, and $\tilde{N}_{(j)}^{(n)} \supseteq \tilde{N}_{(i)}^{(m)}$.

*Proof.*   (1) We have $E^2 = (p_{mi}^{-1},i,m)(p_{mi}^{-1},i,m) = (p_{mi}^{-1},i,m) = E$. Further, $\varphi : R \to E\hat{R}E$ defined by $\varphi(r) = (rp_{mi}^{-1},i,m)$ is an algebra homomorphism because

$$\varphi(r_1r_2) = (r_1r_2p_{mi}^{-1},i,m) = (r_1p_{mi}^{-1},i,m)(r_2p_{mi}^{-1},i,m) = \varphi(r_1)\varphi(r_2)$$

Clearly, $\varphi$ maps $R$ onto $E\hat{R}E = \hat{R}_{(i)}^{(m)}$ and is one-to-one.

(2) Observe that $\tilde{N}_{(i)}^{(m)}p_{mi} = \varphi^{-1}(N_{(i)}^{(m)})$. Since $N_{(i)}^{(m)} = N \cap \hat{R}_{(i)}^{(m)}$ is a right ideal of $\hat{R}_{(i)}^{(m)}$, then $\tilde{N}_{(i)}^{(m)}p_{mi}$ is a right ideal of $R$. Thus, $\tilde{N}_{(i)}^{(m)}$ is a right ideal of $R$ because $p_{mi}$ is a unit in $R$.

(3) A similar argument shows that $\tilde{N}_{(i)}^{(m)}$ is a two-sided ideal of $R$. Now, for $r \in \tilde{N}_{(i)}^{(m)}$, we have $(r,j,n) = (p_{mi}^{-1},j,m)(r,i,m)(p_{mi}^{-1},i,n) \in N$ because $(r,i,m) \in N$. Hence, (3) follows.

Let $(i,m)$, $(j,n) \in I \times M$ be such that $p_{mi}$, $p_{nj}$ are units in $R$. From Lemma 9 it follows that for any ideal $N$ of $\hat{R}$ we have $\tilde{N}_{(i)}^{(m)} = \tilde{N}_{(j)}^{(n)}$. This allows us to associate with any ideal $N$ of $\hat{R}$ an ideal $\mathfrak{I}(N)$ of $R$ defined as $\tilde{N}_{(i)}^{(m)}$, where $i \in I$, $m \in M$ are arbitrary such that $p_{mi}$ is a unit of $R$. This ideal is called the ideal of $R$ induced by $N$. From the proof of Lemma 9, it follows that $\mathfrak{M}(\mathfrak{I}(N),I,M,P) \subseteq N$.

Let $\hat{R} = \mathfrak{M}(R,I,M,P)$ be a Munn algebra over $R$; that is, any row and any column of $P$ contains a unit of $R$. If $r \in R$, $i,j \in I$, $m,n \in M$, then there exist $k \in I$, $l \in M$, such that $p_{li},p_{mk}$ are units in $R$. Therefore, $(r,j,n) = (p_{li}^{-1},j,l)(r,i,m)(p_{mk}^{-1},k,n)$. Thus, the reasoning of Lemma 9 allows us to improve the definition of an induced ideal of $R$ as follows.

**Corollary 10**  Let $N$ be an ideal of a Munn algebra $\hat{R}$. Then, for any $i \in I$, $m \in M$, we have $\mathfrak{I}(N) = \tilde{N}_{(i)}^{(m)} = \{r \in R \mid (r,i,m) \in N\}$.

**Lemma 11** Let $J$ be an ideal of $R$, and let $N$ be an ideal of an algebra $\hat{R} = \mathfrak{M}(R,I,M,P)$ of matrix type over $R$. Then,

(1) $N \subseteq \mathfrak{B}(\mathfrak{I}(N))$.

(2) $J \subseteq \mathfrak{I}(\mathfrak{B}(J))$.

(3) $\mathfrak{I}(N) = 0$ if and only if $N \subseteq \mathfrak{B}(0)$.

*Proof.* (1) Let $X \in N$. Let $i,j \in I$, $m,n \in M$ be such that $p_{mi}$ is a unit in $R$. Then $(1,i,n) \circ P \circ X \circ P \circ (1,j,m) = (1,i,n)X(1,j,m) \in N$, and it is of the form $(r,i,m)$, where $r$ is the $(n,j)$th entry of the matrix $P \circ X \circ P$. From the definition of $\mathfrak{I}(N)$, it follows that $r \in \mathfrak{I}(N)$ and, since $j \in I$, $n \in M$ are arbitrary, then $P \circ X \circ P$ lies over $\mathfrak{I}(N)$. Thus, $X \in B(\mathfrak{I}(J))$, and (1) follows.

(2) From Lemma 7 we know that $\mathfrak{B}(J) \supseteq \mathfrak{M}(J,I,M,P)$. Since $\mathfrak{I}(\mathfrak{M}(J,I,M,P)) = J$ and $\mathfrak{I}$ is an order-preserving mapping of $T(\hat{R})$ into $T(R)$, then $\mathfrak{I}(\mathfrak{B}(J)) \supseteq J$.

(3) If $\mathfrak{I}(N) = 0$, then by (1) $N \subseteq \mathfrak{B}(0)$. Assume that $N \subseteq \mathfrak{B}(0)$. Then, $\hat{R}N\hat{R} = \hat{R} \circ (P \circ N \circ P) \circ \hat{R} = 0$. Since $N_{(i)}^{(m)} \subseteq \hat{R}N\hat{R}$ for any $i \in I$, $m \in M$, by Lemma 9, then $\mathfrak{I}(N) = 0$.

We now summarize the basic facts on the mappings $\mathfrak{B}, \mathfrak{I}$.

**Theorem 12** Let $\hat{R} = \mathfrak{M}(R,I,M,P)$ be an algebra of matrix type. Then,

(1) $\mathfrak{I}$ is a complete lattice homomorphism of $T(\hat{R})$ onto $T(R)$.

(2) $\mathfrak{B}$ is a $\wedge$-complete semilattice embedding of $T(R)$ into $T(\hat{R})$.

(3) $\mathfrak{I}\mathfrak{B}$ is the identity mapping on $T(R)$.

(4) $\mathfrak{B}\mathfrak{I}$ is the identity mapping on $\mathfrak{B}(T(R))$.

(5) For any $J \in T(R)$, $\mathfrak{B}(J)$ is maximal among all ideals $N$ of $\hat{R}$ with the property $\mathfrak{I}(N) = J$.

*Proof.* (1) Let $N_\alpha$, $\alpha \in A$, be a set of ideals of $\hat{R}$, and let $p_{mi}$, $i \in I$, $m \in M$, be a unit in $R$. Then $r \in \mathfrak{I}(\bigcap_{\alpha \in A} N_\alpha)$ if and only if $(r,i,m) \in \bigcap_{\alpha \in A} N_\alpha$, the latter being equivalent to the fact that $r \in \bigcap_{\alpha \in A} \mathfrak{I}(N_\alpha)$. Hence, $\mathfrak{I}(\bigcap_{\alpha \in A} N_\alpha) = \bigcap_{\alpha \in A} \mathfrak{I}(N_\alpha)$.

Since $\mathfrak{I}(\sum_{\alpha \in A} N_\alpha)$ is an ideal of $R$ that contains $\mathfrak{I}(N_\alpha)$, $\alpha \in A$, then it is clear that $\sum_{\alpha \in A} \mathfrak{I}(N_\alpha) \subseteq \mathfrak{I}(\sum_{\alpha \in A} N_\alpha)$. Let $r \in \mathfrak{I}(\sum_{\alpha \in A} N_\alpha)$. Then $(r,i,m) = X_1 + \cdots + X_t$ for some $t \geq 1$, and some $X_k \in N_{\alpha_k}$, $\alpha_k \in A$, $k = 1$, $\ldots$, $t$. Now,

$$(r,i,m) = (p_{mi}^{-1},i,m)(r,i,m)(p_{mi}^{-1},i,m) = \sum_{k=1}^{t}(p_{mi}^{-1},i,m)X_k(p_{mi}^{-1},i,m).$$

Since any $(p_{mi}^{-1}, i, m) X_k (p_{mi}^{-1}, i, m) \in N_{\alpha_k} \cap \hat{R}_{(i)}^{(m)}$, then it must be of the form $(r_k, i, m)$ for some $r_k \in \mathfrak{I}(N_{\alpha_k})$. Thus, $r = \sum_{k=1}^{t} r_k \in \sum_{\alpha \in A} \mathfrak{I}(N_\alpha)$. Therefore, $\mathfrak{I}(\sum_{\alpha \in A} N_\alpha) \subseteq \sum_{\alpha \in A} \mathfrak{I}(N_\alpha)$, and the equality follows.

That $\mathfrak{I}$ maps $\mathcal{T}(\hat{R})$ onto $\mathcal{T}(R)$ will be established once (3) is proved.

(2) $\mathfrak{B}$ is a $\wedge$-complete semilattice homomorphism by Lemma 7, and so (2) will follow from (3).

(3) Let $J$ be an ideal of $R$. Consider the natural homomorphism $\varphi : R \to R/J$ and the induced homomorphism $\hat{\varphi} : \hat{R} \to \hat{R}' = \mathfrak{M}(R/J, I, M, \varphi(P))$. Then $\hat{\varphi}(\mathfrak{B}(J)) = \{\hat{\varphi}(X) \mid X \in \hat{R}, \ P \circ X \circ P \text{ is over } J\} = \{\varphi(X) \mid X \in \hat{R}, \ \hat{\varphi}(P \circ X \circ P) = 0\} \subseteq \{Y \in \hat{R}' \mid \varphi(P) \circ Y \circ \varphi(P) = 0\} = \mathfrak{B}_{\hat{R}}'(0)$. It is clear that $\varphi(\mathfrak{I}(N)) = \mathfrak{I}(\hat{\varphi}(N))$ for any ideal $N$ of $\hat{R}$. Therefore, $\varphi(\mathfrak{I}(\mathfrak{B}(J))) = \mathfrak{I}(\hat{\varphi}(\mathfrak{B}(J))) \subseteq \mathfrak{I}(\mathfrak{B}_{\hat{R}}'(0))$. From Lemma 11 it follows that $\mathfrak{I}(\mathfrak{B}_{\hat{R}}'(0)) = 0$, so that $\varphi(\mathfrak{I}(\mathfrak{B}(J))) = 0$. This means that $\mathfrak{I}(\mathfrak{B}(J)) \subseteq J$. Hence, $\mathfrak{I}(\mathfrak{B}(J)) = J$ by Lemma 11, which establishes (3).

(4) by (3) $\mathfrak{B}\mathfrak{I}\mathfrak{B} = \mathfrak{B}$, so that (4) follows.

(5) From (3) we know that $\mathfrak{I}(\mathfrak{B}(J)) = J$. Assume that $N \supseteq \mathfrak{B}(J)$ is an ideal of $\hat{R}$ satisfying $\mathfrak{I}(N) = J$. Then $\mathfrak{B}(\mathfrak{I}(N)) = \mathfrak{B}(J)$ and, since $N \subseteq \mathfrak{B}(\mathfrak{I}(N))$ by Lemma 11, we get $N \subseteq \mathfrak{B}(J)$. Hence, $N = \mathfrak{B}(J)$ as desired.

The mappings $\mathfrak{I}$, $\mathfrak{B}$ establish a one-to-one correspondence between some important classes of ideals of $R$ and $\hat{R}$. This is a consequence of the following observation.

**Lemma 13** Let $N$ be a semiprime ideal of $\hat{R}$. Then there exists an ideal $J$ of $R$ such that $\mathfrak{B}(J) = N$.

*Proof.* From Lemma 7 we know that $\mathfrak{B}(\mathfrak{I}(N))^3 \subseteq \mathfrak{M}(\mathfrak{I}(N), I, M, P) \subseteq N$. Hence, the hypothesis on $N$ implies that $\mathfrak{B}(\mathfrak{I}(N)) \subseteq N$. From Lemma 11, it then follows that $\mathfrak{B}(\mathfrak{I}(N)) = N$.

**Proposition 14** Let $\hat{R}$ be an algebra of matrix type over $R$. Then the mappings $\mathfrak{I}$, $\mathfrak{B}$ establish a one-to-one correspondence between the sets of maximal, prime, and semiprime ideals of $\hat{R}$ and $R$.

*Proof.* In view of Theorem 12 and Lemma 13, it is enough to show that, for an ideal $J$ of $R$ and an ideal $N$ of $\hat{R}$ such that $N = \mathfrak{B}(J)$, $J = \mathfrak{I}(N)$, $J$ is a maximal (prime, semiprime) ideal of $R$ if and only if $N$ is a maximal (prime, semiprime, respectively) ideal of $\hat{R}$. The case of maximal ideals is a direct consequence of assertions (3) and (4) of Theorem 12 and the fact that $\mathfrak{I}$, $\mathfrak{B}$ are order-preserving mappings.

Assume that $J$ is a prime ideal of $R$ and that $N_1$, $N_2$ are ideals of $R$ such that $N \subseteq N_1$, $N_2$ and $N_1N_2 \subseteq N$. If $r_1 \in \mathfrak{I}(N_1)$, $r_2 \in \mathfrak{I}(N_2)$, then $(r_1,i,m) \in N_1$, $(r_2,i,m) \in N_2$, where $i \in I$, $m \in M$, are such that $p_{mi}$ is a unit in $R$. Thus, $(r_1,i,m)(p_{mi}^{-1}r_2,i,m) = (r_1r_2,i,m) \in N_1N_2$ because $p_{mi}^{-1}r_2 \in \mathfrak{I}(N_2)$. Hence, $r_1r_2 \in \mathfrak{I}(N_1N_2)$. It follows that $\mathfrak{I}(N_1)\mathfrak{I}(N_2) \subseteq \mathfrak{I}(N_1N_2)$. Now, $\mathfrak{I}(N_1N_2) \subseteq \mathfrak{I}(N)$, and so the primeness of $J$ implies that $\mathfrak{I}(N_k) \subseteq \mathfrak{I}(N) = J$ for some $k = 1, 2$. Then, by Lemma 11 and the assumption on $J$, $N$, we have $N_k \subseteq \mathfrak{B}(\mathfrak{I}(N_k)) \subseteq \mathfrak{B}(\mathfrak{I}(N)) = \mathfrak{B}(J) = N$. Hence, $N = N_k$, proving that $N$ is a prime ideal of $\hat{R}$. The same argument with $N_1 = N_2$ shows that $N$ is semiprime whenever $J$ is also.

Now, assume that $J_1, J_2$ are ideals of $R$ such that $J \subseteq J_1, J_2$ and $J_1J_2 \subseteq J$. Then, for $X_1, Y_1 \in \mathfrak{B}(J_1)$, $X_2, Y_2 \in \mathfrak{B}(J_2)$, the matrix $P \circ (X_1X_2Y_1Y_2) \circ P = P \circ X_1 \circ P \circ X_2 \circ P \circ Y_1 \circ P \circ Y_2 \circ P$ lies over $J_1J_2$ because $P \circ X_1 \circ P$ lies over $J_1$ and $P \circ Y_2 \circ P$ lies over $J_2$. Thus, $(\mathfrak{B}(J_1)\mathfrak{B}(J_2))^2 \subseteq \mathfrak{B}(J_1J_2) \subseteq \mathfrak{B}(J) = N$. If $N$ is prime, then $\mathfrak{B}(J_k) \subseteq \mathfrak{B}(J)$ for some $k = 1,2$, and, hence, by Theorem 12, $J_k = \mathfrak{I}(\mathfrak{B}(J_k)) \subseteq \mathfrak{I}(\mathfrak{B}(J)) = \mathfrak{I}(N) = J$. This shows that $J$ is prime. If $N$ is semiprime, then, with $J_1 = J_2$, we get $\mathfrak{B}(J_1)^4 \subseteq N$. Thus, $\mathfrak{B}(J_1) \subseteq N$ and, as above, $J$ must be semiprime.

As a consequence, we derive a description of the prime radical $\mathcal{B}(\hat{R})$ of $\hat{R}$.

**Corollary 15**   $\mathcal{B}(\hat{R}) = \mathfrak{B}(\mathcal{B}(R)) = \{X \in \hat{R} \mid P \circ X \circ P \text{ lies over } \mathcal{B}(R)\}$

*Proof.*   This follows from Proposition 14 and the fact that $\mathcal{B}(R')$ is the least semiprime ideal of $R'$ for any algebra $R'$.

We will briefly describe a connection between the classes of modules over $\hat{R}$ and $R$, which exploits the mapping $\mathfrak{B}$ considered above.

Let $V$ be a right $R$-module. Then there is a natural induced $M_M^{col}(R)$-module structure on $V^M$. Thus, $V^M$ may be regarded as an $\hat{R}$-module through the homomorphism of algebras $\varphi: \hat{R} \to M_M^{col}(R)$ defined by $\varphi(X) = P \circ X$. This module will be denoted by $V^M(P)$. In other words, the action of $\hat{R}$ on $V^M(P)$ is given by $vX = v \circ P \circ X$ for $v \in V^M(P)$, $X \in \hat{R}$.

Let $V_0^M(P) = \{v \in V^M(P) \mid v\hat{R} = 0\}$. Then $V_0^M(P)$ is an $\hat{R}$-submodule of $V^M(P)$ and $V_0^M(P) = \{v \in V^M(P) \mid v \circ P \circ X = 0 \text{ for all } X \in \hat{R}\} = \{v \in V^M(P) \mid v \circ P = 0\}$ because $l_R(R) = 0$.

**Lemma 16**   Let $V$ be a right $R$-module. Then,

(1)   $\text{ann}_{\hat{R}}(V^M(P)/V_0^M(P)) = \mathfrak{B}(\text{ann}_R(V)) = \{X \in \hat{R} \mid P \circ X \circ P \text{ lies over } \text{ann}_R V\}$.

(2)   If $V$ is an irreducible $R$-module, then $V^M(P)/V_0^M(P)$ is an irreducible $\hat{R}$-module.

(3)   The mapping $W \to W^M(P)$ is an embedding of the lattice of $R$-submodules of $V$ into the lattice of $\hat{R}$-submodules of $V^M(P)$.

*Proof.*   (1) We have

$$
\begin{aligned}
\mathrm{ann}_{\hat{R}}\,(V^M(P)/V_0^M(P)) &= \{X \in \hat{R} \mid V^M(P)X \subseteq V_0^M(P)\} \\
&= \{X \in \hat{R} \mid v \circ P \circ X \in V_0^M(P) \text{ for all } v \in V^M(P)\} \\
&= \{X \in \hat{R} \mid v \circ P \circ X \circ P = 0 \text{ for all } v \in V^M(P)\} \\
&= \{X \in \hat{R} \mid P \circ X \circ P \text{ lies over } \mathrm{ann}_R(V)\} = \mathfrak{B}(\mathrm{ann}_R(V))
\end{aligned}
$$

(2) Let $v = (v_m)_{m \in M} \in V^M(P) \backslash V_0^M(P)$. Then $v \circ P$ is a nonzero element of $V^I$, so there exists $j \in I$ such that the $j$th entry $y$ of $v \circ P$ is nonzero. Let $u = (u_m) \in V^M(P)$. Since $V$ is an irreducible $R$-module, then, for any $m \in M$, there exists $r_m \in R$ such that $yr_m = u_m$. Let $X \in \hat{R}$ be defined as the matrix with zeros everywhere except the $j$th row, where it has $r_m$ at the $(j,m)$th entry, for all $m \in M$. It is easy to see that $u = v \circ P \circ X = vX \in v\hat{R}$. This proves (2).

(3) is straightforward.

If $T$ is a ring with a nonzero idempotent $e$, and if $V$ is a right $T$-module, then $Ve$ is a right $eTe$-module that carries some important properties of $V$. In particular, $Ve$ is irreducible or faithful whenever $V$ is also. In the special case of algebras of the matrix type, we derive the following result.

**Lemma 17**   Let $V$ be a right $\hat{R}/\mathfrak{B}(0)$-module. If $p_{mi}$ is a unit of $R$, for some $m \in M$, $i \in I$, then $E = (p_{mi}^{-1}, i, m)$ is an idempotent of $\hat{R}$ and $VE$ is a right $R$-module. Moreover,

(1)   $VE$ is an irreducible $R$-module whenever $V$ is an irreducible $\hat{R}/\mathfrak{B}(0)$-module.

(2)   $VE$ is a faithful $R$-module whenever $V$ is a faithful $\hat{R}/\mathfrak{B}(0)$-module.

*Proof.*   We may treat $V$ as a right $\hat{R}$-module with $V\mathfrak{B}(0) = 0$. From Lemma 9 we know that $E = E^2$ and $E\hat{R}E \cong R$, so that $VE \subseteq V$ may be regarded as an $R$-module. Thus, (1) is a direct consequence of the foregoing remark. If $\mathrm{ann}_{\hat{R}} V = \mathfrak{B}(0)$, and $VE(EXE) = 0$ for some $X \in \hat{R}$, then $EXE \in \mathfrak{B}(0)$. Thus, $EXE = E \circ (P \circ EXE \circ P) \circ E = 0$, which proves (2).

**Corollary 18**   Let $N$ be an ideal of an algebra of matrix type $\hat{R}$. Then $N$ is right primitive if and only if $N = \mathfrak{B}(J)$ for a right primitive ideal $J$ of $R$ and, in this case, the division rings associated to the right primitive rings $\hat{R}/N$ and $R/J$ are isomorphic. Moreover, $\mathcal{J}(\hat{R}) = \mathfrak{B}(\mathcal{J}(R))$.

*Proof.* From Lemma 13 it follows that any right primitive ideal of $\hat{R}$ must be of the form $\mathfrak{B}(J)$ for an ideal $J$ of $R$. Since the kernel of the homomorphism $\hat{\varphi} : \hat{R} \to \hat{R}' = \mathfrak{M}(R/J, I, M, \varphi(P))$ determined by the natural homomorphism $\varphi : R \to R/J$ lies in $\mathfrak{B}(J)$ by Lemma 7, then, passing to $\hat{R}'$, we may assume that $J = 0$. From Lemmas 16 and 17 it follows that $R$ is right primitive if and only if $\hat{R}/\mathfrak{B}(0) = \hat{R}/\mathfrak{B}(J)$ is right primitive. Further, since $\mathfrak{B}(0)^3 = 0$, then $\mathfrak{B}(0) \cap E\hat{R}E = 0$, where $E$ is chosen as in Lemma 17. Thus, $R \cong E\hat{R}E \cong E\hat{R}E/(E\hat{R}E \cap \mathfrak{B}(0)) \cong E'(\hat{R}/\mathfrak{B}(0))E'$ for an idempotent $E'$ of $\hat{R}/\mathfrak{B}(0)$. It is well known that the commuting division rings of $R$ and $\hat{R}/\mathfrak{B}(0)$ must be isomorphic in this case; see [93]. Finally,

$$\mathcal{J}(\hat{R}) = \cap \{N \mid N \text{ right primitive ideal of } \hat{R}\}$$
$$= \cap \{\mathfrak{B}(J) \mid J \text{ right primitive ideal of } R\}$$

the latter equal to $B(\mathcal{J}(R))$ by Lemma 7.

We now consider a special case that is of interest when we look at representations of Munn algebras. Assume that $\hat{R} = \mathfrak{M}(R, I, M, P)$ is a Munn algebra over a simple artinian algebra $R$. Let $R = M_r(D)$ for some $r \geq 1$ and a division algebra $D$. If $T$, $Z$ are nonempty sets and $X = (x_{tz})_{t \in T, z \in Z}$ is a $T \times Z$ matrix over $R$, then we write $\bar{P}$ for the $(T.r) \times (Z.r)$ matrix obtained from $P$ by erasing the matrix brackets of all entries $x_{tz}$ of $X$. Here $T.r$, $Z.r$ denote the disjoint unions of $r$ copies of sets of cardinality $|T|$, $|Z|$, respectively. Similarly, treating elements of $(D^r)^T$ as $\{1\} \times T$ matrices over $D^r$, an element $\bar{v} \in D^{T.r}$ is associated with any element $v \in (D^r)^T$. With this notation, we have the following result.

**Lemma 19** The algebras $\mathfrak{M}(M_r(D), I, M, P)$, $\mathfrak{M}(D, I.r, M.r, \bar{P})$ are isomorphic.

*Proof.* Let $X$, $Y \in \mathfrak{M}(M_r(D), I, M, P)$. It is easy to see that $\bar{X}\bar{Y} = \bar{X} \circ \bar{P} \circ \bar{Y} = \overline{X \circ P \circ Y} = \overline{XY}$, so that $X \to \bar{X}$ determines the desired isomorphism of algebras.

We define the rank of an $M \times I$ matrix $P$ over $M_r(D)$ and write $\rho(P)$ as $\dim_D \varphi_P((D^r)^M)$, where $\varphi_P$ is the homomorphism of left $D$-spaces $(D^r)^M \to (D^r)^I$ determined by $P$, that is, $\varphi_P(v) = v \circ P$. Note that, if $r = 1$ and $M = I$ are finite, then this notion coincides with the rank used in Chapter 1, so that our notation for the rank is consistent with that used before. Observe that under the natural isomorphisms of $D$-spaces $(D^r)^M \cong D^{M.r}$, $(D^r)^I \cong D^{I.r}$, $\varphi_P$ corresponds to the isomorphism $\varphi_{\bar{P}} : D^{M.r} \to D^{I.r}$ given by $\varphi_{\bar{P}}(w) = w \circ \bar{P}$. Thus, $\dim_D \varphi_{\bar{P}}(D^{M.r}) = \dim_D \varphi_P((D^r)^M)$. Since the former is equal to $\dim_D \text{row}(\bar{P})$, then, as a direct consequence, we get the following corollary.

**Corollary 20**  $\rho(P) = \rho(\bar{P}) = \dim_D \text{row}(\bar{P})$.

The following linear algebra result is well known.

**Lemma 21**  Let $P$ be an $M \times I$ matrix over a division algebra $D$. Then $\dim_D \text{col}(P) = \dim_D \text{row}(P) = \sup\{k \mid$ there exists a $k \times k$ submatrix of $P$ that is invertible in $M_k(D)\}$.

This shows that $\rho(P)$ may be equivalently defined as the dimension of the image of $(D^r)^I$ under the homomorphism of right $D$-spaces defined dually to $\varphi_P$ by the action of $P$ on the left.

**Lemma 22**  Let $\hat{D} = \mathfrak{M}(D,I,M,P)$ be a Munn algebra over a division algebra $D$. Assume that $\rho(P) = t < \infty$. Then, $\hat{D}/\mathfrak{B}(0) \cong M_t(D)$.

*Proof.* Since $\rho(P) = \dim_D \text{col}(P)$, then there exist right $D$-independent columns $P^{(i_1)}, \dots, P^{(i_t)}$ of $P$, $i_1, \dots, i_t \in I$, such that $\text{col}(P) = \sum_{j=1}^{t} P^{(i_j)}D$. Put $J = \{i_1, \dots, i_t\}$, and let $\varphi : \hat{D}_{(J)}^{(M)} \to M_{|J|}(D)$ be the homomorphism defined by $\varphi(X) = X \circ P_{MJ}$. In view of Lemma 21, the matrix $P_{MJ}$ has $t$ rows that are left $D$-independent. Thus, from Lemma 3 it follows that $\varphi(\hat{D}_{(J)}^{(M)})$ has dimension $t^2$ as a left vector space over $D$. Therefore, $\varphi$ maps $\hat{D}_{(J)}^{(M)}$ onto $M_{|J|}(D)$ and hence, again, Lemma 3 yields

(a) $$\hat{D}_{(J)}^{(M)}/l(\hat{D}_{(J)}^{(M)}) \cong \varphi(\hat{D}_{(J)}^{(M)}) \cong M_{|J|}(D) \cong M_t(D)$$

On the other hand, from Lemma 2 (see Remark 6), it follows that $r(\hat{D}) = r_{\hat{D}}(\hat{D}_{(J)}^{(M)})$ and, consequently, $r(\hat{D}) \cap \hat{D}_{(J)}^{(M)} = r(\hat{D}_{(J)}^{(M)})$. Since, by Lemma 5, $\hat{D} = \hat{D}_{(J)}^{(M)} + r(\hat{D})$, we come to

$$\hat{D}/r(\hat{D}) = (\hat{D}_{(J)}^{(M)} + r(\hat{D}))/r(\hat{D}) \cong \hat{D}_{(J)}^{(M)}/(r(\hat{D}) \cap \hat{D}_{(J)}^{(M)}) = \hat{D}_{(J)}^{(M)}/r(\hat{D}_{(J)}^{(M)})$$

Thus, from Lemma 8 it follows that

$$\hat{D}/\mathfrak{B}_{\hat{D}}(0) \cong (\hat{D}/r(\hat{D}))/l(\hat{D}/r(\hat{D})) = (\hat{D}_{(J)}^{(M)}/r(\hat{D}_{(J)}^{(M)}))/l(\hat{D}_{(J)}^{(M)}/r(\hat{D}_{(J)}^{(M)}))$$

Applying Lemma 8 once again with respect to the algebra $\hat{D}_{(J)}^{(M)}$, we get

$$\hat{D}/\mathfrak{B}_{\hat{D}}(0) \cong (\hat{D}_{(J)}^{(M)}/l(\hat{D}_{(J)}^{(M)}))/r(\hat{D}_{(J)}^{(M)}/l(\hat{D}_{(J)}^{(M)}))$$

The latter algebra is, in view of (a), isomorphic to $M_t(D)$. This completes the proof of the lemma.

Now we state the final result providing a quantitative version of the statement of Corollary 18 in the case in which $R$ is a simple artinian algebra.

**Proposition 23**   Let $\hat{R} = \mathfrak{M}(R,I,M,P)$ be a Munn algebra over a simple artinian algebra $R = M_r(D)$, $r \geq 1$, for a division algebra $D$. Then $\hat{R}/\mathfrak{B}(0) \cong M_t(D)$, provided that $t = \rho(P) < \infty$.

*Proof.*   From Lemma 19 we know that $\hat{R} \cong \mathfrak{M}(D,I.r,M.r,\bar{P})$. Thus, Lemma 22 and Corollary 20 imply that $\hat{R}/\mathfrak{B}(0) \cong M_{\rho(\bar{P})}(D) = M_{\rho(P)}(D)$.

We continue with an important auxiliary result providing necessary and sufficient conditions for a Munn algebra to have an identity. The following well-known linear algebra argument will be used; see[26], Theorem 5.11.

**Lemma 24**   Let $R$ be a finite dimensional $K$-algebra, and let $P$ be an $m \times n$ matrix over $R$ for some integers $m, n \geq 1$. If $m > n$, then $X \circ P = 0$ for a nonzero $n \times m$ matrix $X$ over $R$. If $n > m$, then $P \circ Y = 0$ for a nonzero $n \times m$ matrix $Y$ over $R$.

**Proposition 25**   Let $R$ be an algebra with a nonzero finite dimensional homomorphic image. Then the following conditions are equivalent for any algebra of matrix type $\hat{R} = \mathfrak{M}(R,I,M,P)$:

(1)   The algebra $\hat{R}$ has an identity.
(2)   $I,M$ are finite sets of the same cardinality, and $P$ is an invertible matrix in $M_{|I|}(R)$.

Moreover, if (1) and (2) hold, then $\hat{R} \cong M_{|I|}(R)$.

*Proof.*   Assume that $\hat{R}$ has an identity $E$. Then there exist finite subsets $J \subseteq I$, $N \subseteq M$ such that $E \in \hat{R}_{(J)}^{(N)}$. Let $i \in I$. We know that, for any $m \in M$,

(b)                $(1,i,m) = E(1,i,m) = E \circ P \circ (1,i,m) \in \hat{R}_{(J)}^{(M)}$

This implies that $i \in J$, and so $I = J$ is a finite set. Moreover, from (b) it follows that the $i$th column of the matrix $E \circ P$ consists of zeros except the $(i,i)$th entry, which is equal to 1. Therefore, $E \circ P$ is the $I \times I$ identity matrix. Similarly, the fact that $E$ is a right identity of $\hat{R}$ implies that $M = N$ is a finite set and that the $M \times M$ matrix $P \circ E$ is the identity matrix. To establish (2), it is enough to show that $|I| = |M|$ since then $E$ is the inverse of $P$ in the algebra $M_{|I|}(R)$. Let $\varphi : R \to R'$ be a homomorphism onto a nonzero finite dimensional algebra $R'$, and let $\hat{\varphi} : \hat{R} \to \hat{R}' = \mathfrak{M}(R',I,M,\varphi(P))$ be the induced homomorphism. Then $\hat{R}'$ has an identity $\varphi(E)$ and is finite dimensional since $I$, $M$ are finite sets.

Now, $X = \varphi(E) \circ \varphi(P) \circ X = X \circ \varphi(P) \circ \varphi(E)$ for all $X \in \hat{R}'$, and from Lemma 24 it follows that $|I| = |M|$. This proves (2).

If (2) holds, then the homomorphism $\psi : \hat{R} \to M_{|I|}(R)$ given by $\varphi(X) = X \circ P$ is an isomorphism because $P$ is invertible in $M_{|I|}(R)$. This proves (1), as well as the remaining assertion.

It may be shown that Proposition 25 does not hold for arbitrary algebras $R$. This is because, in general, the assumption that $E \circ P$, $P \circ E$ are the identity $I \times I$, respectively, $M \times M$ matrices does not imply that $P$ is a square invertible matrix; see [28]. On the other hand, it is the case under some hypotheses on $R$ weaker than the existence of finite dimensional nonzero homomorphic images. However, the above statement is sufficient to derive the following consequence for semigroup algebras.

**Corollary 26**  Let $S = \mathfrak{M}^0(G, I, M, P)$ be a semigroup of matrix type. Then the following conditions are equivalent:

(1)  The algebra $K_0[S]$ has an identity.
(2)  $I, M$ are finite sets of the same cardinality, and $P$ is an invertible matrix in $M_{|I|}(K[G])$.

Moreover, if (1) and (2) hold, then $K_0[S] \cong M_{|I|}(K[G])$, and $S$ is a completely 0-simple semigroup.

If $S$ is an inverse semigroup, then (1) and (2) are equivalent to the fact that $S$ has finitely many idempotents.

*Proof.*  Since the augmentation homomorphism maps $K[G]$ onto $K$, then Proposition 25, in view of Lemma 1, establishes the equivalence of (1) and (2) and the fact that $K_0[S] \cong M_{|I|}(K[G])$. Moreover, the invertibility of $P$ implies that $P$ has no zero columns or rows, so that $S$ is completely 0-simple.

Assume now that $S$ is inverse. Then $S$ is the so-called Brandt semigroup, and $S \cong \mathfrak{M}^0(G, I, I, Q)$, where $Q$ is the identity $I \times I$ matrix; see [26], Theorem 3.9. Hence, by Lemma 4 in Chapter 1, $S$ has finitely many idempotents if and only if $I$ is a finite set. This completes the proof.

We close this chapter with an application of the above result to a study of the structure of algebras of certain not necessarily completely 0-simple semigroups. Let $p$ be a prime or zero, and let $F$ be the prime field of characteristic $p$. We say that a semigroup $S$ is strongly $p$-semisimple if all the principal factor algebras $F_0[S_t]$, $t \in S$, have unity elements. (Note that, if $\mathrm{ch}(K) = p$, then $K_0[S_t] \cong K \otimes F_0[S_t]$ has a unity if and only if $F_0[S]$ has a unity.) It is clear that $S$ has no nonzero nilpotent principal factors in this case, so that $S$ is a semisimple semigroup. From Corollary 26 it follows that every inverse semigroup with finitely many idempotents is strongly $p$-semisimple for every (prime or zero) $p$.

**Corollary 27**  Let $S$ be a nonzero strongly $p$-semisimple semigroup with finitely many principal factors, and let $\mathrm{ch}(K) = p$ (not necessarily zero). Then,

$$K_0[S] \cong K_0[T_1] \oplus \cdots \oplus K_0[T_r]$$

for some $r$, $n_i \geq 1$, and nonzero principal factors $T_i$ of $S$. Moreover, if $S$ is completely semisimple, then,

$$K_0[S] \cong M_{n_1}(K[G_1]) \oplus \cdots \oplus M_{n_r}(K[G_r])$$

where $G_i$ are maximal subgroups of $S$.

*Proof.*  Let $T$ be a minimal nonzero ideal of $S$. Then, by the hypothesis on $S$, $K_0[T]$ is an algebra with unity, so that Corollary 8 in Chapter 4 implies that $K_0[S] \cong K_0[T] \oplus K_0[S/T]$. Thus, the former assertion follows by induction on the number of principal factors of $S$. The latter is then a consequence of Corollary 26.

If the hypothesis on the finiteness of the set of the principal factors is removed in Corollary 27, one can still prove a useful result on the connections between $K[S]$ and its principal factor algebras. We state this result in a more general form, covering the case of semilattice decompositions discussed in Chapter 6. Recall that we put $S/I = S$, and $K[I] = 0$ for $I = \varnothing$.

**Proposition 28**  Let $S$ be a semigroup with a family of subsets $I_\alpha \subset J_\alpha$, $\alpha \in \mathcal{A}$, satisfying the conditions:

(i)   $J_\alpha$ are ideals of $S$, $I_\alpha$ are ideals of $S$ if nonempty, and $S = \bigcup_{\alpha \in \mathcal{A}}(J_\alpha \setminus I_\alpha)$.

(ii)  If $J_\alpha \cap (J_\beta \setminus I_\beta) \neq \varnothing$ for $\alpha$, $\beta \in \mathcal{A}$, then $J_\alpha \supseteq J_\beta$.

Then $K_0[S]$ is a subdirect product of the algebras $K_0[S/T_\alpha]$, where $T_\alpha = \bigcup_{\gamma \in C} J_\gamma$ and $C = \{\gamma \in \mathcal{A} \mid J_\gamma \not\supseteq J_\alpha\}$.
Moreover,

(iii) For every $\alpha \in \mathcal{A}$, $K_0[J_\alpha/I_\alpha]$ is an algebra with unity.

then $K_0[S]$ is a subdirect product of all $K_0[J_\alpha/I_\alpha]$, $\alpha \in \mathcal{A}$. In particular, if $S$ is a strongly $\mathrm{ch}(K)$-semisimple semigroup, then $K_0[S]$ is a subdirect product of the principal factor algebras $K_0[S_t]$, $t \in S$.

*Proof.* We can assume that $S$ has a zero element $\theta$, adjoining it to $S$ if necessary. Clearly, $T_\alpha$ is an ideal of $S$, or it is empty. Let $0 \neq a \in K_0[S]$. Since $S$ must be a disjoint union of $J_\beta \backslash I_\beta$, $\beta \in \mathcal{A}$, then there exists $\alpha \in \mathcal{A}$ such that $J_\alpha$ is minimal among all ideals $J_\delta$, $\delta \in \mathcal{A}$, satisfying $(J_\delta \backslash I_\delta) \cap \mathrm{supp}(a) \neq \varnothing$. Thus, $a \notin K[T_\alpha]$. Therefore, $\bigcap_{\alpha \in \mathcal{A}} K[T_\alpha] = 0$, and the first assertion follows.

Assume that (iii) holds. Then every $K_0[J_\alpha/I_\alpha]$ is a direct summand of the algebra $K_0[S/I_\alpha]$. Let $\tau_\alpha : K_0[S] \to K_0[S/I_\alpha]$ be the natural homomorphism, and let $e_\alpha \in K[J_\alpha]$ be such that $\tau_\alpha(e_\alpha)$ is the unity of $K_0[J_\alpha/I_\alpha]$. There exists a natural homomorphism $\phi_\alpha : K_0[S/I_\alpha] \to K_0[J_\alpha/I_\alpha]$ such that $\phi_\alpha(x) = \tau_\alpha(e_\alpha)x$. Let $M_\alpha$ denote the kernel of $\phi_\alpha \tau_\alpha$ and put $M = \bigcap_{\alpha \in \mathcal{A}} M_\alpha$. It is enough to show that $M = 0$.

Suppose that $y \in M$, $y \neq 0$. For each $t \in \mathrm{supp}(y)$, choose $\alpha_t \in \mathcal{A}$ such that $t \in J_{\alpha_t} \backslash I_{\alpha_t}$. Let $s \in \mathrm{supp}(y)$ be such that $J_{\alpha_s}$ is maximal among all $J_{\alpha_t}$, $t \in \mathrm{supp}(y)$. Put $\beta = \alpha_s$ and write $y = u + z$ where $\mathrm{supp}(u) \subseteq J_\beta \backslash I_\beta$ and $\mathrm{supp}(z) \cap (J_\beta \backslash I_\beta) = \varnothing$. Since $e_\beta y \in M$, then $\phi_\beta \tau_\beta(e_\beta y) = 0$. The choice of $s$ and the definition of $z$ imply, in view of condition (ii), that $e_\beta z \in K[I_\beta]$. Hence, $\tau_\beta(e_\beta z) = 0$, and

$$0 = \phi_\beta \tau_\beta(e_\beta y) = \phi_\beta \tau_\beta(e_\beta u + e_\beta z)$$
$$= \phi_\beta \tau_\beta(e_\beta u) = \phi_\beta \tau_\beta(e_\beta)\phi_\beta \tau_\beta(u) = \phi_\beta \tau_\beta(u)$$

On the other hand, since $s \in \mathrm{supp}(u)$, then $u \neq 0$, and so $\phi_\beta \tau_\beta(u) \neq 0$, a contradiction. This shows that $M = 0$.

Assume now that $S$ is strongly $\mathrm{ch}(K)$-semisimple. Then the family of the principal ideals $S^1 t S^1$, $t \in S$, of $S$, and the subsets of nongenerators of $S^1 t S^1$, satisfy the hypothesis. Here $J_\alpha/I_\alpha$ runs through all principal factors of $S$. The result follows.

## Comments on Chapter 5

Algebras studied in this chapter are also called Rees algebras. We follow the terminology of Clifford and Preston [26], motivated by the fact that the first application of this object (in the context of semigroup algebras of finite semigroups and their representations) was given by Munn in [154]. Most of the results of this section are based on the paper of McAlister [147]. Some connections with the general problems in ring theory were discussed by Petrich in [208]. In particular, Munn algebras appeared in a characterization of simple rings with minimal one-sided ideals as those algebras of matrix type over a division algebra $D$, the sandwich matrix of which has rows and columns left, respectively right, $D$-independent; see

[208], Theorem II.2.8. General radicals of Munn algebras over an arbitrary ring $R$ were studied by Weissglass in [267].

We note that the problem of relating the properties of the semigroup algebra of a completely 0-simple semigroup $S$ with the properties of the group algebra of the maximal subgroup $G$ of $S$ was in fact also the major subject, or the tool, in the papers of Domanov [46], Hall [82], Lallement and Petrich [133], and McAlister [146]. For a survey of the results in terms of representations of $S$ and $G$, we refer to [148]; see also [26]. A substantial work has been independently done by Ponizovskii, most of whose early papers are hardly available to the reader. We refer to [212], [215], [219], and [221] for some of his results and references on this topic.

# 6

# Gradations

In this chapter, we discuss a wide class of semigroup-graded semigroups and algebras. The aim is to give some general techniques coming from gradations, which are especially useful for the theory of semigroup algebras. We refer to [168] as a general reference on graded algebras, especially those of basic importance—graded by groups.

Let $T$ be a semigroup. A $K$-algebra $R$ is $T$-graded if the additive group of $R$ is the direct sum of some subgroups $R(t)$, $t \in T$, and $R(s)R(t) \subseteq R(st)$ for every $s, t \in T$. The sets $R(s)$ are called the (homogeneous) components of $R$. Clearly, any semigroup algebra $R = K[S]$ is $S$-graded with the components defined by $R(s) = Ks$ for $s \in S$. Similarly, a semigroup $S$ is said to be $T$-graded if $S$ is a disjoint union of some subsets $S(t)$, $t \in T$, and $S(s)S(t) \subseteq S(st)$ for every $s, t \in T$ such that $S(s), S(t) \neq \varnothing$. In this case, the algebra $K[S]$ has an induced $T$-gradation with the component $R(t)$, $t \in T$, defined as the $K$-subspace $KS(t)$ spanned by $S(t)$ if $S(t) \neq \varnothing$, and $R(t) = 0$

otherwise. Moreover, the map $\phi : S \to T$ given by $\phi(s) = t$, where $s \in S(t)$, is a semigroup homomorphism. On the other hand, every homomorphism $\psi : S \to T$ determines a $T$-gradation on $S$ by $S(t) = \{s \in S \mid \psi(s) = t\}$.

It is sometimes convenient to consider the following more general situation. We say that a semigroup $S$ with zero $\theta$ is $T_0$-graded if $S = \bigcup_{t \in T} S_0(t)$ for some subsets $S_0(t)$ of $S$ such that $S_0(s)S_0(t) \subseteq S_0(st)$ for all $s, t \in T$, with $S_0(s), S_0(t) \neq \varnothing$, and $S_0(s) \cap S_0(t) = \theta$ if $s \neq t$.

A subset $X$ of a $T$-graded algebra $R$ is said to be graded (or homogeneous) if $X = \sum_{t \in T} (X \cap R(t))$. Further $\mathcal{L}_T(R), \mathcal{R}_T(R)$ denote the lattices of all graded left, respectively right, ideals of $R$. For a semigroup $S$, $\mathcal{LI}(S)$, $\mathcal{RI}(S)$ are the lattices of left, and right, ideals of $S$.

**Remark 1** Let $R$ be an arbitrary $T$-graded $K$-algebra, and let $H(R)$ denote the set of all homogeneous elements of $R$, that is, $H(R) = \bigcup_{t \in T} R(t)$. Then $H(R)$, with the multiplication coming from $R$, is a semigroup with zero. Clearly, $H(R)$ is $T_0$-graded with the components $H(R)(t) = R(t)$, $t \in T$. Moreover, it is easy to see that, if a nonempty subset $J$ of $R$ is a graded right ideal of $R$, then $J \cap H(R)$ is a right ideal of $H(R)$. Conversely, a right ideal $I$ of $H(R)$ leads to a $T$-graded right ideal $\sum_{t \in T} I \cap R(t)$ of $R$. Thus, the lattice of graded right ideals of $R$ may be identified with the lattice of right ideals of $H(R)$.

On the other hand, when passing from $H(R)$ to the semigroup algebra $K_0[H(R)]$, we get an embedding of the lattice of right ideals of $H(R)$ into the lattice of $T$-graded right ideals of $K_0[H(R)]$, given by $I \to K_0[I]$ which, in general, may not be an isomorphism. (Here $K_0[H(R)]$ is $T$-graded with components coming from the components of $R$.) We also note that $\mathcal{R}_T(K_0[H(R)])$ maps onto $\mathcal{R}_T(R)$ because there is a natural homomorphism of algebras $K_0[H(R)] \to R$ that respects gradation. Thus, we get a chain of natural homomorphisms:

$$R \longrightarrow H(R) \longrightarrow K_0[H(R)] \longrightarrow R$$

such that every graded ideal $J$ of $R$ is acted on in the following way:

$$J \longrightarrow J \cap H(R) \longrightarrow K_0[J \cap H(R)] \longrightarrow J.$$

This establishes a strong connection between the theory of graded semigroups and graded algebras.

We give some important examples of gradations.

**Example 2** (i) Let $\Omega$ be a semilattice, that is, a commutative semigroup consisting of idempotents. If $S$ is a $\Omega$-graded semigroup, and $S(\alpha) \neq \varnothing$ for every $\alpha \in \Omega$, then every $S(\alpha)$ is a subsemigroup of $S$. $S$ is then called a semilattice $\Omega$ of its subsemigroups $S(\alpha)$, $\alpha \in \Omega$. The corresponding homomorphism $S \to \Omega$ maps $S$ onto $\Omega$. The resulting congruence on $S$ is called a semilattice congruence on $S$.

It is easy to check, and well known, that for every semigroup $S$ there exists the least semilattice congruence $\eta$ on $S$. Moreover, if $S$ is commutative, then $(s,t) \in \eta$ if and only if there exist $x,y \in S$ and $n,m \geq 1$ such that $sx = t^m, ty = s^n$; see [26], Section 4.3. The components of the natural $S/\eta$-grading on $S$ are called the archimedean components of $S$.

(ii) Let $S$ be a subsemigroup of a group $G$. If $H$ is a normal subgroup of $G$, then put $S(gH) = S \cap gH$ for every $g \in G$. It is clear that $S$ is a disjoint union of $S(gH)$, $gH \in G/H$, and so this defines a $G/H$-gradation on $S$.

We note that every contracted semigroup algebra $K_0[S]$ is $S$-graded in an obvious way. A more interesting example of a similar type is given below.

**Example 3**   Let $\hat{R} = \mathfrak{M}(R,I,M,P)$ be a Munn algebra over a $K$-algebra $R$ with unity. Let $S$ be the semigroup with zero of matrix type $\mathfrak{M}^0(\{1\}^0, I, M, P')$, where $p'_{mi} = 1$ if $p_{mi} \neq 0$, and $p'_{mi} = \theta$ if $p_{mi} = 0$ for $i \in I$, $m \in M$. If $s = (1,i,m) \in S$, then we put $\hat{R}(s) = \hat{R}^{(m)}_{(i)}$, $\hat{R}\{\theta\} = 0$. It is clear that this defines an $S$-gradation on $\hat{R}$.

We state a useful extension of the situation described in Lemma 7 of Chapter 4. If allows us to use the ideal structure of $T$ when studying $T$-graded algebras, as in the case of semigroup algebras. The proof is straightforward and will be omitted.

**Lemma 4**   Let $T$ be a semigroup, and let $R$ be a $T$-graded $K$-algebra. If $I$ is an ideal of $T$, then $J = \sum_{t \in I} R(t)$ is an ideal of $R$, and $R/J$ is a $(T/I)$-graded algebra with the components $(R/J)(t) = R(t)$ for $t \in T \setminus I$, $(R/J)(\theta) = 0$.

Recall that a semigroup $S$ is called separative if, for every $s,t \in S$, the following conditions are satisfied:

$$s^2 = st, t^2 = ts \Longrightarrow s = t \quad \text{(left separativity)}$$
$$s^2 = ts, t^2 = st \Longrightarrow s = t \quad \text{(right separativity)}$$

The notion of separativity provides us with the following important reduction procedure arising from gradations described in Example 2 (i). Recall that every semilattice $\Omega$ is ordered by the relation $\leq$ defined for $\alpha, \beta \in \Omega$ by $\alpha \geq \beta$ if $\alpha\beta = \beta$. The groups of fractions considered below are the Ore groups of fractions, (see [26] and [31]), which will be discussed in more detail in Chapter 7. Using some of the results from [207], we summarize the key results on semilattices of cancellative semigroups.

**Theorem 5**   A semigroup $S$ is separative if and only if $S$ is graded by a semilattice $\Omega$ such that every $S(\alpha)$, $\alpha \in \Omega$, is a cancellative semigroup. Moreover, if every $S(\alpha)$ has a group of right fractions $G_\alpha$, then there exists a semigroup $G$ that is graded by $\Omega$ with $G(\alpha) = G_\alpha$ for every $\alpha \in \Omega$, and an embedding $\phi : S \to G$ such that $\phi(s) = s$ for every $s \in S$.

*Proof.*   The first assertion is explicit in [207], Theorem II.6.4. If $S(\alpha)$ is an ideal of a cancellative semigroup $T_\alpha$, then, for every $u, v \in T_\alpha$, $s \in S(\alpha)$, we have $us, vs \in S(\alpha)$. The fact that $S(\alpha)$ has a group of right fractions $G_\alpha$ implies that $usS(\alpha) \cap vsS(\alpha) \neq \varnothing$, so that $uT_\alpha \cap vT_\alpha \neq \varnothing$. Therefore $T_\alpha$ has a group of right fractions. Moreover, $uv^{-1} = (us)(vs)^{-1} \in S(\alpha)S(\alpha)^{-1} = G_\alpha$ and, hence, $T_\alpha T_\alpha^{-1} = G_\alpha$. Now, from [207], Theorem III.7.2 and Corollary III.5.15, it follows that for all $\alpha, \beta \in \Omega$, such that $\alpha \geq \beta$, there exist homomorphisms $\phi_{\alpha,\beta} : S(\alpha) \to G(\beta)$ satisfying the following conditions:

(1)   $\phi_{\alpha,\alpha}$ is the identity mapping on $S(\alpha)$.
(2)   $\phi_{\alpha,\alpha\beta}(S(\alpha))\phi_{\beta,\alpha\beta}(S(\beta)) \subseteq S(\alpha\beta)$.
(3)   If $\alpha, \beta \geq \gamma$, then for all $s \in S(\alpha)$, $t \in S(\beta)$, we have

$$\phi_{\alpha\beta,\gamma}(\phi_{\alpha,\alpha\beta}(s)\phi_{\beta,\alpha\beta}(t)) = \phi_{\alpha,\gamma}(s)\phi_{\beta,\gamma}(t)$$

Moreover, in this case, $st = \phi_{\alpha,\alpha\beta}(s)\phi_{\beta,\alpha\beta}(t)$. From the universal property of the group of fractions (see [31], Section 0.8), it follows that every $\phi_{\alpha,\beta}$ extends to a group homomorphism $\bar{\phi}_{\alpha,\beta} : G_\alpha \to G_\beta$, so that $\bar{\phi}_{\alpha,\beta}(st^{-1}) = \phi_{\alpha,\beta}(s)\phi_{\alpha,\beta}(t)^{-1}$ for $s, t \in S(\alpha)$. Thus, [207], Theorem II.7.2, implies that $\bigcup_{\alpha \in \Omega} G_\alpha$ is a semigroup under multiplication given by $gh = \phi_{\alpha,\alpha\beta}(g)\phi_{\beta,\alpha\beta}(h)$ for $g \in G_\alpha$, $h \in G_\beta$, and the result follows.

We note that the multiplication in the semigroup $G$ constructed above may be explicitly defined in terms of $S$ as follows. Let $s_1, t_1 \in S(\alpha)$, $s_2, t_2 \in S(\beta)$. Take any $z \in S(\alpha\beta)$. Then $zs_2, zt_1 \in S(\alpha\beta)$, and so there exist elements $x, y \in S(\alpha\beta)$ such that $(zs_2)x = (zt_1)y$. The cancellativity in $S(\alpha\beta)$ implies that $s_2 x = t_1 y$. Now, $\phi_{\beta,\alpha\beta}(s_2)x = \phi_{\alpha,\alpha\beta}(t_1)y$, and so

$$
\begin{aligned}
\bar{\phi}_{\alpha,\alpha\beta}&(s_1 t_1^{-1})\bar{\phi}_{\beta,\alpha\beta}(s_2 t_2^{-1}) \\
&= \phi_{\alpha,\alpha\beta}(s_1)\left(\phi_{\alpha,\alpha\beta}(t_1)^{-1}\phi_{\beta,\alpha\beta}(s_2)\right)\phi_{\beta,\alpha\beta}(t_2)^{-1} \\
&= \phi_{\alpha,\alpha\beta}(s_1)(yx^{-1})\phi_{\beta,\alpha\beta}(t_2)^{-1} = (\phi_{\alpha,\alpha\beta}(s_1)y)(\phi_{\beta,\alpha\beta}(t_2)x)^{-1} \\
&= (s_1 y)(t_2 x)^{-1}
\end{aligned}
$$

(Here the inverses are taken in the appropriate groups $G_\gamma$.) Consequently, the product $(s_1 t_1^{-1})(s_2 t_2^{-1})$ may be defined as $(s_1 y)(t_2 x)^{-1}$ because $s_1 y \in S(\alpha\beta)$, $t_2 x \in S(\alpha\beta)$.

If $\rho$ is a congruence on $S$, then $\rho$ is called separative if $S/\rho$ is a separative semigroup. In the case of commutative semigroups, there exists a nice criterion for separativity, see [26], Theorem 4.14.

**Lemma 6**  Let $S$ be a commutative semigroup. Then the binary relation $\xi = \{(s,t) \in S \times S \mid st^n = t^{n+1}, s^n t = s^{n+1} \text{ for some } n \geq 1\}$ is the least separative congruence on $S$.

Certain extension of the least separative congruence will be useful in the case of semigroup algebras of positive characteristics. If $p$ is a prime number, then $S$ is called $p$-separative if, for every $s, t \in S$, the condition $s^p = t^p$ implies that $s = t$. A congruence $\rho$ on $S$ is called $p$-separative if $S/\rho$ is a $p$-separative semigroup.

**Proposition 7**  Let $S$ be a commutative semigroup. Then, for every prime $p$, the following conditions hold:

(i)   The binary relation $\xi_p$ defined on $S$ by $\xi_p = \{(s,t) \in S \times S \mid s^{p^n} = t^{p^n}$ for some $n \geq 1\}$ is the least $p$-separative congruence on $S$.

(ii)  $\xi_p \geq \xi$.

(iii) $S/\xi_p$ is graded by a semilattice, with all components being cancellative semigroups, the groups of fractions of which have no elements of order $p$.

*Proof.*  (i) Assume that $s^{p^n} = t^{p^n}$, $t^{p^m} = u^{p^m}$ for some $s, t, u \in S$, $n, m \geq 1$. Then $s^{p^{n+m}} = (s^{p^n})^{p^m} = (t^{p^n})^{p^m} = (t^{p^m})^{p^n} = (u^{p^m})^{p^n} = u^{p^{m+n}}$. Moreover, for every $x \in S$, we have $(sx)^{p^n} = s^{p^n} x^{p^n} = t^{p^n} x^{p^n} = (tx)^{p^n}$ because $S$ is commutative. It follows that $\xi_p$ is a congruence on $S$. It is clear that $S/\xi_p$ is a $p$-separative semigroup since $(s^{p^n}, t^{p^n}) \in \xi_p$ implies that $(s,t) \in \xi_p$ for every $s, t \in S$, $n \geq 1$. It is also straightforward that, if $(s,t) \in \xi_p$, then $(s,t)$ lies in every $p$-separative congruence on $S$. This establishes (i).

(ii) Assume that $(s,t) \in \xi$, that is, $st^n = t^{n+1}$, $ts^n = s^{n+1}$ for some $n \geq 1$. If $m \geq n$, then we have

$$s^m t^m = s^{m-1}(st^n)t^{m-n} = s^{m-1} t^{n+1} t^{m-n} = s^{m-2}(st^n)t^{m-n+1} = \cdots$$
$$= (st^n)t^{2m-n-1} = t^{n+1} t^{2m-n-1} = t^{2m}$$

Similarly, using the equality $ts^n = s^{n+1}$, one sees that $s^m t^m = s^{2m}$. Therefore, there exists $k \geq 1$ such that $s^{p^k} = t^{p^k}$, which shows that $(s,t) \in \xi_p$.

(iii) Since $S/\xi_p$ is a separative semigroup by (ii), then (iii) follows from Theorem 5 and the fact that the group of fractions of a cancellative $p$-separative semigroup has no elements of order $p$.

We continue with a useful general result on semigroup algebras of semilattices of semigroups.

**Proposition 8** Let $S$ be a semilattice $\Omega$ of semigroups $S(\alpha), \alpha \in \Omega$. Then,

(i) $K_0[S]$ is a subdirect product of all algebras $K_0[S/T_\alpha]$, where $T_\alpha = \bigcup_{\beta \ngeq \alpha} S(\beta)$.

(ii) If every $K[S(\alpha)]$ has a unity, then $K[S]$ is a subdirect product of all $K[S(\alpha)]$, $\alpha \in \Omega$.

*Proof.* Clearly, every $T_\alpha$ is an ideal of $S$ if $T_\alpha \neq \varnothing$. Define $I_\alpha = \bigcup_{\beta < \alpha} S(\beta)$, $J_\alpha = \bigcup_{\beta \leq \alpha} S(\beta)$. Then, $I_\alpha \subseteq J_\alpha$, $J_\alpha$ is an ideal of $S$, $I_\alpha$ is an ideal of $S$ if nonempty, and $K_0[J_\alpha/I_\alpha] \cong K[S(\alpha)]$. If $I_\alpha = \varnothing$, then $\alpha = \theta_\Omega$ and $K_0[J_\alpha/I_\alpha] \cong K_0[S_\alpha]$. The hypotheses of Proposition 28 in Chapter 5 are thus satisfied, and the result follows.

We will now present certain aspects of the gradations discussed in Example 2 (ii). Some results and techniques coming from the theory of smash products will be needed. Let $S$ be a semigroup with zero that is $T_0$-graded for a semigroup $T$. Our original definition does not require any restriction on $S, T$. By $s_*$ we denote the image of $s \in S$, $s \neq \theta_S$, under the resulting mapping $S \setminus \{\theta_S\} \to T$, so that $S(t) = \{s \in S \mid s_* = t\} \cup \{\theta_S\}$ for every $t \in T$. In the set $(S \times T) \cup \{\theta\}$ define a multiplication by the rules

$$(s_1,t_1)(s_2,t_2) = \begin{cases} (s_1s_2,t_2) & \text{if } t_1 = (s_2)_*t_2 \\ \theta & \text{if } t_1 \neq (s_2)_*t_2 \end{cases}$$

$$(s_1,t_1)\theta = \theta(s_1,t_1) = \theta$$

It is easy to see that this defines a semigroup structure on $(S \times T) \cup \{\theta\}$. In fact, let $s_1,s_2,s_3 \in S$, $t_1,t_2,t_3 \in T$. Then,

$$[(s_1,t_1)(s_2,t_2)](s_3,t_3) = (s_1s_2,t_2)(s_3,t_3) = (s_1s_2s_3,t_3)$$

if $t_1 = (s_2)_*t_2$ and $t_2 = (s_3)_*t_3$, and this product is equal to zero if this is not the case. On the other hand,

$$(s_1,t_1)[(s_2,t_2)(s_3,t_3)] = (s_1,t_1)(s_2s_3,t_3) = (s_1s_2s_3,t_3)$$

if $t_2 = (s_3)_*t_3$ and $t_1 = (s_2s_3)_*t_3$, and this product is equal to zero if this is not the case. Thus, if $t_2 \neq (s_3)_*t_3$, then both products are equal to zero. If $t_2 = (s_3)_*t_3$, then $(s_2s_3)_*t_3 = (s_2)_*(s_3)_*t_3 = (s_2)_*t_2$, and we also get the equality.

The above-defined semigroup, factored out by the ideal $\{(s,t) \mid s = \theta_s\} \cup \{\theta\}$, will be denoted by $S\#T$ and called the smash product of $S$ and $T$.

We will deal mainly with the case in which $S$ is a monoid and $T = G$ is a group. In this case, there is a very convenient way of representing the semigroup $S\#G$ as a semigroup of matrices, which will be presented below.

If $A$ is a $G$-graded algebra, then one may also define the smash product algebra $A\#G$ as the free left $A$-module on generators $g$, $g \in G$, with the multiplication defined by the rule

$$(a_1 g_1)(a_2 g_2) = a_1 a_2(g)g_2$$

where $g = g_1 g_2^{-1}$ and $a_2(g)$ is the $g$th component of $a_2$ in $A$. If $0 \neq a \in H(A)$, then we write $a_*$ for $g \in G$ such that $a \in R(g)$. If $a = 0$, then, let $a_*$ be the identity of $G$. It is easy to see that the above operation restricted to $a_i \in H(A)$ coincides with that defined earlier for the case of semigroup smash products. It may then be easily verified that, for every $G_0$-graded semigroup $S$, the algebra $K_0[S]\#G$ is isomorphic to $K_0[S\#G]$.

For the sake of simplicity, we will further assume that $A$ has an identity. Let $M_G(A)$ denote the set of all $G \times G$ matrices over $A$ that are row-finite and column-finite (that is, any element of $M_G(A)$ has finitely many nonzero entries in each row and in each column). Clearly, $M_G(A)$ is an algebra under the natural addition and multiplication of matrices. Consider the following subsets of $M_G(A)$

$$\bar{A} = \{\bar{a} \mid a \in A\} \qquad \hat{G} = \{\hat{g} \mid g \in G\} \qquad P = \{p_g \mid g \in G\} \cup \{0\}$$

where:

For every homogeneous element $c \in A$, $\bar{c}$ stands for the matrix with $c$ in the $(x, c_*^{-1}x)$-entry for all $x \in G$ and zeros elsewhere, and $\bar{a} = \sum_{g \in G} \overline{a(g)}$ for an arbitrary $a \in A$ (in other words, the $(g,h)$th entry of $\bar{a}$ is $\bar{a}(gh^{-1})$, $g, h \in G$).

$\hat{g}$ is the matrix with 1 in the $(x, xg)$-entry for every $x \in G$ and zeros elsewhere.

$p_g$ has 1 in the $(g,g)$-entry and zeros elsewhere.

By an analogy to the notation used in Chapter 5, we will write $p_g = (1,g,g)$ while, for $c \in H(A)$, $g \in G$, the elements $\bar{c}, \hat{g}$ will be denoted by $[c,x,c_*^{-1}x]_{x \in G}$, $[1,x,xg]_{x \in G}$, or $[c,x,c_*^{-1}x]_x$, $[1,x,xg]_x$. It is clear that

$$\bar{c}\bar{b} = [c,x,c_*^{-1}x]_x [b,y,b_*^{-1}y]_y = [cb,x,b_*^{-1}c_*^{-1}x]_x = \overline{cb}$$

for every $c, b \in H(A)$. Since $a \to \bar{a}$ is additive, then it establishes an isomorphism of algebras $A \to \bar{A}$. Similarly, $g \to \hat{g}$ establishes a group isomorphism $G \to \hat{G}$. Clearly, $P$ is a semigroup of "diagonal" idempotents.

Moreover, it is easy to see that, for every $c \in H(A)$, $g, h \in G$, we have the following multiplication rules:

(a)
$$\tilde{g}\bar{c} = [1,x,xg]_x[c,y,c_*^{-1}y]_y = [c,x,c_*^{-1}xg]_x$$
$$= [c,y,c_*^{-1}y]_y[1,x,xg]_x = \bar{c}\tilde{g}$$
$$\bar{c}p_h = [c,y,c_*^{-1}y]_y(1,h,h) = (c,c_*h,h)$$
$$p_h\bar{c} = (1,h,h)[c,y,c_*^{-1}y]_y = (c,h,c_*^{-1}h)$$
$$\tilde{g}p_h = [1,x,xg]_x(1,h,h) = (1,hg^{-1},h)$$
$$p_h\tilde{g} = (1,h,h)[1,x,xg]_x = (1,h,hg)$$

We will be interested in the subalgebra $R$ of $M_G(A)$ generated by $\bar{A}, \tilde{G}, P$. By $B$ we denote the subalgebra of $R$ generated by $\bar{A}, P$. Since, by (a), $(\bar{c}p_g)(\bar{b}p_h) = \overline{cb}p_h$ if $b \in A(gh^{-1})$ and 0 otherwise, then the fact that $\bar{A}P$ has a $K$-basis consisting of elements of the form $\bar{c}p_g$ implies that $cg \to cp_g$ determines an isomorphism of algebras $A\#G \cong \bar{A}P$. Consider two cases:

   (1) $G$ is finite. Then $\sum_{x \in G} p_x$ is the identity of $M_G(A)$. Hence, $\bar{a} = \sum_{x \in G} p_x\bar{a} = \sum_{x \in G}(a,x,a_*^{-1}x)$, $\tilde{g} = \sum_{x \in G} p_x\tilde{g} = \sum_{x \in G}(1,x,xg)$, so that $\bar{A} \subseteq \bar{A}P$ and $\tilde{G}$ is contained in the subspace generated by $P\tilde{G}$. Then the subalgebra $\bar{A}P\tilde{G}$ of $M_G(A)$ coincides with $R$, and $B = \bar{A}P$. Clearly, $R \supseteq B$, and $B$ is an algebra with an identity $\sum_{x \in G} p_x = \bar{1} = \tilde{1}$ (here 1 is used both for the identity of $A$ and of $G$).

   (2) $G$ is infinite. In this case, we see that $R = \bar{A}P\tilde{G} + \bar{A}\tilde{G}$, and $B = \bar{A}P + \bar{A}$. Then $B$ is also an algebra with an identity $\bar{1} = [1,x,x]_x = \tilde{1} \in \tilde{G} \cap \bar{A}$, which is the identity of $M_G(A)$, but $B \neq \bar{A}P$.

   With the above notation, we have the following basic result.

**Lemma 9**  Let $A$ be a $G$-graded algebra with an identity. Then,

   (i)   $\tilde{G}$ is a basis of the left $B$-module $R$.
   (ii)  If $G$ is finite, then $R = M_G(A)$.

Moreover, if $A = K_0[S]$ for a $G_0$-graded monoid $S$, then $P \subseteq \bar{S}P \subseteq \bar{S}P\tilde{G}$ are semigroups such that

   (iii) $\bar{S}P = P\bar{S} = \{(s,x,s_*^{-1}x) \mid s \in S, x \in G\}$ and $\bar{S}P\tilde{G} = \tilde{G}\bar{S}P = \{(s,x,y) \mid s \in S, x,y \in G\}$.
   (iv)  The nonzero elements of $\bar{S}P\tilde{G}$ are linearly independent in $M_G(K[S])$, and the nonzero elements of $\bar{S}P\tilde{G} \cup \bar{S}\tilde{G}$ are linearly independent if $G$ is infinite.

*Proof.*  If $c \in H(A)$, $g, h \in G$, then we have

(b)
$$\tilde{g}\bar{c}p_h = [1,x,xg]_x(c,c_*h,h) = (c,c_*hg^{-1},h)$$
$$\bar{c}p_h\tilde{g} = (c,c_*h,h)[1,x,xg]_x = (c,c_*h,hg)$$

Now, for $b \in H(A)$, $u,v \in G$, we get $(c,x,c_*^{-1}xy)(b,u,b_*^{-1}uv) = (cb,x,b_*^{-1}uv)$
if $c_*^{-1}xy = u$, and 0 otherwise. This implies that the set $\overline{H(A)P\tilde{G}} = \{(a,x,y) \mid$
$a \in H(A), x,y \in G\}$ is closed under multiplication. $\overline{H(A)P} = \{(a,a_*x,x) \mid$
$a \in H(A), x \in G\}$ is also a semigroup. Hence, $R$ is the subspace generated
by the set $\check{Z}P\tilde{G} \cup \check{Z}\tilde{G} = (\check{Z}P \cup \check{Z})\tilde{G}$, where $Z$ is any $K$-basis of $A$ consisting
of homogeneous elements, while $B$ is the subspace generated by $\check{Z}P \cup \check{Z}$.

Now, by (b), every element $\bar{c}p_x\tilde{y}$ of $\check{Z}P\tilde{G}$ is uniquely determined by the
triple $c,x,y$, and has only one nonzero entry, which is taken from the basis
$Z$ of $A$. Thus (i) follows in the case where $G$ is finite, because then $R,B$
are the subspaces generated by $\check{Z}P\tilde{G}$, $\check{Z}P$, respectively.

If $G$ is infinite, then it is easy to see that every nontrivial linear
combination of elements of the form $[c,x,c_*^{-1}xg]_k \in \check{Z}\tilde{G}$ (see (a)) with
fixed $c \in \check{Z}$ is a matrix with infinitely many nonzero entries, all of which are
in $Kc$. Hence, every nontrivial combination of elements of $\check{Z}\tilde{G}$ has infinitely
many nonzero entries. Since the elements of $\check{Z}P\tilde{G}$ have one nonzero entry,
then the preceding implies that the set $\check{Z}P\tilde{G} \cup \check{Z}\tilde{G}$ is independent.

(ii) Assume that $G$ is finite. We have seen that, for every $c \in H(A)$,
$x,y \in G$, the element $(c,x,y)$ lies in $R$. It is then clear that $R = M_G(A)$.

Finally, for a $G_0$-graded monoid $S$ and $A = K_0[S]$, the remaining
assertions follow as in (i) with $Z = S \setminus \{\theta\}$.

We state an immediate consequence of the above result, which shows
that, for a $G_0$-graded monoid $S$, the resulting subalgebra $R \subseteq M_G(K[S])$
is a contracted semigroup algebra.

**Corollary 10**  Let $S$ be a $G_0$-graded monoid, and let $A = K_0[S]$. Then,

(i)  $B \cong K_0[T]$, where $T = \check{S}P$ if $G$ is finite and $T = \check{S}P \cup \check{S}$ if $G$ is infinite.
(ii)  $R \cong K_0[U]$, where $U = \check{S}P\tilde{G}$ if $G$ is finite and $U = \check{S}P\tilde{G} \cup \check{S}\tilde{G}$ if
     $G$ is infinite.

**Remark 11**  Observe that, in any case, $R$ is a skew group algebra $B * \tilde{G}$ of
the group $\tilde{G}$ over the algebra $B$. This is a consequence of the following facts:

(i)  $\tilde{G}$ lies in the group of units of $R$.
(ii)  $\tilde{G}$ is a basis of the left $B$-module $R$.
(iii)  $\tilde{G}$ acts as a group of automorphisms of $B$ by conjugation because,
     in view of (a),

$$\tilde{g}^{-1}\bar{a}\tilde{g} = \bar{a}$$

and

$$\tilde{g}^{-1}\bar{a}p_h\tilde{g} = \bar{a}\tilde{g}^{-1}p_h\tilde{g} = \bar{a}[1,x,xg^{-1}]_k(1,h,hg) = \bar{a}(1,hg,hg) = \bar{a}p_{hg}$$

Let $A^G$ be the set of all column vectors over $A$ with entries indexed by $G$ ($G$ being ordered as in the construction of $M_G(A)$), and with only finitely many nonzero entries. Then $A^G$ is a left $M_G(A)$-module, where the action is defined by the natural multiplication.

If $a \in A$, then we write $(a)_g$ for an element of $A^G$, the $g$th entry of which is $a$, the other entries being zero. Moreover, $\underline{a}$ denotes the element of $A^G$, the $g$th entry of which is equal to $a(g)$ for all $g \in G$. Thus, if $a \in H(A)$, then $\underline{a} = (a)_{a_*}$. If $X$ is a subset of $A$, then put $\underline{X} = \{\underline{x} \mid x \in X\}$.

We note that the action of $R$ on $A^G$ is determined for $g, h \in G$, $a \in H(A)$, $b \in A$, by the following rules:

$$
\text{(c)} \qquad p_g \underline{a} = \begin{cases} 0 & \text{if } g \neq a_* \\ a & \text{if } g = a_* \end{cases}
$$
$$
\bar{b}\underline{a} = \underline{ba}
$$
$$
\tilde{g}\underline{a} = (a)_h \qquad \text{where } h = a_* g^{-1}
$$

Recall that, if $A$ is a $G$-graded algebra, then a left $A$-module $M$ is graded if $M = \oplus_{g \in G} M(g)$ as an abelian group such that $A(g)M(h) \subseteq M(gh)$ for every $g, h \in G$. Moreover, the graded Jacobson radical $\mathcal{J}_G(N)$ of any left $A$-module $N$ is defined as the intersection of the maximal graded submodules or $N$, of $\mathcal{J}_G(N) = 0$ if $N$ has no maximal graded submodules.

**Remark 12**  It is well known that $\mathcal{J}_G(A)$, defined as the graded Jacobson radical of the left $A$-module $A$, is a graded ideal of $A$. Moreover, $\mathcal{J}_G(A)$ is the largest graded ideal of $A$ such that $\mathcal{J}_G(A) \cap A(1) \subseteq \mathcal{J}(A(1))$; see [168], I.7. Now $I = \sum_{g \in G} \mathcal{J}(A) \cap A(g)$ is a graded ideal of $A$ such that $I \cap A(1) = \mathcal{J}(A) \cap A(1)$. Moreover, $\mathcal{J}(A) \cap A(1) \subseteq \mathcal{J}(A(1))$ because $A(1)$ is a direct summand of the left $A(1)$-module $A$. Consequently, $I \subseteq \mathcal{J}_G(A)$, which also leads to $\mathcal{J}(A) \cap A(1) \subseteq \mathcal{J}_G(A) \cap A(1) \subseteq \mathcal{J}(A(1))$.

The following result shows the interplay between the graded and the ordinary structures of the defined objects, and it is crucial for applications of the above embedding of $A\#G$ into $M_G(A)$.

**Lemma 13**  Let $A$ be an algebra with an identity graded by a group $G$. Then,

(i)   For every $g \in G$, $\tilde{g}\underline{A}$ is a left $B$-submodule of $A^G$ which is isomorphic to $\bar{A}$ as a left $\bar{A}$-module, and $A^G \cong \oplus_{g \in G} \tilde{g}\underline{A}$ as left $B$-modules. Moreover $A^G$ is a graded $R$-module with components $\tilde{g}\underline{A}$, $g \in G$.

(ii)  Every submodule $M$ of the left $B$-module $\tilde{g}\underline{A}$, $g \in G$, is of the form $\tilde{g}\underline{N}$ where $N$ is a graded left ideal of $A$, and every graded left ideal of $A$ determines a $B$-submodule of $\tilde{g}\underline{A}$ in this way.

(iii)   Every submodule $L$ of the left $R$-module $A^G$ is of the form $Q^G$ for a left ideal $Q$ of $A$, and every left ideal $P$ of $A$ leads to an $R$-submodule $P^G$ of $A^G$.

(iv)   If $G$ is finite, then $\mathcal{J}(B) = \overline{\mathcal{J}_G(A)}P$, and $A^G \cong B$ as left $B$-modules

*Proof.* (i) From Remark 11 and (c), it follows that $B\tilde{g}\underline{A} = \tilde{g}B\underline{A} = \tilde{g}\underline{A}$, so that $\tilde{g}\underline{A}$ is a left $B$-module. Moreover, the map $\bar{A} \to \tilde{g}\underline{A}$ given by $\bar{a} \to \tilde{g}\underline{a}$ is a homomorphism of $\bar{A}$-modules because $\tilde{G}$ commutes with $\bar{A}$. Since $\tilde{g}$ is invertible in $R$, then this is an isomorphism. From (c) it also follows that every element of $A^G$ can be uniquely written as a sum of elements of the form $\tilde{g}\underline{a}$ where $g \in G$, $a \in H(A)$. Therefore, $A^G \cong \oplus_{g \in G}\tilde{g}\underline{A}$ as left $B$-modules. Clearly $\tilde{g}\underline{A}$ are components of $A^G$ as a $G$-graded $R$-module.

   (ii) Let $m = \sum_{x \in G} m(x) \in A$, where $m(x)$ are the $G$-components of $m$ as an element of the $G$-graded algebra $A$, be such that $\tilde{g}\underline{m} \in M$. Then, for all $x \in G$, $\tilde{g}\underline{m(xg)} = \tilde{g}p_{xg}\underline{m} = p_x\tilde{g}\underline{m} \in PM \subseteq M$. Therefore $N = \{m \in A \mid \tilde{g}\underline{m} \in M\}$ is a graded subset of $A$. Since by (c) $\tilde{g}a\underline{m(x)} = \tilde{g}\bar{a}\underline{m(x)} = \bar{a}\tilde{g}\underline{m(x)} \in M$ for $a \in A$, then $N$ is a left ideal of $A$. Moreover, $\tilde{g}\underline{N} = M$. The remaining assertion is clear.

   (iii) Let $v \in L$, and let $v^g$ be the $g$th entry of the element $v$ in $A^G$. Then, for every $h \in G$, $a \in H(A)$, we have

$$(av^g)_h = (a,h,a_*^{-1}h)[1,x,xh^{-1}a_*g]_x v = p_h \bar{a}\bar{f}v \in Rv \subseteq L$$

where $f = h^{-1}a_*g$. Therefore the set $Q = \{b \in A \mid (b)_g \in L\}$ is a left ideal of $A$ that does not depend on $g$, and so we have $L = Q^G$. The latter assertion is clear.

   (iv) We know that $B = \bar{A}P$ because $G$ is finite. For every $g \in G$, $a \in H(A)$, define $\phi(\bar{a}p_g) = (a)_{a_*g}$. It is easy to see that this extends to a mapping $\phi : B \to A^G$, which is a linear isomorphism. If $b \in H(A)$, $h \in G$, then,

$$(\bar{b}p_h)(\bar{a}p_g) = (b,b_*h,h)(a,a_*g,g) = \begin{cases} (ba,b_*a_*g,g) & \text{if } h = a_*g \\ 0 & \text{if } h \neq a_*g \end{cases}$$

$$(\bar{b}p_h)(a)_{a_*g} = (b,b_*h,h)(a)_{a_*g} = \begin{cases} (ba)_{b_*a_*g} & \text{if } h = a_*g \\ 0 & \text{if } h \neq a_*g \end{cases}$$

If follows that $\phi$ is an isomorphism of $B$-modules $B$ and $A^G$. This and (ii) imply that

$$\phi(\mathcal{J}(B)) = \mathcal{J}(_BA^G) = \oplus_{g \in G}\mathcal{J}(_B\tilde{g}\underline{A}) = \oplus_{g \in G}\tilde{g}\,\underline{\mathcal{J}_G(A)} = \mathcal{J}_G(A)^G$$

From the definition of $\phi$, we see that $\phi(\overline{\mathcal{J}_G(A)}P) = \mathcal{J}_G(A)^G$. Therefore, $\overline{\mathcal{J}_G(A)}P = \mathcal{J}(B)$.

To derive the first important consequence of the above lemma, we need a result, which is stated below. This is a special case of the theorem on the behavior of the radical under normalizing extensions; see [203], Theorem 7.2.5. In fact, for every skew group ring $C * G$ of a finite group $G$ over a ring $C$ with an identity, $G$ is a normalizing basis of $C * G$ over $C$. Moreover, the assumption on the invertibility of $|G|$ in $C$ implies that $C * G$ meets the hypothesis of [203], Theorem 7.2.5 (see Lemma 7.2.2.).

**Lemma 14** Let $C$ be a ring with an identity, and let $G$ be a finite group of automorphisms of $C$. Then,

(i)   $\mathcal{J}(C * G)^{|G|} \subseteq \mathcal{J}(C) * G \subseteq \mathcal{J}(C * G)$.
(ii)  $\mathcal{J}(C * G) = \mathcal{J}(C) * G$ if $|G|$ is invertible in $C$.

Recall that the grading on a $G$-graded algebra $A$ is nondegenerated if, for every $g \in G$, $0 \neq a \in A(g)$, we have $aA(g^{-1}) \neq 0$, and $A(g^{-1})a \neq 0$. In this case, every nonzero homogeneous ideal $I$ of $A$ intersects nontrivially all components of $A$. In fact, if $0 \neq a \in I \cap A(g)$ for some $g \in G$, then $0 \neq aA(g^{-1}h) \subseteq I \cap A(h)$ for $h \in G$.

If $A$ has no identity and the components $A(g)$, $g \in G$, are $K$-subspaces of $A$, then the standard extension $A^1$ of $A$ to a $K$-algebra with an identity is $G$-graded with the components $A(g)$, $1 \neq g \in G$, and $A(1)^1$. Moreover, since $\mathcal{J}_G(A)$ is a subspace of $A$, then $\mathcal{J}_G(A)^1$ is $G$-graded, so we can use the foregoing results with respect to this algebra. (If $A(g)$ are not subspaces, one can use the unitary Z-extension of $A$, because the smash product construction works for arbitrary $G$-graded rings.) Since $\mathcal{J}(A^1) = \mathcal{J}(A)$, it also follows that the assertions on $\mathcal{J}(A)$ in the following result, first established by Cohen and Montgomery in [27], are valid in the nonunitary case as well.

**Corollary 15** Let $A$ be an algebra with an identity graded by a finite group $G$. Then $\mathcal{J}(A)^{|G|} \subseteq \mathcal{J}_G(A) = \sum_{g \in G} \mathcal{J}(A) \cap A(g) \subseteq \mathcal{J}(A)$, and $\mathcal{J}(A) = \mathcal{J}_G(A)$ if $|G| \neq 0$ in $K$. Moreover, if the $G$-grading on $A$ is nondegenerated and $\mathcal{J}(A(1)) = 0$, then $\mathcal{J}(A)^{|G|} = 0$, and $\mathcal{J}(A) = 0$ if $|G| \neq 0$ in $K$.

*Proof.* Since $B\tilde{G}$ is the skew group ring of $\tilde{G}$ over $B$ by Remark 11, and $B\tilde{G} = M_G(A)$ by Lemma 9, then from Lemma 14 we know that

$$M_G(\mathcal{J}(A)^{|G|}) = \mathcal{J}(M_G(A))^{|G|} = \mathcal{J}(B\tilde{G})^{|G|} \subseteq \mathcal{J}(B)\tilde{G} \subseteq \mathcal{J}(B\tilde{G})$$
$$= \mathcal{J}(M_G(A)) = M_G(\mathcal{J}(A))$$

and

$$\mathcal{J}(B\tilde{G}) = \mathcal{J}(B)\tilde{G} \text{ if } |G| \neq 0 \text{ in } K$$

Now, Lemma 13 (iv) implies that $\mathcal{J}(B)\tilde{G} = \overline{\mathcal{J}_G(A)}P\tilde{G}$, the latter being equal to $M_G(\mathcal{J}_G(A))$ by Lemma 9 (ii) applied to $J_G(A)^1$. Consequently, $\mathcal{J}(A)^{|G|} \subseteq \mathcal{J}_G(A) \subseteq \mathcal{J}(A)$, and $\mathcal{J}(A) = \mathcal{J}_G(A)$ if $|G| \neq 0$ in $K$. Moreover, the fact that $\mathcal{J}_G(A)$ is a graded ideal of $A$ implies then that $\mathcal{J}_G(A) \subseteq \sum_{g \in G} \mathcal{J}(A) \cap A(g)$. Therefore, from Remark 12 it follows that $\mathcal{J}_G(A) = \sum_{g \in G} \mathcal{J}(A) \cap A(g)$. The equality $\mathcal{J}(A(1)) = 0$ implies that $\mathcal{J}(A) \cap A(1) \subseteq \mathcal{J}(A(1)) = 0$. Hence, by the remark preceding the corollary, $\mathcal{J}_G(A) = 0$, provided that the grading is nondegenerated. Thus, the remaining assertion follows.

As the second main application of the smash product techniques, we will give a result on the noetherian property of graded algebras. As above, an auxiliary result on skew group rings is needed. We refer to an extended version of this fact, which may be found in [168]. Recall that a $G$-graded algebra $C$ is strongly graded if $C(g)C(h) = C(gh)$ for all $g,h \in G$.

**Lemma 16** Let $C$ be an algebra with an identity strongly graded by a polycyclic-by-finite group $G$. Assume that $M$ is a graded left $C$-module. Then $M$ is noetherian if and only if the identity component $M(1)$ is a noetherian left $C(1)$-module.

The following result was obtained by Quinn [231]; see [23].

**Corollary 17** Let $C$ be an algebra with an identity graded by a polycyclic-by-finite group $G$. Then $A$ is left noetherian if and only if $A$ is graded left noetherian, that is, $A$ satisfies a.c.c. on graded left ideals.

*Proof.* The necessity is clear. Thus, assume that $A$ is graded left noetherian. We use the notation of Lemma 13. From Lemma 13 (ii) it follows that the left $B$-module $\underline{A}$ is noetherian. Lemma 16, with $C = R$, $M = A^G$, $C(1) = B$, $M(1) = \underline{A}$, implies that $A^G$ is a noetherian left $R$-module. Thus, from Lemma 13 (iii) it follows that $A$ is left noetherian.

# Comments on Chapter 6

Results on general semigroups that are graded by semilattices, referred to as semilattice decompositions, may be found in the monographs of Howie [90], and Petrich [207]. The latter discusses in detail the connections with the subdirect product decompositions. The assertion of Theorem 5 generalizes the very well-known case of commutative semigroups; see [26], Section 4.3. The case of an extension of this class of semigroups has been recently considered by Nordahl [172]. The notion of $p$-separative semigroups was introduced by Munn in [159] in the context of the radical of commutative semigroup algebras, but $p$-separative congruences arose implicitly earlier when the nil radical was studied; see [64] and our Chapter 21. We note that Gilmer [64] uses the term "$p$-torsion free" for our "$p$-separative." Also, separative semigroups are there referred to as "free of asymptotic torsion." The name comes from the fact that they are exactly semigroups satisfying the condition: if $s, t \in S$ are such that $s^n = t^n$ for all $n$ exceeding some integer $k$, then $s = t$. The classical terminology is motivated by the fact that every two elements of $S$ can be separated by a homomorphism into the multiplicative subgroup of a field of characteristic zero; see Chapter 21.

Rings graded by a semilattice were first considered by Weissglass [269] and called supplementary semilattice sums of rings. We refer to the papers of Chick and Gardner [22], and Kelarev and Volkov [112], for some general recent results and bibliography on this class of rings. Some specific problems on semigroup algebras of semigroups being semilattices of semigroups from certain classes have been extensively studied. The basic bibliography on this topic is given in Comments on Chapter 21. Finiteness conditions for rings graded by semilattices with applications to semigroup rings were discussed by Wauters in [266]. We note that Proposition 8 was essentially obtained in [22] in the setting of semilattice-graded rings, but some aspects of this result were exploited earlier by many authors, starting with a paper of Teply et al. [258].

The smash product techniques were applied to connect group gradations and actions of groups on rings via so-called duality theorems by Cohen and Montgomery [27]. One of these theorems, asserting that $(A\#G) * G \cong M_G(A)$ for a ring $A$ graded by a finite group $G$, is in fact established in Lemma 9 and Remark 11. The second result asserts that $(A * G)\#G \cong M_G(A)$ whenever $G$ is a finite group of automorphisms of a ring $A$, and $A * G$ is viewed as a $G$-graded ring with components $Ag, g \in G$. These results, together with an extension to the case of infinite groups, have lead to several theorems on the radicals, prime ideals, and dimensions of graded rings. We refer to the papers of Cohen and Montgomery [27], Quinn [231], and Chin and Quinn [23], and also to a further extension due to Beattie [9], for the main results of this type. Our exposition is based on [231].

# Part II

# Semigroup Algebras of Cancellative Semigroups

In Part I, we considered a class of semigroups strongly related to groups, which arise as principal factors of a given semigroup $S$. Cancellative semigroups form another important class of semigroups close to groups. Observe, however, that any cancellative semigroup $S$ that is $\pi$-regular must be a group. In fact, for any $s \in S$, there exists $n \geq 1$ such that $s^n$ is a regular element of $S$; that is, $s^n t, us^n$ are idempotents for some $t, u \in S$. Since $S$ is cancellative, then $s^n t, us^n$ must be the identity of $S$, so that $s$ is invertible in $S$, and $S$ is a group. Therefore, the generalizations of the class of groups considered in this part lead to a quite different direction—different properties and specific methods will be involved. Instead of relating $S$ to some groups arising from the principal factors of $S$, which is not possible for cancellative semigroups, the natural way is to consider the group generated by $S$, if it exists. Unfortunately, it is well known that there exist cancellative semigroups not embeddable in groups. Examples of this type were first constructed by Malcev; see [142]. In the case where $S \subseteq G$ for a group $G$, the properties of the algebra $K[S]$ are easier to handle through the relations with the group algebra $K[G]$. As will be apparent, the best case arises if $S$ has a group of fractions. Here, by a (right, two-sided) group of fractions, we always mean the classical Ore group of fractions; see [29], [26], § 1.10. While there are examples of subsemigroups of groups that do not have groups of fractions, many important ring theoretic assumptions of $K[S]$ imply that $S$ has a group of fractions. Thus, the relationship between $K[S]$ and $K[SS^{-1}]$ will turn out to be crucial for applications in Part IV. This is, in fact, a consequence of Proposition 8 in Chapter 1, which shows how any irreducible skew linear representation of a semigroup $S$ determines, and is dependent on (see Lemma 21), some cancellative semigroups arising from $S$. Let us note one more argument showing the role of cancellative semigroups in the theory of general semigroup algebras: that any member of an important class of so-called separative semigroups is, in fact, composed of cancellative components; see Theorem 5 in Chapter 6.

After discussing groups of fractions in Chapter 7, we turn to the problem of characterizing cancellative semigroup algebras of polynomial growth. Thus, we present the recent result of Grigorchuk extending the celebrated Gromov's theorem on groups of polynomial growth. Next, we develop an analog of a powerful tool in group algebras, so called $\Delta$-methods, for cancellative semigroups. Then, two useful classes of unique-product (u.p.) and two-unique-product (t.u.p.) (cancellative) semigroups are discussed. Algebras of polycyclic-by-finite groups form one of the most important and extensively studied classes of group algebras. The last chapter is devoted to subsemigroups of such groups.

# 7

# Groups of Fractions

We first determine some conditions under which a cancellative semigroup $S$ has a group $G$ of right fractions. It is well known that this holds if and only if $S$ satisfies the right Ore condition:

$$\text{For every } s, t \in S \qquad sS \cap tS \neq \varnothing$$

Then $G$ is unique, up to isomorphism, and may be identified with $SS^{-1}$. If $S$ satisfies also the left Ore condition, defined symmetrically, then $G = SS^{-1} = S^{-1}S$ is called the group of fractions of $S$. We refer to [26], § 1.10, [29], Chapter 12, and also to [31], Chapter 8, for a presentation of these standard results.

**Lemma 1** Assume that $S$ is a cancellative semigroup with no noncommutative free subsemigroups. Then $S$ has a two-sided group of fractions.

*Proof.* Let $s, t \in S$. Since $\langle s, t \rangle$ is not a noncommutative free semigroup, then there exist two distinct words $w(x,y)$, $v(x,y)$ in letters $x$, $y$ such that $w(s,t) = v(s,t)$. Since $S$ is left cancellative, then we may assume that $w(x,y)$, $v(x,y)$ have no common initial segment. If $w(x,y)$, $v(x,y)$ are both nonempty words, then the equality $w(s,t) = v(s,t)$ implies that $sS \cap tS \neq \varnothing$. Assume that $v(x,y)$ is the empty word. Then $w(s,t)$ is the identity of $\langle s, t \rangle$. If $x$ is the initial letter of $w(x,y)$, then $w(s,t) = su$ for some $u \in \langle s, t \rangle^1$, and so $t = w(s,t)t \in sS$. Thus, $sS \cap tS \neq \varnothing$.

Similarly, it is shown that $sS \cap tS \neq \varnothing$ if $y$ is the initial letter of $w(x,y)$ or if $w(x,y)$ is the empty word. Therefore, $S$ satisfies the right Ore condition. A symmetric argument establishes the left Ore condition for $S$.

Note that an algebra analog of this result is also true; see [29], Proposition 12.1.3. This establishes the existence of a division ring of quotients of a domain that has no free noncommutative subalgebras. More generally, Irving and Small proved in [92] that every semiprime algebra with a.c.c. on left annihilators is left and right Goldie whenever it does not contain free noncommutative subalgebras.

We now consider an important class of semigroups satisfying the hypothesis of Lemma 1. The following definition and Theorems 2 and 3 are due to Malcev [143].

Let $x$, $y$, $w_1$, $w_2$ ... be elements of a semigroup $S$. Consider the sequence of elements defined inductively as follows:

$$x_0 = x, \qquad y_0 = y$$
$$x_{n+1} = x_n w_{n+1} y_n \qquad y_{n+1} = y_n w_{n+1} x_n \qquad \text{for } n \geq 0$$

We say that the identity $X_n = Y_n$ is satisfied in $S$ if $x_n = y_n$ for all $x$, $y$, $w_1$, $w_2$, ... $\in S$.

**Theorem 2**  Let $n \geq 1$. Then the following conditions are equivalent for a group $G$:

(1)  $G$ is a nilpotent group of class $n$.
(2)  $n$ is the least positive integer for which the identity $X_n = Y_n$ is satisfied in $G$.

*Proof.*  If $n = 1$, then $xw_1y = yw_1x$ for any $x, y, w_1 \in G$, implies that $xy = yx$. Thus (1), (2) are equivalent to the commutativity of $G$. Assume that the assertion holds for some $n > 1$. For any $x \in G$, denote by $\bar{x}$ the image of $x$ in $G/Z(G)$. If $G$ is nilpotent of class $n + 1$, then $G/Z(G)$ is nilpotent of class $n$ and, by the induction hypothesis $\bar{x}_n = \bar{y}_n$, for any $x$, $y$, $w_1$, ...,

$w_{n+1} \in G$. Thus, $y_n = x_n z$ for some $z \in Z(G)$, and so

$$x_{n+1} = x_n w_{n+1} y_n = x_n w_{n+1} x_n z = x_n z w_{n+1} x_n = y_n w_{n+1} x_n = y_{n+1}$$

proving that $X_{n+1} = Y_{n+1}$ is satisfied in $G$.

Now, assume that $X_{n+1} = Y_{n+1}$ is satisfied in $G$. Then, for any $x$, $y$, $w_1, \ldots, w_{n+1} \in G$, we have $x_n w_{n+1} y_n = y_n w_{n+1} x_n$, and so $x_n y_n = y_n x_n$. This implies that

$$y_n^{-1} x_n w_{n+1} = w_{n+1} x_n y_n^{-1} = w_{n+1} y_n^{-1} x_n$$

that is, $y_n^{-1} x_n \in Z(G)$. Thus, $G/Z(G)$ satisfies the identity $X_n = Y_n$, and the induction hypothesis implies that $G/Z(G)$ is nilpotent of class $\leq n$. Consequently, $G$ is nilpotent of class $\leq n + 1$. The result follows.

The above result motivates the following definition. A semigroup $S$ is called weakly nilpotent of class $n$ if $S$ satisfies the identity $X_n = Y_n$ and $n$ is the least positive integer with this property. Observe that every nilpotent semigroup, that is, a semigroup $S$ with zero such that $S^m = \theta$ for some $m \geq 1$, satisfies some $X_n = Y_n$, and so it is weakly nilpotent.

**Theorem 3**  Let $S$ be a cancellative weakly nilpotent semigroup of class $n$. Then $S$ has a group of fractions that is nilpotent of class $n$.

*Proof.*  Since the identity $X_n = Y_n$ is satisfied in $S$, then $S$ has no free noncommutative subsemigroups. Consequently, $S$ has a group $G$ of fractions by Lemma 1. If $n = 1$, then $S$ is commutative, and the assertion is trivial. Let $n \geq 1$, and let $x$, $y$, $w_1, \ldots, w_n \in S$. Then, substituting $x_{n-1}, x_{n-1}^2$ in the place of $w_n$ in the equality $x_{n-1} w_n y_{n-1} = y_{n-1} w_n x_{n-1}$, we get $x_{n-1}^2 y_{n-1} = y_{n-1} x_{n-1}^2$ and $x_{n-1}^3 y_{n-1} = y_{n-1} x_{n-1}^3$, respectively. Hence, $y_{n-1} x_{n-1}^3 = x_{n-1} x_{n-1}^2 y_{n-1} = x_{n-1} y_{n-1} x_{n-1}^2$, which yields $x_{n-1}^{-1} y_{n-1} = y_{n-1} x_{n-1}^{-1}$. Further $w_n y_{n-1} x_{n-1}^{-1} = x_{n-1}^{-1} y_{n-1} w_n = y_{n-1} x_{n-1}^{-1} w_n$. Therefore, $y_{n-1} x_{n-1}^{-1} \in Z(S) \subseteq Z(G)$. This implies that the image $\bar{S}$ of $S$ in $G/Z(G)$ satisfies the identity $X_{n-1} = Y_{n-1}$. By the induction hypothesis, $G/Z(G)$ is a nilpotent group of class $\leq n - 1$ because it is a group of fractions of $\bar{S}$. Hence, $G$ is nilpotent of class $\leq n$, and the result follows.

Our next aim is to extend the class of semigroups for which the group of fractions is known to exist. If $X$ is a subset of a group $G$, then, by $\langle\langle X \rangle\rangle$ we denote the subgroup of $G$ generated by $X$.

**Lemma 4**  Let $G$ be a group generated by its subsemigroup $S$. Assume that $H$ is a subgroup of $G$ such that, for every $g \in G$, there exists $n \geq 1$ with $g^n \in H$. Then, for all $g \in G$, $s \in S$,

(i)  $H$ is the subgroup generated by $Ss \cap H$.
(ii)  $S \cap gH \neq \emptyset, S \cap Hg \neq \emptyset$.

*Proof.*   (i) We first show that $\langle\langle S \cap H \rangle\rangle = H$. Let $h \in H$, and let $h = t_1 \ldots t_m$ be the shortest presentation of $h$ as a product of elements of $S \cup S^{-1}$. If $m = 1$, then $h \in S \cap H$ or $h \in S^{-1} \cap H$, so that $h \in \langle\langle S \cap H \rangle\rangle$. Assume that $m > 1$. By the hypothesis on $H$, there exists $n \geq 1$ such that $t_1^n \in H$. Now, $h = t_1^n(t_1^{-n+1}t_2)t_3 \ldots t_m$ and $t_1^n \in \langle\langle S \cap H \rangle\rangle$. Moreover, the minimality of $m$ implies that either $t_1^{-1}, t_2 \in S$ or $t_1^{-1}, t_2 \in S^{-1}$, so that $t_1^{-n+1}t_2$ always lies in $S \cup S^{-1}$. Therefore, the element $\bar{h} = (t_1^{-n+1}t_2)t_3 \ldots t_m$ has a shorter presentation in $S \cup S^{-1}$ than $h$, and $\bar{h} \in H$. Thus, the proof may be completed by induction on $m$.

Now, consider another special case where $H = G$. Then $\langle\langle Ss \rangle\rangle \supseteq (Ss)(Ss)^{-1} = SS^{-1} \supseteq S$. Since $\langle\langle Ss \rangle\rangle$ is a group, then $\langle\langle Ss \rangle\rangle \supseteq \langle\langle S \rangle\rangle = G$, and so $\langle\langle Ss \rangle\rangle = G$.

Finally, in the general case, we know that $s^k \in H$ for some $k \geq 1$. Consequently, $Ss^k \cap H = (S \cap H)s^k$. Therefore $\langle\langle Ss \cap H \rangle\rangle \supseteq \langle\langle Ss^k \cap H \rangle\rangle = \langle\langle (S \cap H)s^k \rangle\rangle$. The first part of the proof implies that $\langle\langle S \cap H \rangle\rangle = H$, so that the second special case above shows that $\langle\langle (S \cap H)s^k \rangle\rangle = \langle\langle S \cap H \rangle\rangle = H$. Hence, we get $\langle\langle Ss \cap H \rangle\rangle \supseteq H$. The converse inclusion is clear.

(ii) Let $H^* = \bigcap_{i=1}^{n} g^i H g^{-i}$ where $n$ is such that $g^n \in H$. Let $x \in G$. For every $i \in \{1, \ldots, n\}$, there exists $y \in G$ such that $x = g^i y g^{-i}$. Then $x^m = g^i y^m g^{-i} \in g^i H g^{-i}$ for some $m \geq 1$. It follows that there exists $k \geq 1$ such that $x^k \in H^*$, so that $H^*$ inherits the hypothesis on $H$. Clearly, $H^*$ is a normal subgroup in the group $F = \langle\langle g, H^* \rangle\rangle$. From (i) we know that $S \cap F$ generates $F$ as a group. The image $\overline{S \cap F}$ of $S \cap F$ under the natural homomorphism $F \to F/H^*$ generates $F/H^*$. Since $F/H^*$ is a torsion group, then $\langle\langle \overline{S \cap F} \rangle\rangle = \overline{S \cap F}$. It follows that $\overline{S \cap F} = F/H^*$. Therefore, $S \cap F$ intersects nontrivially every coset of $H^*$ in $F$. In particular, $S \cap gH^* \neq \emptyset$. Thus, $S \cap gH \neq \emptyset$ and, also, $S \cap Hg \neq \emptyset$. This completes the proof.

We apply the above lemma to the case of groups of fractions.

**Lemma 5**  Let $S$ be a subsemigroup of a group $G$. Assume that $H$ is a subgroup of $G$ such that, for every $g \in G$, there exists $n \geq 1$ with $g^n \in H$. Then,

(i)  If $G = SS^{-1}$, then $H = (S \cap H)(S \cap H)^{-1}$.
(ii) If $S \cap H$ has a group of right fractions, then $S$ has a group of right fractions, and $SS^{-1} = S(S \cap H)^{-1}$.

*Proof.*  (i) We know that, for every $s,t \in S \cap H$, there exist $x,y \in S$ such that $sx = ty$. Then $sx^n = t(yx^{n-1})$, where $x \geq 1$ is such that $x^n \in H$. Consequently, $x^n, yx^{n-1} \in S \cap H$, showing that $S \cap H$ satisfies the right Ore condition. From Lemma 4 it follows that $\langle\langle S \cap H \rangle\rangle = H$. Therefore, the group $(S \cap H)(S \cap H)^{-1}$ coincides with $H$.

(ii) Let $s,t \in S$. The hypothesis implies that there exist $n$, $m \geq 1$ such that $s^m, t^n \in H$. Then $s^m(S \cap H) \cap t^n(S \cap H) \neq \varnothing$ because $S \cap H$ satisfies the right Ore condition. Hence, $sS \cap tS \neq \varnothing$, and so $S$ has a group of right fractions. Now, $st^{-1} = st^{m-1}t^{-m} \in St^{-m} \subseteq S(S \cap H)^{-1}$. Thus, $SS^{-1} \subseteq S(S \cap H)^{-1}$, and the result follows.

**Lemma 6**   Let $G$ be the group of right fractions of its subsemigroup $S$. Then,

(i)   For every right (left) ideal $T$ of $S$, $G$ is the group of right fractions of $T$. Moreover, if $SS^{-1} = S^{-1}S$, then $TT^{-1} = T^{-1}T$.

(ii)   For every $s \in S$, $G$ is the group of right fractions of $sSs$.

*Proof.*  (i) Let $s,t \in T$. Since $S$ satisfies the right Ore condition, then there exist $x,y \in S$ such that $s^2x = t^2y$. Then $s(sxs) = t(tys)$ and $sxs, tys \in T$. Consequently, $T$ has a group of right fractions $TT^{-1} \subseteq G$. On the other hand, from Lemma 4 it follows that $\langle\langle T \rangle\rangle = G$, so that we get $TT^{-1} = G$. If we also assume that $SS^{-1} = S^{-1}S$, which means that $G$ is the two-sided group of fractions of $S$, then the foregoing and its left-right symmetric analog imply that $TT^{-1} = SS^{-1} = S^{-1}S = T^{-1}T$.

(ii) If $s \in S$, then $(sSs)(sSs)^{-1} = sSS^{-1}s^{-1} = SS^{-1}$ because $SS^{-1}$ is a group containing $s$.

If $G$ is a group, then the set $\Delta(G) = \{g \in G \mid [G:C_G(g)] < \infty\}$ is called the $FC$-center of $G$. If it coincides with $G$, then $G$ is called an $FC$-group. Note that the condition $[G:C_G(g)] < \infty$ is equivalent to the fact that $g$ has finitely many conjugates in $G$.

We will often use the following basic result on $FC$-groups.

**Lemma 7**   Let $G$ be an $FC$-group. Then,

(i)   $G/Z(G)$ is a periodic group, and it is finite if $G$ is finitely generated.

(ii)   If $G$ has an abelian subgroup of finite index, then $G/Z(G)$ is finite.

*Proof.*  (i) Let $g \in G$. Since $[G:C_G(g)] < \infty$, then there exist $n \geq 1$ and elements $x_1, \ldots, x_n \in G$ such that $G = \bigcup_{i=1}^n C_G(g)x_i$. Put $H = C_G(g) \cap \bigcap_{i=1}^n C_G(x_i)$. Since $[G:C_G(x_i)] < \infty$ for any $i$, then also $[G:H] < \infty$. Thus, there exists $m \geq 1$ such that $g^m \in H$. Then $g^m$ commutes with all

$x_i$, and so $g^m \in Z(G)$. This shows that $G/Z(G)$ is a periodic group. Now, if $G$ is finitely generated, then $Z(G)$ coincides with the intersection of the centralizers of the generators of $G$. Hence, $Z(G)$ has finite index in $G$.

(ii) If $G = \bigcup_{i=1}^{n} Ag_i$, $g_i \in G$, and $A$ is an abelian group, then, as above, $A \cap \bigcap_{i=1}^{n} C_G(g_i)$ has finite index in $G$. Clearly, this subgroup lies in $Z(G)$, so that $[G : Z(G)] < \infty$.

**Corollary 8** Let $S$ be a subsemigroup of a group $G$ such that $G/\Delta(G)$ is a torsion group. Then $S$ has a two-sided group of fractions, $Z(S) \neq \varnothing$, and $SS^{-1} = S(S \cap Z(\Delta(G)))^{-1}$. Moreover, if $G = \Delta(G)$, then $SS^{-1} = SZ(S)^{-1}$ and the commutator subgroup of the group $SS^{-1}$ coincides with the subsemigroup of $SS^{-1}$ generated by the set $\{sts^{-1}t^{-1} \mid s,t \in S\}$.

*Proof.* From Lemma 7 it follows that for every $s,t \in S$ there exists $n \geq 1$ such that $s^n t^n = t^n s^n$ so that $\langle s, t \rangle$ is not a free noncommutative semigroup. Thus, Lemma 1 implies that $S$ has a group of fractions $H \subseteq G$. Replacing $G$ by $H$, we can assume that $G = SS^{-1}$. From Lemmas 5 and 7 it then follows that $G = S(Z(\Delta(G)) \cap S)^{-1}$ because $G/Z(\Delta(G))$ is a torsion group. If $G = \Delta(G)$, then $Z(\Delta(G)) \cap S \subseteq Z(S)$, so that $G = SZ(S)^{-1}$. Let $u,z \in Z(S)$, $s,t \in S$. Then $(su^{-1})(tz^{-1})(su^{-1})^{-1}(tz^{-1})^{-1} = sts^{-1}t^{-1}$. Further, $(sts^{-1}t^{-1})^{-1} = tst^{-1}s^{-1}$. It follows that $\langle \{sts^{-1}t^{-1} \mid s,t \in S\} \rangle$ is a group containing the commutator subgroup of $SS^{-1}$. Consequently, these two groups must coincide.

**Proposition 9** Let $G$ be a group generated by its subsemigroup $S$. Assume that $Sg \cap W \subseteq \bigcup_{i=1}^{m} \bigcup_{j=1}^{n_i} G_i g_{ij}$ for some $m, n_i \geq 1$, elements $g, g_{ij} \in G$, and subgroups $W, G_1, \ldots, G_m$ of $G$ with $[G : W] < \infty$. Then,

(i)  There exists $r$ such that $[G : G_r] < \infty$.
(ii) If $G_r$ is the only subgroup in the set $\{G_1, \ldots, G_m\}$ of finite index in $G$ and $G = W$, then $G = \langle\langle S \rangle\rangle = \bigcup_{j=1}^{n_r} G_r g_{rj}$.

*Proof.* We can assume that all cosets $G_i g_{ij}$ are distinct.

(i) We will proceed in two steps.

Step 1: $W = G$, $g$ is the identity of $G$. Observe that, for any $s \in S$, the semigroup $Ss$ satisfies $\langle\langle Ss \rangle\rangle = G$, and $Ss$ embeds into a finite union of cosets of the subgroups $G_1, \ldots, G_m$. Thus, replacing, if necessary, $S$ by some $Ss$, $s \in S$, and omitting some of the cosets, we can assume that, if $t \in S$, then $St$ is not contained in any union of finitely many cosets of the subgroups $G_{i_1}, \ldots, G_{i_{m-1}}$ for $i_1, \ldots, i_{m-1} \in \{1, \ldots, m\}$ and, if $St \subseteq \bigcup_{i=1}^{m} \bigcup_{j=1}^{k_i} G_i u_{ij}$,

$u_{ij} \in G$, then $k_1 \geq n_1$. Let $s \in S$. Then,

$$Ss \subseteq G_1g_{11}s \cup \cdots \cup G_1g_{1n_1}s \cup \bigcup_{i=2}^{m}\bigcup_{j=1}^{n_i} G_ig_{ij}s$$

The minimality of $n_1$ implies that, for every $j \in \{1,\ldots,n_1\}$, $Ss \cap G_1g_{1j}s$ is not contained in a union of finitely many cosets of the subgroups $G_2$, ..., $G_m$. Thus, the fact that

$$Ss \subseteq S \subseteq G_1g_{11} \cup \cdots \cup G_1g_{1n_1} \cup \bigcup_{i=2}^{m}\bigcup_{j=1}^{n_i} G_ig_{ij}$$

implies that $Ss \cap G_1g_{1j}s$ intersects nontrivially some $G_1g_{1k}$, $k \leq n_1$. This means that $G_1g_{1j}s = G_1g_{1k}$. Since $G_1g_{1j} \neq G_1g_{1i}$ for $i \neq j$, then we get $\bigcup_{j=1}^{n_1} G_1g_{1j} = \bigcup_{j=1}^{n_1} G_1g_{1j}s$ and, consequently, $\bigcup_{j=1}^{n_1} G_1g_{1j} = \bigcup_{j=1}^{n_1} G_1g_{1j}s^{-1}$. It follows that $\bigcup_{j=1}^{n_1} G_1g_{1j}$ is fixed under right multiplication by the elements of $\langle\langle S \rangle\rangle = G$. Therefore, $G = \bigcup_{j=1}^{n_1} G_1g_{1j}$, which establishes the assertion in this case.

Step 2: The general case is as follows. Renumbering elements $g_{ij}$ and subgroups $G_i$, we can write $Sg \cap W \subseteq \bigcup_{i=1}^{l} H_ih_i$ for some $l \geq 1$, $h_i \in G$. From Lemma 4 we know that $S \cap gW \neq \varnothing$, so that there exists $t \in S$ with $gW = tW$. Then $g = tw$ for some $w \in W$. This implies that

$$Sg \cap W = Stw \cap W = (St \cap W)w$$

and so

$$St \cap W \subseteq \bigcup_{i=1}^{l} H_ih_iw^{-1}$$

Since $[G : W] < \infty$, then $[H_i : (H_i \cap W)] < \infty$ for every $i$ and, hence, $St \cap W$ is contained in a union of finitely many cosets of the subgroups $H_1 \cap W$, ..., $H_l \cap W$ in $W$. Now, Lemma 4 implies that $St \cap W$ generates $W$ as a group, and the first part of the proof (applied to $St \cap W$ in place of $S$ and $W$ in place of $G$) implies that $[W : (H_r \cap W)] < \infty$ for some $r$. It then follows that $[G : (H_r \cap W)] < \infty$ and, consequently, $[G : H_r] < \infty$. This completes the proof of (i).

(ii) Assume now that

(a)
$$S \subseteq \bigcup_{i=1}^{l} Zz_i \cup \bigcup_{i=1}^{n} H_ih_i$$

for some subgroups $Z$, $H_1$, ..., $H_n$ of $G$ such that $[G : Z] < \infty$ and $[G : H_i] = \infty$, $i = 1, ..., n$. If some coset $Zx$, $x \in G$, does not appear in this presentation, then

$$S \cap Zx \subseteq \bigcup_{i=1}^{n} H_i h_i$$

Since $x$ can be chosen from $S$ by Lemma 4, then $(S \cap Z)x \subseteq S \cap Zx$, and so $S \cap Z \subseteq \bigcup_{i=1}^{n} H_i h_i x^{-1}$. This contradicts (i) in view of the hypothesis on the indices of the $H_i$ and $Z$. It follows that all cosets of $Z$ appear in (a), and so $G = \bigcup_{i=1}^{l} Z z_i$, which establishes (ii).

It may be shown that under the assumptions of Proposition 9, the cosets of the $G_i$ that are of infinite index in $G$ always can be omitted. Moreover, $[G : G_r]$ is bounded by a function of $n = \sum_{i=1}^{m} n_i$.

We can now extend the assertion of Theorem 3 to a class of semigroups crucial for the considerations in Chapter 8. Let $T$ be a subsemigroup of $S$. We say that $T$ has finite index in $S$ if there exists a finite subset $F$ of $S$ such that, for every $s \in S$, there exists $f \in F$ with $sf \in T$. This notion extends that of a subgroup of finite index, as the following result shows.

**Corollary 10**  (1) Let $H$ be a subgroup of finite index in a group $G$. If $S \subseteq G$ is a subsemigroup that generates $G$ as a group, then $S \cap H$ is a subsemigroup of finite index in $S$.
(2) Assume that $T$ is a subsemigroup of finite index in $S$, and let $H$ be a group of right fractions of $T$. Then $S$ has a group of right fractions $G \supseteq H$ and $[G : H] < \infty$.

*Proof.*  (1) By Lemma 4 we can choose a finite set $F \subseteq S$ of coset representatives for $H$ in $G$. Then, for every $s \in S$, there exists $f \in F$ with $sf \in H$. Hence, $sf \in S \cap H$ and (1) follows.
(2) Let $F \subseteq S$ be a finite subset chosen for $T$ as in the above definition. If $s_1, s_2 \in S$, then $s_1 f_1, s_2 f_2 \in T$ for some $f_1, f_2 \in F$. Thus, $s_1 f_1 T \cap s_2 f_2 T \neq \varnothing$ by the assumption on $T$, and so $S$ has a group $G \supseteq H$ of right fractions. Since $S \subseteq TF^{-1} \subseteq HF^{-1}$, then Proposition 9 implies that $[G : H] < \infty$.

Our next result follows directly from Corollary 10 and Theorem 3.

**Theorem 11**  The following conditions are equivalent for a cancellative semigroup $S$:

(i)   $S$ has a weakly nilpotent subsemigroup $T$ of finite index.
(ii)  $S$ has a group of fractions that is nilpotent-by-finite.

Semigroups characterized in the above theorem will also be called almost weakly nilpotent semigroups.

We list two other important cases in which $S$ has a group of right fractions.

**Proposition 12** Let $S$ be a cancellative semigroup such that either of the following holds:

(i) $S$ has the a.c.c. on right ideals.
(ii) $K[S]$ has a finite right Goldie dimension.

Then $S$ has a group of right fractions.

*Proof.* Let $s, t \in S$. Suppose that $sS \cap tS = \varnothing$. Then $s^i tS \cap s^j tS = \varnothing$ for every $i, j \geq 1, i \neq j$, because $S$ is cancellative. Now, $\bigcup_{i=1}^{n} s^i tS, n \geq 1$, is an ascending chain of right ideals of $S$, which contradicts (i). Further, $\sum_{i=1}^{\infty} K[s^i tS]$ is a direct sum of right ideals $K[s^i tS]$ of $K[S]$, which contradicts (ii). Thus, the right Ore condition is satisfied in $S$, and the result follows.

We start presenting advantages coming from the fact that $S$ has a group of (right) fractions. The general aim is to connect problems concerning the algebra $K[S]$ to their analogs for $K[SS^{-1}]$ and then to exploit the group algebra results.

The following basic result is well known; see [203], Lemma 10.2.13.

**Lemma 13** Let $T$ be a right Ore subset of a ring $R$. Then,

(i) For every $a_1, \ldots, a_n \in RT^{-1}$, there exists $t \in T$ such that $a_i t \in R$, $i = 1, \ldots, n$.
(ii) For every right ideal $I$ of $RT^{-1}$, we have $(I \cap R)RT^{-1} = I$.

Moreover, if $Z$ is a right Ore subset of a cancellative semigroup $S$, then $Z$ is a right Ore subset in $K[S]$ and $K[S]Z^{-1} = K[SZ^{-1}]$.

Our first result is an easy observation on the behavior of primeness and semiprimeness under localizations.

**Lemma 14** Let $B$ be a ring that is the localization of its subring $A$ with respect to a right Ore subset $Z$. Then,

(i) $B$ is prime (semiprime) whenever $A$ is prime (semiprime, respectively).
(ii) If $Z \subseteq Z(B)$, then the converse holds.

*Proof.* (i) is a consequence of the fact that every nonzero ideal of $B$ nontrivially intersects $A$; see Lemma 13.

(ii) If $0 \neq I, J$ are ideals of $A$ with $IJ = 0$, then $(BI + I)(JB + J) = 0$ and $BI + I, JB + J$ are ideals of $B$. For example, $B(BI + I)B \subseteq (BI + I)B = (BI + I)AZ^{-1} \subseteq Z^{-1}(BI + I) \subseteq BI + I$. The result follows.

We now state a result that will be crucial for studying the relationship between the properties of $K[S]$ and $K[SS^{-1}]$.

**Lemma 15**  Let $G$ be a group of right fractions of its subsemigroup $S$. Then,

(i)   For any right ideals $R_1 \subseteq R_2$ of $K[G]$, we have $R_1 \cap K[S] \subseteq R_2 \cap K[S]$.

(ii)  If $P$ is a prime ideal of $K[G]$ and $K[G]/P$ is a Goldie ring, then $P \cap K[S]$ is a prime ideal of $K[S]$.

(iii) If all prime homomorphic images of $K[G]$ are Goldie rings, then $\mathcal{B}(K[S]) = \mathcal{B}(K[G]) \cap K[S]$.

*Proof.* (i) is a direct consequence of Lemma 13.

(ii) Since $K[G]$ is a localization of $K[S]$, then $K[G]/P$ is a localization of $K[S]/(P \cap K[S])$. Thus, the hypothesis implies that the latter is an order in a prime artinian ring of quotients of $K[G]/P$. By [86], Theorem 7.2.3, $K[S]/(P \cap K[S])$ is a prime ring.

(iii) Since the Baer radical property is hereditary on subrings (see [93], Theorem VIII.2.3), and $\mathcal{B}(K[G]) \cap K[S]$ is an ideal of $K[S]$, then $\mathcal{B}(K[G]) \cap K[S] \subseteq \mathcal{B}(K[S])$. From (ii) it follows that $\mathcal{B}(K[G]) \cap K[S] = \bigcap_P (P \cap K[S])$, where $P$ runs over the set of primes of $K[G]$, is a semiprime ideal of $K[S]$. Thus, it contains $\mathcal{B}(K[S])$, and (iii) follows.

We now turn to the problem of finding conditions under which $K[S]$ is prime or semiprime. First, we summarize the group algebra results (see [203], Section II.2).

**Lemma 16**  Let $G$ be a group. Then the following conditions are equivalent:

(i)   $K[G]$ is prime (semiprime, respectively).

(ii)  $G$ has no nontrivial finite normal subgroups (ch$(K) = 0$, or ch$(K) = p > 0$ and $G$ has no finite normal subgroups of order divisible by $p$).

(iii) The $FC$-center of $G$ is torsion-free abelian (ch$(K) = 0$, or ch$(K) = p > 0$ and the $FC$-center of $G$ has no $p$-torsion).

(iv)  $Z(K[G])$ is prime (semiprime).

We recall from [203] that a subset $T$ of a group $G$ is large if, for all subgroups $W$ of finite index in $G$, $T \cap W$ cannot be covered by a finite union of right cosets of subgroups of $G$ of infinite index. We say that $T$ is very large if $T$ and all its right translates $Tg$, $g \in G$, are large. The following result of Passman is crucial for applications of very large subsets; see [203], Section II.2.

**Lemma 17** Let $T$ be a very large subset of a group $G$. If $\sum_{i=1}^{n} a_i x b_i = 0$ for some fixed $n \geq 1$, $a_i, b_i \in K[G]$, and every $x \in T$, then this identity holds for every $x \in G$.

Since our Proposition 9 may be reformulated to say that $S$ is a very large subset of the group $G = \langle\langle S \rangle\rangle$, then the following is straightforward.

**Corollary 18** Assume that $G = \langle\langle S \rangle\rangle$. If an identity $\sum_{i=1}^{n} a_i x b_i = 0$, $n \geq 1$, $a_i, b_i \in G$, holds for $x \in S$, then it holds for every $x \in G$.

Now, we state the first main application of the above reasoning. The result is essentially due to Okniński [186].

**Theorem 19** Let $G$ be the group generated by its subsemigroup $S$. Then,

(i) If $K[G]$ is prime (semiprime), then $K[S]$ is prime (semiprime, respectively).

(ii) Assume that $G$ is the group of right fractions of $S$. Then the following conditions are equivalent:

    (1) $K[S]$ is prime (semiprime).

    (2) $K[G]$ is prime (semiprime).

    (3) $G$ has no nontrivial finite normal subgroups (ch($K$) = 0, or ch($K$) = $p > 0$ and $G$ has no finite normal subgroups of order divisible by $p$).

*Proof.* (i) Assume that $IJ = 0$ for some nonzero ideals $I, J$ of $K[S]$. Then $IsJ = 0$ for every $s \in S$ and, by Corollary 18, $IgJ = 0$ for every $g \in G$. Thus, $(K[G]IK[G])(K[G]JK[G]) = 0$, and so the primeness (semiprimeness) of $K[G]$ implies the respective condition on $K[S]$.

(ii) Since $K[G]$ is a localization of $K[S]$ by Lemma 13, then the equivalence of (1) and (2) follows from (i) and Lemma 14. Thus, the assertion follows from Lemma 16.

A commutative semigroup $S$ is called torsion-free if, for every $s, t \in S$, we have, $s = t$ whenever $s^n = t^n$ for some $n \geq 1$. If $S$ is a commutative cancellative semigroup and $(st^{-1})^n$ is the identity of the group $SS^{-1}$, then clearly $s^n = t^n$. Thus, $S$ is torsion-free if and only if $SS^{-1}$ is a torsion-free group. Now, Corollary 8, Lemma 14 (or Theorem 19), and Lemma 16 yield the following consequence.

**Corollary 20** Let $S$ be a subsemigroup of an $FC$-group. Then $K[S]$ is prime if and only if $S$ is a commutative torsion-free semigroup.

We close this chapter with an important technical result extending the assertions of Lemma 15. The hypotheses of Lemma 21 will be met when considering linear representations of certain semigroups in Part IV. In fact, in view of Proposition 8 in Chapter 1, such representations will lead to the situation described below.

**Lemma 21**  Let $J$ be a completely 0-simple ideal of a semigroup $U$, and let $H \neq \theta_J$ be a subgroup of $J$. Assume that $S$ is a subsemigroup of $U$ such that $H$ is the group of right fractions of $S \cap H$. Then,

(i)   For every right ideal $I$ of $K[H]$, there exists a right ideal $I_S$ of $K_0[S]$ such that $I_S \cap K[H] = I \cap K_0[S]$ and $(I_S \cap K[H])K[H] = I$. Moreover, if $J$ is a right ideal of $K[H]$ such that $J \supseteq I$, then the ideal $J_S$ may be chosen so that $J_S \supseteq I_S$.

   Assume further that every prime homomorphic image of $K[H]$ is a Goldie ring. Then,

(ii)  for every prime ideal $Q$ of $K[H]$, then there exists a prime ideal $Q^S$ of $K_0[S]$ such that $Q^S \cap K[H] = Q \cap K_0[S]$. Moreover, if $P$ is a prime ideal of $K[H]$ with $P \supseteq Q$, then the ideal $P^S$ may be chosen so that $P^S \supseteq Q^S$.

*Proof.*  (i) Let $\bar{I}$ be the right ideal of $K_0[U]$ generated by $I$. Since, by Lemma 4 (vi) in Chapter 1, $HU = HeJ$, where $e$ is the identity of $H$, then $\bar{I}$ is the right ideal of $K_0[eJ]$ generated by $I$. Moreover, $\bar{I}$ is a $K$-subspace because $I$ is a $K$-subspace. From Lemma 21 and Example 22, both in Chapter 4, it then follows that $\bar{I} \cap K[H] = I$. Now, put $I_S = \bar{I} \cap K_0[S]$. Since $\bar{I}$ is a right ideal of $K_0[U]$, then $I_S$ is a right ideal of $K_0[S]$. Further, $I_S \cap K[H] = \bar{I} \cap K_0[S] \cap K[H] = I \cap K_0[S]$. Finally, $(I \cap K_0[S])K[H] = (I \cap K[S \cap H])K[H] = I$ by Lemma 13, and (i) follows.

   If $J$ has a zero element, then we denote it by $\theta$.

   (ii) Let $Q'$ be the ideal of $K_0[S]$ generated by $Q \cap K_0[S]$. If $s \in S \cap H$, then there exists $h \in H$ such that $sh = hs$ is the identity $e$ of $H$. Hence, $eS^1e = hsS^1sh \subseteq h(H \cup \theta)h \subseteq H \cup \theta$. Then

$$Q \cap K_0[S] \subseteq Q' \cap K_0[H] = K_0[S^1](Q \cap K_0[S])K[S^1] \cap K[H]$$
$$\subseteq eK_0[S^1]e(Q \cap K_0[S])eK_0[S^1]e \subseteq K[H](Q \cap K_0[S])K[H] \subseteq Q$$

Consequently, $Q' \cap K[H] = Q \cap K_0[S]$ and Zorn's lemma may be used to find an ideal $Q^S$ of $K_0[S]$ that is maximal with respect to the property $Q^S \cap K[H] = Q \cap K_0[S]$. We claim that $Q^S$ is a prime ideal of $K_0[S]$. Let $J_1, J_2$ be ideals of $K_0[S]$ with $J_1, J_2 \supseteq Q^S$. Then $J_i \cap K[H] \supseteq Q \cap K_0[S]$ for $i = 1,2$. If $J_1 J_2 \subseteq Q^S$, then $(J_1 \cap K[H])(J_2 \cap K[H]) \subseteq Q^S \cap K[H] = Q \cap K_0[S]$.

The primeness of $Q$ implies, in view of Lemma 15, that $Q \cap K_0[S]$ is a prime ideal of $K[S \cap H]$ (we use the additional hypothesis on the prime factors of $K[H]$ here). Thus, $J_i \cap K[H] \subseteq Q \cap K_0[S]$ for some $i \in \{1, 2\}$, a contradiction. This shows that $Q^S$ is a prime ideal of $K_0[S]$.

Observe that $eQ^S e = h(sQ^S s)h \subseteq h(Q^S \cap K[H])h \subseteq hQh \subseteq Q$. Therefore,

$$
\begin{aligned}
P \cap K_0[S] &\subseteq (Q^S + K_0[S^1](P \cap K_0[S])K_0[S^1]) \cap K[H] \\
&\subseteq e(Q^S + K_0[S^1]PK_0[S^1])e \cap K[H \cap S] \\
&= (eQ^S e + eK_0[S^1]ePeK_0[S^1]e) \cap K_0[S] \\
&\subseteq (Q + P) \cap K_0[S] \subseteq P \cap K_0[S]
\end{aligned}
$$

and so $(Q^S + K_0[S^1](P \cap K_0[S])K_0[S^1]) \cap K[H] = P \cap K_0[S]$. Hence, by Zorn's lemma, there exists an ideal $P^S$ of $K_0[S]$ that is maximal with respect to the properties $Q^S \subseteq P^S$, $P^S \cap K[H] = P \cap K_0[S]$. As above, one shows that $P^S$ is a prime ideal of $K_0[S]$. Clearly, $Q^S \subseteq P^S$ because $Q \cap K_0[S] \subseteq P \cap K_0[S]$. This completes the proof of (ii).

## Comments on Chapter 7

Some of the auxiliary lemmas used in this chapter are "folklore" among semigroup and ring theorists. This is also the case with respect to the basic Lemma 1 which, however, may be found in the paper of Makar–Limanov [140]. The same author established the existence of noncommutative free subsemigroups in all division algebras over nondenumerable fields [141]. Some other aspects of the absence of free noncommutative subsemigroups in $S$ have also been studied. In particular, Shneerson [248], [249] proved that if the number $n$ of defining relations in a finitely presented semigroup $S$ is less than the number $m$ of generators, then $S$ has no free noncommutative subsemigroups if and only if $S$ satisfies a nontrivial identity. Moreover, this is never the case if $n < m - 1$. For some other results on this topic, including Adjan's contribution, we refer to the survey paper of Shevrin and Volkov [247]. The case of subsemigroups of polycyclic-by-finite groups is discussed in Chapter 11. A discussion of various aspects of groups of quotients is given in [31]. The problem of embeddability of semigroups in groups is surveyed in [26]. For a discussion of some necessary and sufficient conditions for $K[S]$ to satisfy the Ore condition, we refer to a recent paper of Engel and Makar–Limanov [53a]. Weakly nilpotent semigroups were further studied by Lallement in [131]. He proved that regular semigroups of this type exactly are semilattices of nilpotent groups of bounded nilpotency class. Moreover, finitely generated weakly nilpotent regular semigroups are residually finite. The terminology "weakly nilpotent" may seem unfortunate. We have to make a distinction, however, between this class and the class of semigroups satisfying $S^n = \theta$ for some $n \geq 1$ and called nilpotent (semigroups). This choice is motivated by the much greater use of the latter class throughout the text, and it is usually exploited in the context of the corresponding nilpotent contracted semigroup algebras. We refer to [203], [259] for the basic facts on $FC$-groups and their role in the theory of group algebras. Finally, we note that Proposition 9 is an extended version of a result obtained in [186].

# 8

# Semigroups of Polynomial Growth

Let $S$ be a finitely generated semigroup with a generating set $\{s_1, \ldots, s_n\}$. By $d_S(m)$ we denote the number of elements in $S$ that may be presented as products of at most $m$ generators $s_i$. $S$ is said to have polynomial growth if there exist non-negative numbers $C, d$ such that

$$d_S(m) \leq Cm^d \qquad \text{for all } m \geq 1$$

In this case, the limit $d_S = \limsup_{m \to \infty} \log_m d_S(m)$ is finite, is not dependent on the choice of the generating set, and is called the exponent (or degree) of the growth of $S$. Clearly, $d_S = 0$ exactly when $S$ is a finite semigroup. We refer to [121] for basic results on the growth of groups and algebras.

Groups of polynomial growth were described by Gromov in [80]. Namely, a finitely generated group $G$ has polynomial growth if and only if it is nilpotent-by-finite, that is, $G$ has a nilpotent subgroup $N$ of finite index. Moreover, in this case, $N$ is a finitely generated group, and the exponents of growth of $N$ and $G$ are equal; see [121], § 11. Therefore, the growth of $G$ may be computed using Bass's formula for finitely generated nilpotent groups. Namely,

$$d_G = d_N = \sum_{j=1}^{p} jr_j$$

where $N = N_1 \supset N_2 \supset \cdots \supset N_{p+1} = \{1\}$ is the lower central series for $N$, and $r_j$ is the torsion-free rank of the quotient $N_j/N_{j+1}$.

In this chapter, we present an extension of Gromov's theorem to the class of cancellative semigroups. It is due to Grigorchuk [78].

We start with the following simple observation.

**Lemma 1**   Let $S$ be a cancellative semigroup of polynomial growth. Then $S$ has no free noncommutative subsemigroups and $S$ has a group of fractions.

*Proof.*   If $S$ has a free noncommutative semigroup $T = \langle x, y \rangle$, then, choosing a generating set of $S$ containing $x$, $y$, we see that $d_S(m) \geq 2^m$ for every $m \geq 1$. This contradiction, in view of Lemma 1 in Chapter 7, proves the assertion.

We will further assume that $S$ has a group of fractions $G$. We fix a generating set $\{s_1, \ldots, s_n\}$ of $S$. Then $G$ is generated as a semigroup by the set $\{s_1, \ldots, s_n, s_1^{-1}, \ldots, s_n^{-1}\}$. If $w$ is a word in the alphabet $s_1, \ldots, s_n$, then the length of $w$ is denoted by $l(w)$. The length of the shortest presentation of an element $s \in S$ in the generators $s_1, \ldots, s_n$ is denoted by $\partial_S(s)$. Similarly, for every $g \in G$, $\partial_G(g)$ stands for the length of the shortest presentation of $g$ in $s_1, \ldots, s_n, s^{-1}, \ldots, s_n^{-1}$.

Our first aim is to determine some quantitative relations between the presentations of an element $g$ of $G$ in the generators $s_1, \ldots, s_n, s_1^{-1}, \ldots, s_n^{-1}$, and its presentations as $st^{-1}$ with $s, t \in S$.

Let $g \in G$. If $\partial_G(g) = m$, then $g = s_{i_1}^{\epsilon_1} s_{i_2}^{\epsilon_2} \ldots s_{i_m}^{\epsilon_m}$, where $\epsilon_j \in \{-1, 1\}$ and $i_j \in \{1, \ldots, n\}$. This determines a presentation

(a)                    $$g = u_1 u_2^{-1} u_3 u_4^{-1} \ldots u_{4k-3} u_{4k-2}^{-1} u_{4k-1} u_{4k}^{-1}$$

where all $u_j$ are words in $s_1, \ldots, s_n$ with $\sum_{j=1}^{4k} l(u_i) = m$. If we assume that $u_j$ are all nonempty words, except possibly $u_1$, $u_{4k-2}$, $u_{4k-1}$, $u_{4k}$, then this presentation is unique. We call the words $u_j$, $j = 1, \ldots, 4k$, the components of this presentation, while all $u_{4j+1} u_{4j+2}^{-1} u_{4j+3} u_{4j+4}^{-1}$ are referred to as quadruples occurring in (a). It is clear that $k$ (= the number of quadruples) does not exceed $m/4 + 1$.

We first consider the quadruple $u_1u_2^{-1}u_3u_4^{-1}$. Since $S$ satisfies the right Ore condition, then there exist $x, y \in S$ such that $u_2^{-1}u_3 = xy^{-1}$. Then,

$$u_1u_2^{-1}u_3u_4^{-1} = u_1xy^{-1}u_4^{-1} = v_1v_2^{-1}$$

where $v_1 = u_1x$, $v_2 = u_4y$ are elements of $S$. We will call the above transformation of a quadruple a convolution. We proceed with the convolutions for the remaining quadruples in (a). This leads to a new presentation of $g$ as a product of quadruples,

(b) $$g = v_1v_2^{-1}v_3v_4^{-1}\ldots v_{4l-3}v_{4l-2}^{-1}v_{4l-1}v_{4l}^{-1}$$

Here the number of quadruples $l$ does not exceed $\frac{1}{2}[(m/4) + 1] + 1$. Presentation (b) will be called a one-step convolution of (a).

Applying the one-step convolution to $g$ successively $j$ times, $j \geq 1$, we come to a presentation of $g$ with the number of quadruples at most

$$\underbrace{\frac{1}{2}\left\{\cdots\frac{1}{2}\left[\frac{1}{2}\left(\frac{m}{4}+1\right)+1\right]+\cdots+1\right\}}_{j}+1$$

The latter does not exceed $(m/2^{j+2}) + 1 + \frac{1}{2} + \frac{1}{4} + \cdots + \frac{1}{2^j} \leq (m/2^{j+2}) + 2$. Thus, after at most $[\![\log_2 m]\!] - 1$ one-step convolutions, we come to a presentation with at most two quadruples. (Here $[\![x]\!]$ denotes the maximal integer not exceeding $x$.) Therefore, after at most $[\![\log_2 m]\!] + 1$ steps, we come to

(c) $$g = w_1w_2^{-1}$$

where $w_1, w_2 \in S$. This presentation of $g$ will be called a convolution of (a). We will also need the following simple observation.

**Lemma 2** Let $G = G_1 \supseteq G_2 \supseteq \cdots$ be the lower central series of a group $G$. Then,

(1) If $s \in G$, $t \in G_k$ for some $k \geq 1$, then the elements $[s,t]^j$, $[s^j, t]$, $[s, t^j]$ lie in the same coset of $G_{k+2}$ in $G$.

(2) If $G$ is generated by a subset $A_1 = \{s_1, \ldots, s_n\}$, and $A_{k+1} = \{[s,t] \mid s \in A_1, t \in A_k\}$ for $k \geq 1$, then the image of $A_{k+1}$ in $G_{k+1}/G_{k+2}$ generates $G_{k+1}/G_{k+2}$ as a group for all $k \geq 0$.

*Proof.* Passing to the group $G/G_{k+2}$, we can assume that $G_{k+2} = \{1\}$, that is, $G_{k+1} \subseteq Z(G)$. For every $j \geq 1$, $s, t, s_1, \ldots, s_j, t_1, \ldots, t_j \in G$, we have

$$[s, t_1 \ldots t_j] = st_1 \ldots t_j s^{-1}(t_1 \ldots t_j)^{-1} = st_1 s^{-1}[s, t_2 \ldots t_j]t_1^{-1}$$
$$[s_1 \ldots s_j, t] = s_1 \ldots s_j t(s_1 \ldots s_j)^{-1}t^{-1} = s_1[s_2 \ldots s_j, t]ts_1^{-1}t^{-1}$$

(1) We proceed by induction on $j$. If $j > 1$, then the induction hypothesis implies that

$$[s^{j-1}, t] = [s, t^{j-1}] = [s, t]^{j-1} \in Z(G)$$

Hence, putting $t_1 = \cdots = t_j = t, s_1 = \cdots = s_j = s$, we get

$$[s^j, t] = [s, t][s^{j-1}, t] = [s, t]^j$$
$$[s, t^j] = [s, t][s, t^{j-1}] = [s, t]^j$$

as desired.

(2) We proceed by induction on $k$ here. The case where $k = 0$ is clear. Let $[s, z] \in G_{k+1}, s \in G, z \in G_k$ for some $k \geq 1$. By the induction hypothesis, $z \in \langle\!\langle A_k \rangle\!\rangle G_{k+1}$. Since $G_{k+1} \subseteq Z(G)$, then $[s, z] = [s, t]$ for some $t \in \langle\!\langle A_k \rangle\!\rangle$. Then, using the above identities, we deduce that $[s, z] \in \langle \{ [x, y] \mid x \in A_1 \cup A_1^{-1}, y \in A_k \cup A_k^{-1} \} \rangle$. If $x \in A_1, y \in A_k$, then

$$[x^{-1}, y] = (x^{-1}[x, y]x)^{-1} = [x, y]^{-1}$$
$$[x, y^{-1}] = (y^{-1}[x, y]y)^{-1} = [x, y]^{-1}$$
$$[x^{-1}, y^{-1}] = (xy)^{-1}[x, y](xy) = [x, y]$$

because $[x, y] \in Z(G)$. Therefore, $[s, z] \in \langle\!\langle \{ [x, y] \mid x \in A_1, y \in A_k \} \rangle\!\rangle = \langle\!\langle A_{k+1} \rangle\!\rangle$, and so $G_{k+1} = \langle\!\langle A_{k+1} \rangle\!\rangle$.

As a notational convenience, $d_S(\lambda)$ is also defined for every number $\lambda$ by $d_S(\lambda) = d_S([\![\lambda]\!])$.

**Theorem 3**   Let $S$ be a finitely generated cancellative semigroup. Then the following conditions are equivalent:

(i)   $S$ has polynomial growth.
(ii)  $S$ has a group of fractions that is nilpotent-by-finite.
(iii) $S$ has a weakly nilpotent subsemigroup of finite index.

Moreover, if this is the case, then the exponents of growth of $S$ and $G$ coincide.

*Proof.*   The equivalence of (ii) and (iii) was established in Theorem 11 in Chapter 7. The implication (ii) $\Rightarrow$ (i) is clear because the group of fractions of $S$ has polynomial growth by Gromov's theorem.

Thus, assume that (i) holds. From Lemma 1 we know that $S$ has a group of fractions $G$. Let $S = \langle s_1, \ldots, s_n \rangle$, and view $G$ as a semigroup with generators $s_1, \ldots, s_n, s_1^{-1}, \ldots, s_n^{-1}$.

We know that $d_S(m) \leq Cm^d$, $m = 1, 2, \ldots$, for some numbers $C > 0$, $d \geq 0$. If $d = 0$, then $S$ is finite, so that $G = S$. Thus, we can further assume that $d > 0$.

We will prove that there exist numbers $M, N$ such that, for every $m \geq N$ and $g \in G$ with $\partial_G(g) \leq m$, there exists a convolution (c) of the presentation (a) of $g$ such that $l(w_i) \leq m^M$ for $i = 1, 2$. Then,

$$d_G(m) = | \{g \in G \mid \partial_G(g) \leq m\} | \leq (d_S(m^M))^2$$
$$\leq (C(m^M)^d)^2 = C^2 m^{2Md}$$

which shows that $G$ has polynomial growth and establishes the equivalence of (i) and (ii) via Gromov's theorem.

First, choose a number $\lambda$ such that

$$1 < \lambda < \min \left\{ 2^{1/d}, \frac{3}{2} \right\}$$

Then $1 - d \log_2 \lambda > 0$, and so there exists a number $D$ such that

$$D - Dd \log_2(\lambda) - \log_\lambda 2 > 9 \log_\lambda \frac{1}{\lambda - 1}$$

Note also that $\log_\lambda(1/(\lambda - 1)) > 1$. We define the following subsets of the set of natural numbers:

$$\mathcal{E} = \mathcal{E}_m = \{i \mid d_S(\lambda^{i+1}) \geq 2d_S(\lambda^i), i = 1, 2, \ldots, [\![D \log_2 m]\!]\}$$
$$\mathcal{F} = \mathcal{F}_m = \{i \mid d_S(\lambda^{i+1}) < 2d_S(\lambda^i), i \geq [\![\log_\lambda m]\!], i \leq [\![D \log_2 m]\!]\}$$

Then $d_S(\lambda^{[\![D \log_2 m]\!]+1}) \geq 2^{|\mathcal{E}|}$, which implies in view of the bound for $d_S(m)$ that

$$C \lambda^{d(D \log_2 m + 1)} \geq 2^{|\mathcal{E}|}$$

This, in turn, implies that

$$dD \log_2 \lambda \log_2 m + \log_2(C\lambda^d) \geq |\mathcal{E}|$$

Since $\mathcal{F}, \mathcal{E}$ are disjoint sets, then,

$$|\mathcal{F}| \geq |\mathcal{A}| - |\mathcal{E}|$$

where $\mathcal{A} = \{z \in \mathbb{N} \mid [\![\log_\lambda m]\!] \leq z \leq [\![D \log_2 m]\!]\}$. This implies that

$$|\mathcal{F}| \geq D \log_2 m - \log_\lambda m - |\mathcal{E}|$$
$$\geq D \log_2 m - \log_\lambda m - dD \log_2 \lambda \log_2 m - \log_2(C\lambda^d)$$
$$= (D - \log_\lambda 2 - dD \log_2 \lambda) \log_2 m - \log_2(C\lambda^d)$$

Then, by the choice of $D$, we get

$$|\mathcal{F}| \geq 9 \log_\lambda \frac{1}{\lambda - 1} \log_2 m - \log_2(C\lambda^d)$$

Since $C$, $\lambda$, $d$ do not depend on $m$, and $\log_\lambda(1/(\lambda - 1)) > 0$, then there exists a number $N > 4$ such that, for all $m \geq N$, we have

$$|\mathcal{F}| \geq 8\log_\lambda \frac{1}{\lambda - 1}\log_2 m$$

From the fact that the inequality $2 \, (\llbracket y \rrbracket + 1)(x + 1) + 4 < 8yx$ holds for every $y > 2$, $x > 1$, it follows (with $x = \log_\lambda(1/(\lambda - 1))$, $y = \log_2 m$) that a subset $\mathcal{H}$ of $\mathcal{F}$ may be chosen so that

(d)                                  $\mathcal{H} = \{i_1, i_2, \ldots, i_{2r}\}$

where $r = \llbracket \log_2 m \rrbracket + 1$ and $i_1 > \log_\lambda m + \log_\lambda(1/(\lambda - 1)), i_{j+1} - i_j > \log_\lambda(1/(\lambda - 1))$ for $j = 1, \ldots, 2r - 1$.

Now, let $g$ be an element of $G$ with a presentation (a), and assume that $m > N$. Every component $u_i$, $i = 1, \ldots, 4k$, satisfies $l(u_i) \leq m$ because $\sum_{i=1}^{4k} l(u_i) = m$. We will show that there exists a one-step convolution of the presentation (a) such that the lengths of the components in the resulting presentation of $g$ do not exceed $\lambda^{i_2}$. Let

$$u_{4i+1}u_{4i+2}^{-1}u_{4i+3}u_{4i+4}^{-1}$$

be the $i + 1$th quadruple in (a), $0 \leq i \leq k - 1$. Define subsets $B(\gamma) = \{s \in S \mid \partial_S(s) \leq \gamma\}$ for all positive numbers $\gamma$. If $u_{4i+2}B(\lambda^{i_1}) \cap u_{4i+3}B(\lambda^{i_1}) = \varnothing$, then,

$$d_S(\lambda^{i_1} + m) = |B(\lambda^{i_1} + m)| \geq 2|B(\lambda^{i_1})| = 2d_S(\lambda^{i_1})$$

because $l(u_{4i+2})$, $l(u_{4i+3}) \leq m$. On the other hand, (d) implies that $\lambda^{i_1} > m(1/(\lambda - 1))$ so that $\lambda^{i_1+1} > \lambda^{i_1} + m$. Therefore, we get

$$d_S(\lambda^{i_1+1}) \geq d_S(\lambda^{i_1} + m) \geq 2d_S(\lambda^{i_1})$$

This contradicts the fact that $i_1 \in \mathcal{F}$, and shows that there exist $x_i, y_i \in B(\lambda^{i_1})$ such that

$$u_{4i+2}x_i = u_{4i+3}y_i$$

and so

$$u_{4i+1}u_{4i+2}^{-1}u_{4i+3}u_{4i+4}^{-1} = (u_{4i+1}x_i)(u_{4i+4}y_i)^{-1} = v_{2i+1}^{-1}v_{2i+2}$$

We note that

$$l(v_j) \leq \lambda^{i_1} + m < \lambda^{i_1+1} < \lambda^{i_2} \qquad \text{for } j = 2i + 1, 2i + 2$$

as desired.

Similarly, let $v_{4i+1}v_{4i+2}^{-1}v_{4i+3}v_{4i+4}^{-1}$ be a quadruple in the presentation (b) of $g$. If $v_{4i+2}B(\lambda^{i_3}) \cap v_{4i+3}B(\lambda^{i_3}) = \varnothing$, then the fact that $i_{j+1} - i_j > \log_\lambda(1/(\lambda - 1))$ implies, as above, that

$$d_S(\lambda^{i_3+1}) \geq d_S(\lambda^{i_3} + \lambda^{i_2}) \geq 2d_S(\lambda^{i_3})$$

contradicting the fact that $i_3 \in \mathcal{F}$. Since

$$\lambda^{i_3} + \lambda^{i_2} < \lambda^{i_3+1} < \lambda^{i_4}$$

this means that we can proceed with a one-step convolution of the presentation (b) so that the components of the resulting presentation have lengths not exceeding $\lambda^{i_4}$.

As shown in the comments preceding the theorem, proceeding this way, we reach, in at most $r = [\![\log_2 m]\!] + 1$ steps, a presentation (c) $g = w_1 w_2^{-1}$ with the bounds

$$l(w_j) \leq \lambda^{i_{2r}} \qquad \text{for } j = 1,2$$

Since $i_{2r} \in \mathcal{H}$, then $i_{2r} \leq D \log_2 m$ and, consequently,

$$l(w_j) \leq \lambda^{D \log_2 m} = m^{D \log_2 \lambda} \qquad j = 1,2$$

This shows that every element $g \in G$ with $\partial_G(g) = m$ may be presented as a word in $s_1, \ldots, s_n, s_1^{-1}, \ldots, s_n^{-1}$ of length not exceeding $m^{2D \log_2 \lambda}$. It is also clear that the same bound applies to all elements $h \in G$, with $\partial_G(h) < m$. This proves our original claim (put $M = D \log_2 \lambda$), and completes the proof of the equivalence of (i), (ii), (iii).

It remains to show that the exponents of growth of $S$ and $G$ are equal provided that $G$ is a nilpotent-by-finite group and $G = SS^{-1}$. Clearly, $d_S \leq d_G$. Since $d_G = d_N$ for a nilpotent subgroup $N$ of finite index in $G$, and $(S \cap N)(S \cap N)^{-1} = N$ (see Lemma 5 in Chapter 7), then we can assume that $G$ is a nilpotent group. We know that the identity $X_p = Y_p$ (see Chapter 7) is satisfied in $G$ for some natural number $p$. Let $u_1 u_2^{-1} u_3 u_4^{-1}$ be a quadruple such that

$$\partial_S(u_1) + \partial_S(u_2) + \partial_S(u_3) + \partial_S(u_4) = m$$

Substituting the elements $u_2$, $u_3$ for $x$ and $y$, respectively, and setting $w_1 = w_2 = \cdots = w_p = 1$ in the identity $X_p = Y_p$, we come to $u_2 s = u_3 t$ for some $s, t \in \langle u_2, u_3 \rangle$, which are words of length $\leq 2^p - 1$ in $u_2, u_3$. Thus,

$$\partial_S(s) < 2^p (\partial_S(u_2) + \partial_S(u_3))$$
$$\partial_S(t) < 2^p (\partial_S(u_2) + \partial_S(u_3))$$

Consequently,

$$u_1 u_2^{-1} u_3 u_4^{-1} = (u_1 s)(u_4 t)^{-1}$$

and

(e)  $\partial_S(u_1 s) + \partial_S(u_4 t) \leq 2^{p+1}(\partial_S(u_2) + \partial_S(u_3)) + \partial_S(u_1) + \partial_S(u_4)$
$< 2^{p+1}(\partial_S(u_1) + \partial_S(u_2) + \partial_S(u_3) + \partial_S(u_4)) \leq 2^{p+1}m$

It then follows that the quadruple $u_1 u_2^{-1} u_3 u_4^{-1}$ can be transformed to the form $v_1 v_2^{-1}$, where

$$\partial_S(v_1) + \partial_S(v_2) \le 2^{p+1} m$$

Let $G = G_1 \supset G_2 \supset \cdots \supset G_{p+1} = \{1\}$ be the lower central series of $G$. Inductively, we define subsets $A_k$ of $G$ by

$$A_1 = \{s_1, \ldots, s_n\},$$
$$A_{k+1} = \{[s, t] = s t s^{-1} t^{-1} \mid s \in A_1, t \in A_k\} \text{ for } k = 1, 2, \ldots, p$$

From Lemma 2 it follows that the image of $A_k$ in the quotient $G_k/G_{k+1}$ generates this group. We need the following technical result, providing an estimate on the length of presentations of elements of $G_k$ modulo $G_{k+1}$.

(f)  Let $k$, $m$ be natural numbers, and let $N$ be an integer such that $|N| \le m^k$. Assume that $t \in A_k$. Then there exist an element $t^{(N)} \in G_k$ and elements $u^{(N)}, v^{(N)} \in S$ such that $t^{(N)} (t^N)^{-1} \in G_{k+1}, t^{(N)} = u^{(N)} (v^{(N)})^{-1}$, and $\partial_S(u^{(N)}) + \partial_S(v^{(N)}) \le L_k m$, where $L_k = 2^{3^k (p+1)}$

To prove this, let us first assume that $N \ge 0$. If $k = 1$, then we set $t^{(N)} = t^N$. Since $A_1 = \{s_1, \ldots, s_n\}$, then $\partial_S(t^{(N)}) \le N \le m^k = m$, and the elements $u^{(N)} = t^{2N}$, $v^{(N)} = t^N$ satisfy the desired conditions. We proceed by induction on $k$. Assume that the assertion holds for some $k \ge 1$, and let $w = [s, t] \in A_{k+1}$, where $s \in A_1$, $t \in A_k$. Since $N \le m^k$, then $N = jm + l$, where $0 \le l < m$ and $0 \le j < m^k$. Then, by Lemma 2,

$$w^N = [s, t]^N = [s, t]^l [s, t]^{jm} \equiv [s^l, t][s^m, t^j] \bmod G_{k+2}$$

Using the elements $u^{(1)}$, $v^{(1)}$, $u^{(j)}$, $v^{(j)}$, $t^{(j)}$ existing for $t$ by the induction hypothesis, we can further write

$$[s^l, t][s^m, t^j] \equiv [s^l, t][s^m, t^{(j)}] \bmod G_{k+2} = [s^l, u^{(1)} (v^{(1)})^{-1}][s^m, u^{(j)} (v^{(j)})^{-1}]$$
$$= s^l u^{(1)} (v^{(1)})^{-1} s^{-l} v^{(1)} (u^{(1)})^{-1} s^m u^{(j)} (v^{(j)})^{-1} s^{-m} v^{(j)} (u^{(j)})^{-1}$$
$$= (s^l u^{(1)})(s^l v^{(1)})^{-1} v^{(1)} (u^{(1)})^{-1} (s^m u^{(j)})(s^m v^{(j)})^{-1} v^{(j)} (u^{(j)})^{-1}$$

The latter is a product of two quadruples. Thus, applying three times the convolution to it, we can come, in view of (e), to a presentation

$$w^N \equiv u^{(N)} (v^{(N)})^{-1} \bmod G_{k+2}$$

for some $u^{(N)}, v^{(N)} \in S$, that satisfy

$$\partial_S(u^{(N)}) + \partial_S(v^{(N)})$$
$$\le 2^{3(p+1)} [2 \partial_S(v^{(j)}) + 2 \partial_S(u^{(j)}) + 2 \partial_S(v^{(1)}) + 2 \partial_S(u^{(1)}) + 2 \partial_S(s^l) + 2 \partial_S(s^m)]$$

Then, the fact that $\partial_S(s) = 1$, together with the induction hypothesis on the lengths of $v^{(j)}$, $u^{(j)}$, $v^{(1)}$, $u^{(1)}$, implies that

$$\partial_S(u^{(N)}) + \partial_S(v^{(N)}) \le 2^{3(p+1)}(4 \cdot 2^{3^k(p+1)}m + 2m + 2l)$$
$$\le 2^{3(p+1)}(4 \cdot 2^{3^k(p+1)} + 4)m < 2^{3^{k+1}(p+1)}m$$

This proves assertion (f) in the case where $N \ge 0$. If $N < 0$, then it is enough to define $t^{(N)} = (t^{(-N)})^{-1}$, and $u^{(N)} = v^{(-N)}$, $v^{(N)} = u^{(-N)}$, respectively.

Let $k$ be an integer such that $1 \le k \le p$, and let $r_k$ denote the rank of the group $G_k/G_{k+1}$. Since $A_k$ generates $G_k$ modulo $G_{k+1}$ by Lemma 2, then there exist elements $t_{k,1}, t_{k,2}, \ldots, t_{k,r_k} \in A_k$, the images of which generate a free abelian subgroup of rank $r_k$ in $G_k/G_{k+1}$. For a fixed natural number $m$, consider the set

$$M_k = \{t_{k,1}^{(q_1)}t_{k,2}^{(q_2)}\ldots t_{k,r_k}^{(q_{r_k})} \mid |q_i| \le m^k \text{ for } i = 1,\ldots,r_k\}$$

Then $|M_k| = (2m^k + 1)^{r_k}$ and, by (f), every element of $M_k$ has length at most $r_k L_k m$ in the generators $s_1, \ldots, s_n, s_1^{-1}, \ldots, s_n^{-1}$. Moreover, the mapping,

$$M_1 \times M_2 \times \cdots \times M_p \longrightarrow G \text{ given by } (t_1, t_2, \ldots, t_p) \longrightarrow t_1 t_2 \ldots t_p \text{ for } t_i \in M_i$$

is an injection. In fact, if $t_1 \ldots t_p = t_1' \ldots t_p'$ for some $t_i' \in M_i$, then $t_1 \in t_1' G_2$ and, consequently, $t_1 = t_1'$. Then $t_2 \ldots t_p = t_2' \ldots t_p'$ and, similarly, one shows that, $t_2 = t_2'$, $\ldots$, $t_p = t_p'$. It follows that the set $Q_m = M_1 M_2 \ldots M_p \subseteq G$ contains $\prod_{k=1}^{p}(2m^k + 1)^{r_k}$ elements of length not exceeding

(g)
$$\sum_{k=1}^{p} r_k L_k m = m \sum_{k=1}^{p} 2^{3^k(p+1)} r_k \le mR \cdot 2^{3^p(p+1)}$$

where $R = \sum_{k=1}^{p} r_k$. Since $\sum_{k=1}^{p} r_k k = d_G$ by Bass's formula, then it follows that

$$\delta m^{d_G} \le |Q_m|$$

for some positive integer $\delta$ not dependent on $m$.

By (f) each element of $Q_m$ may be presented as a product of at most $R$ elements of the form $uv^{-1}$ with $u,v \in S$ and $\partial_S(u) + \partial_S(v) \le L_k m$. Therefore, applying convolution to this product at most $R$ times, we can, in view of (e) and the bound (g), represent an arbitrary element of $Q_m$ as a product $xy^{-1}$, where $x,y \in S$ and $\partial_S(x) + \partial_S(y) < R \cdot 2^{(3^p+R)(p+1)}m$. Let $\varepsilon = R \cdot 2^{(3^p+R)(p+1)}$. We have thus shown that, for every natural number $m$,

$$\delta m^{d_G} \le |B(\varepsilon m)B(\varepsilon m)^{-1}|$$

where $B(j)$ is defined as before by $B(j) = \{s \in S \mid \partial_S(s) \le j\}, j \ge 1$. Consider the mapping $\varphi$,

$$T = \{(x,y) \in B(2\varepsilon m) \times B(2\varepsilon m) \mid x = sw,\ y = tw \text{ for some } s,t,w \in B(\varepsilon m)\}$$
$$\xrightarrow{\varphi} B(\varepsilon m)B(\varepsilon m)^{-1}$$

given by $\varphi((x,y)) = xy^{-1} = (st)(tw)^{-1} = st^{-1}$. Since $\varphi$ maps $T$ onto $B(\varepsilon m)B(\varepsilon m)^{-1}$, and the inverse image of every $st^{-1}$ has at least $|B(\varepsilon m)|$ elements, then $|B(2\varepsilon m)|^2 \ge |T| \ge |B(\varepsilon m)| \cdot |B(\varepsilon m)B(\varepsilon m)^{-1}|$. This shows that

(h)  $$\frac{|B(2\varepsilon m)|^2}{|B(\varepsilon m)|} \ge |B(\varepsilon m)B(\varepsilon m)^{-1}| \ge \delta m^{d_G}$$

Suppose that there exists a natural number $t$ such that

$$|B(2\varepsilon m)| > 4^{d_G}|B(\varepsilon m)| \qquad \text{for } m > t$$

Since $|B(m)| = d_S(m) < Cm^{d_G}$ by the hypothesis, then,

$$C(2^q \varepsilon m)^{d_G} \ge |B(2^q \varepsilon m)| > 4^{qd_G}|B(\varepsilon m)| \qquad \text{for } q = 1, 2, \ldots$$

Hence,

$$\left(\frac{1}{2}\right)^{qd_G} > \frac{|B(\varepsilon m)|}{C(\varepsilon m)^{d_G}} \qquad \text{for } q = 1, 2, \ldots$$

a contradiction. This means that there exists a sequence of natural numbers $m_1, m_2, \ldots$ such that

$$|B(2\varepsilon m_j)| \le 4^{d_G}|B(\varepsilon m_j)| \qquad j = 1, 2, \ldots$$

Therefore, by (h),

$$\delta m_j^{d_G} \le \frac{|B(2\varepsilon m_j)|^2}{|B(\varepsilon m_j)|} \le \frac{16^{d_G}|B(\varepsilon m_j)|^2}{|B(\varepsilon m_j)|} = 16^{d_G}|B(\varepsilon m_j)| \qquad j = 1, 2, \ldots$$

It follows that

$$\begin{aligned}
d_S &= \limsup_m \log_m d_S(m) = \limsup_m \log_m |B(m)| \\
&\ge \limsup_j \log_{\varepsilon m_j} |B(\varepsilon m_j)| \ge \limsup_j \log_{\varepsilon m_j} \delta(\frac{m_j}{16})^{d_G} \\
&= d_G
\end{aligned}$$

Therefore, $d_S = d_G$, which completes the proof of the theorem.

## Comments on Chapter 8

The growth function $d_R(m)$ may be defined for an arbitrary finitely generated $K$-algebra $R$. Then, the Gelfand–Kirillov dimension of $R$ is defined as $GK \dim R = \limsup_m \log_m d_R(m)$. If $R = K[S]$ for a semigroup $S$, then $d_S =$

$GK$ $\dim K[S]$. We refer to [121] for an exposition of the main results on the Gelfand–Kirillov dimension. Let us note that if $GK$ $\dim K[S] = d < \infty$, then there exists $N \geq 1$ such that $\log_m d_S(m) < d + 1$ for all $m > N$. Therefore, there exists $c > 0$ such that $d_S(m) \leq cm^{d+1}$ for all $m \geq 1$, which means that the polynomial growth of $S$ is equivalent to the finiteness of $GK$ $\dim K[S]$. Thus, Gromov's theorem and Theorem 3 may be restated in terms of the corresponding semigroup algebras. In the case where $S$ is a group and $d_S < \infty$, it is also known that $d_S = \lim_m \log_m d_S(m)$; see [80], [121]. It is not clear whether the proof of Theorem 3 may be extended to establish the same for all cancellative semigroups. $GK$ $\dim R$ actually is the limit of $\log_m d_R(m)$ for several classes of algebras [121], [149], including commutative and some of their generalizations and algebras of Gelfand–Kirillov dimension not exceeding 2. The equality $GK$ $\dim R = \lim_m \log_m d_R(m)$ has some important consequences. In particular, $GK$ $\dim(R \otimes_K R') = GK$ $\dim R + GK$ $\dim R'$ for any other algebra $R'$ in this case, which is not true in general; see [121], [128]. We note that examples of groups with subexponential but not polynomial growth have been recently given by Grigorchuk in [76], [77]. Finally, more information of $GK$ $\dim K[S]$ for $PI$-algebras $K[S]$ is given in Part IV.

# 9
# $\Delta$-Methods

The class of $FC$-groups, that is, groups any element of which has finitely many conjugates, plays an important role in the theory of group algebras. Among other things, it provides an effective technique (called $\Delta$-methods) for studying this class of algebras and yields a description of the center and of prime and semiprime group algebras; see [203], Sections 4.1, 4.2. This technique was, in fact, implicitly used in the proof of Lemma 17 in Chapter 7. The aim of this chapter is to present an analog of the $FC$-center of a group in the case of an arbitrary cancellative semigroup.

Assume that $S$ is a cancellative semigroup, and let $s \in S$. If, for some $x \in S$, there exists $t \in S$ such that $xs = tx$, then, since $S$ is cancellative, $t$ is uniquely determined and will be denoted by $s^x$. Consider the set $\Delta(S)$ of all elements $s \in S$ for which any $s^x$, $x \in S$, are determined in $S$ and such that the set $\{s^x \mid x \in S\}$ is finite. If $s \in \Delta(S)$, then we put $D_S(s) = \{s^x \mid x \in S\}$. It is clear that, if $S$ embeds into a group $G$, then, for $s \in \Delta(S)$, $D_S(s)$ embeds into the set of conjugates of $s$ in $G$. Moreover, in this case, $s \in \Delta(S)$ if and only if $\{xsx^{-1} \mid x \in S\}$ is a finite subset of $S$, and then $D_S(s) = \{xsx^{-1} \mid x \in S\}$. In particular, if $S$ is a group, then it is clear that $\Delta(S)$ coincides with the $FC$-center of $S$, defined as the set of elements with finite conjugacy classes. For this reason, $\Delta(S)$ will be called the $FC$-center of $S$. We note that $\Delta(S)$ may be defined, and originally was defined by Krempa in [123], alternatively through the conditions given below.

**Lemma 1** Let $S$ be a cancellative semigroup. Then $\Delta(S)$ coincides with the set of all elements $s \in S$ for which there exists a finite subset $X_s$ of $S$ satisfying the following two conditions:

(i)   For every element $x \in S$, there exists an element $t \in S$ such that $xs = tx$.
(ii)  For every element $t \in X_s$, there exists an element $x \in S$ such that $xs = tx$.

Moreover, if $s \in \Delta(S)$, then the set $X_s$ is uniquely determined and is equal to $D_S(s)$.

*Proof.* It is clear that, for $s \in \Delta(S)$, conditions (i) and (ii) are satisfied if we choose $X_s = D_S(s)$. Thus, assume that a set $X_s \subseteq S$ satisfying (i) and (ii) have been chosen for $s \in S$. From (i) it follows that all $s^x$, $x \in S$, are defined in $S$ and $\{s^x \mid s \in S\} \subseteq X_s$. Consequently, $s \in \Delta(S)$ and (ii) implies that $\{s^x \mid x \in S\} = X_s$. The result follows.

We start by listing some basic properties of the correspondence $s \to s^x$ when it is defined.

(a)   If $s, t, x \in S$ and $s^x$, $t^x$ exist in $S$, then $(st)^x$ exists and $(st)^x = s^x t^x$.

In fact, $xs = s^x x$, $xt = t^x x$ imply that $xst = s^x xt = s^x t^x x$ and (a) follows.

(b)   Assume that $s, x, y \in S$ and $s^y$ exists in $S$. If any of the elements $s^{xy}$, $(s^y)^x$ exists in $S$, then the other also exists, and $s^{xy} = (s^y)^x$.

If $s^{xy} \in S$, then we have $s^{xy} xy = xys = xs^y y$, whence $s^{xy} x = xs^y$, and so $s^{xy} = (s^y)^x$. If $(s^y)^x \in S$, then $xys = xs^y y = (s^y)^x xy$, and the assertion also follows.

Now we are ready to establish some fundamental properties of $\Delta(S)$.

**Proposition 2** Let $S$ be a cancellative semigroup such that $\Delta(S) \neq \emptyset$. Then,

(i)   $\Delta(S)$ is a subsemigroup of $S$.

(ii)   If $s \in \Delta(S)$, then, for any $x \in S$, there exists a unique element $s_x \in D_S(s)$ such that $sx = xs_x$, and then $D_S(s) = \{s_x \mid x \in S\}$.

(iii)  If $s \in \Delta(S)$ and $t \in D_S(s)$, then $t \in \Delta(S)$ and $D_S(t) = D_S(s)$.

(iv)  For any fixed $x \in S$, the correspondence $y \to y^x$ is an automorphism of $\Delta(S)$, which maps any $D_S(s)$, $s \in \Delta(S)$, onto itself.

(v)   If $s \in \Delta(S)$, then there exists $n \geq 1$ such that $x^n \in C_S(s)$ for any $x \in S$.

*Proof.* (i) is a direct consequence of the property (a) above because it follows that $\{(st)^x \mid x \in S\} = \{s^x t^x \mid x \in S\} \subseteq D_S(s)D_S(t)$ is a finite set for any $s, t \in S$.

(ii) From the property (b), we know that, for a fixed element $x \in S$, the rule $\phi_x(y) = y^x$ determines a mapping $\phi_x : \Delta(S) \to \Delta(S)$ such that $\phi_x(D_S(s)) \subseteq D_S(s)$ for any $s \in S$. Since $y^x = z^x$, $y, z \in \Delta(S)$, implies that $xy = y^x x = z^x x = xz$, and so $y = z$, then $\phi_x$ is a one-to-one mapping. From the finiteness of $D_S(s)$, it then follows that $\phi_x(D_S(s)) = D_S(s)$. For any $t \in D_S(s)$, define $t_x$ as the element of $D_S(s)$ such that $\phi_x(t_x) = t$. It is clear that $xt_x = tx$, and the cancellativity of $S$ implies that $t_x$ is the only element of $S$ with this property. Moreover, by the definition, we have $\{s_x \mid x \in S\} \subseteq D_S(s)$. Now, replacing $D_S(s)$ by the set $\{s_x \mid s \in S\}$ in the above considerations, and using a symmetric argument involving an analog of the property (b) with respect to the correspondence $y \to y_x$, $y \in \{s_x \mid x \in S\}$, we get the converse inclusion $D_S(s) \subseteq \{s_x \mid x \in S\}$. This proves (ii).

(iii) Fix an element $x \in \Delta(S)$. By (ii), there exists an element $y \in S$ such that $s_y = s^x$. Thus, $s = (s^x)^y$, and (b) implies that, for any $z \in S$, we have $s^z = ((s^x)^y)^z = (s^x)^{zy} \in D_S(s^x)$. Consequently, $D_S(s) \subseteq D_S(s^x)$ and, again by (b), the equality follows.

(iv) By (iii), $\Delta(S)$ is a disjoint union of the subsets $D_S(s)$, $s \in \Delta(S)$. Hence, the fact that $\phi_x$ is a permutation when restricted to any $D_S(s)$, which was established in (ii), implies that $\phi_x$ is a permutation of $\Delta(S)$. This and property (a) show that $\phi_x$ is an automorphism of $\Delta(S)$.

(v) Let $s \in \Delta(S)$, and let $m = |D_S(s)|$. For any $x \in S$, consider the set $\{s_0, s_1, \ldots, s_m\}$, where $s_0 = s$, $s_k = s^{x^k}$ for $k = 1, \ldots, m$. Since $s_k \in D_S(s)$ for all $k = 0, 1, \ldots, m$, then there exist integers $i, j$ such that $0 \leq i < j \leq m$ and $s_i = s_j$. Using the property (b), we get $x^i s_{j-i} = x^{j-i}(x^i s_{j-i}) = x^{j-i}(s_j x^i) = x^{j-i}(s_i x^i) = x^{j-i} x^i s = x^j s$. Thus, $s = s_{j-i}$, which means that $x^{j-i} s = s x^{j-i}$. Since $0 < j - i \leq j \leq m$, then (v) follows with $n = m!$

**Remark 3**  The assertion (ii) of the above proposition shows that the sets of "right conjugates" and "left conjugates" of an element $s \in \Delta(S)$, by elements of $S$, coincide. It then follows easily that the definition of $\Delta(S)$ is left-right symmetric. In other words, $\Delta(S)$ may be equivalently defined as the set of all elements $s \in S$ such that an element $s_x \in S$ satisfying $sx = xs_x$ exists in $S$ for any $x \in S$ and $\{s_x \mid x \in S\}$ is a finite subset of $S$.

In the case where $S$ is a subsemigroup of a group, $\Delta(S)$ may be easily connected with the $FC$-center of the subgroup generated by $S$.

**Lemma 4**  Assume that $G$ is a group generated by its subsemigroup $S$. Then, for every $s \in \Delta(S)$, we have $D_G(s) = D_S(s)$ and $\Delta(S) = \{s \in \Delta(G) \mid xsx^{-1} \in S$ for all $x \in S\} = \{s \in \Delta(G) \mid gsg^{-1} \in S$ for all $g \in G\}$.

*Proof.*  If $s \in \Delta(S)$, then, by Proposition 2, $\{xsx^{-1} \mid x \in S\} = \{s^x \mid x \in S\} = \{s_x \mid x \in S\} = \{x^{-1}sx \mid x \in S\} = D_S(s)$. Thus, $xD_S(s)x^{-1} = D_S(s) = x^{-1}D_S(s)x$ for any $x \in S$. Since $S$ generates $G$ as a group, then $gD_S(s)g^{-1} = D_S(s)$ for any $g \in G$, and the result follows.

It may happen that $\Delta(S)$ is an empty set; for example, take a free noncommutative semigroup $S$. However, it is easy to avoid this problem by adjoining an identity to $S$ and using the following observation.

**Lemma 5**  Let $S$ be a cancellative semigroup. Then $\Delta(S^1) = \Delta(S)^1$.

*Proof.*  If $s \in \Delta(S^1)$ is not the identity of $S^1$, then, for any $x \in S$, $s^x$ is not the identity of $S^1$ because $y \to y^x$ determines an automorphism of $\Delta(S^1)$ by Proposition 2. Hence, $s^x \in S$, and so $s \in \Delta(S)$. It follows that $\Delta(S^1) \subseteq \Delta(S)^1$. The converse inclusion is clear.

Assertion (ii) in Proposition 2 and the definition of $\Delta(S)$ result in the following immediate consequence.

**Corollary 6**  Let $S$ be a cancellative semigroup with $\Delta(S) \neq \varnothing$. Then $\Delta(S)$ is a right and left Ore subset in $S$, and in $\Delta(S)$, too. Further, $\Delta(S)^{-1}S = S\Delta(S)^{-1}$ and $\Delta(S)^{-1}\Delta(S) = \Delta(S)\Delta(S)^{-1}$, the latter being a group.

**Example 7**  Let $G$ be an $FC$-group, and let $S$ be a subsemigroup of $G$ such that $SS^{-1} = G$ and $xSx^{-1}$ is not contained in $S$ for some $x \in G$. Then $\Delta(S) \neq S = S \cap \Delta(G)$ by Lemma 4. However, from Corollary 8 in Chapter 7, it follows that $\Delta(G) = \Delta(S\Delta(S)^{-1}) = \Delta(SZ(S)^{-1})$. Specifically, $G$ may be chosen as the direct product of the permutation group $S_3$ and a cyclic infinite group $\langle\langle x \rangle\rangle$, while $S = S_3\langle x \rangle \cup \{\tau, e\}$, where $\tau \in S_3$ is an element of order two and $e$ the identity of $G$.

The above example shows that $\Delta(S)$ may not carry the whole information on $\Delta(G)$ if $G$ is a group of fractions of $S$. Further, it suggests that $\Delta(S\Delta(S)^{-1})$ may be the right object to study in the general case. Thus, for any cancellative semigroup $S$, we define $\hat{S} = S\Delta(S)^{-1}$ provided that $\Delta(S) \neq \emptyset$. Since an Ore localization of a cancellative semigroup always is cancellative (see [31], § 0.8), then we may also define $\hat{\Delta} = \Delta(S\Delta(S)^{-1})$. With this notation, we prove the following crucial result.

**Proposition 8** Let $S$ be a cancellative semigroup such that $\Delta(S) \neq \emptyset$. Then,

   (i)   $\hat{S}$ is a cancellative semigroup.
  (ii)  $D_{\hat{S}}(s) = D_S(s)$ for any $s \in \Delta(S)$.
 (iii)  $\Delta(S)\Delta(S)^{-1}$ is an $FC$-group, $Z(S) \neq \emptyset$, and $\Delta(S)\Delta(S)^{-1}$
      $= \Delta(S)Z(S)^{-1} \subseteq \hat{\Delta}$.
 (iv)  $\hat{S} = SZ(S)^{-1}$.
  (v)  $\hat{\Delta}$ is an $FC$-group.

*Proof.* (i) holds by the foregoing remark.

Let $s \in \Delta(S)$. It is easy to see that, for every $y \in \Delta(S)$, the element $s^{y^{-1}}$ exists in $\Delta(S)$ and it is equal to $s_y$. Hence, by the property (b), $(s)^{y^{-1}}$ is defined in $\hat{S}$ and it is equal to $(s_y)^x$. Since $(s_y)^x \in D_S(s)$ by Proposition 2, then, we get

$$D_S(s) = \{s^u \mid u \in S\} = \{s^u \mid u \in S\Delta(S)^{-1}\} = D_{\hat{S}}(s)$$

which proves (ii).

From Corollary 6 we know that $\Delta(S)\Delta(S)^{-1}$ is a group. Let $s, t \in \Delta(S)$, and fix an element $x \in S$. Then $t^x \in \Delta(S)$ by Proposition 2. Hence, $t, t^x$ are invertible in $\Delta(S)\Delta(S)^{-1}$ and the fact that $xt = t^x x$ implies that $(t^x)^{-1}x = xt^{-1}$. In other words, $(t^x)^{-1} = (t^{-1})^x$. From the property (a) it follows that $(st^{-1})^x$ exists in $\Delta(S)\Delta(S)^{-1}$ and is equal to $s^x(t^x)^{-1}$. Therefore,

$$\{(st^{-1})^x \mid x \in \Delta(S)\} = \{s^x \mid x \in \Delta(S)\}\{(t^x)^{-1} \mid x \in \Delta(S)\}$$
$$= D_{\Delta(S)}(s)D_{\Delta(S)}(t)^{-1}$$

is a finite subset of $\Delta(S)\Delta(S)^{-1}$. Since, as in Lemma 4, we see that $\{(st^{-1})^g \mid g \in \Delta(S)\Delta(S)^{-1}\} = \{(st^{-1})^x \mid x \in \Delta(S)\}$, then $\Delta(S)\Delta(S)^{-1}$ is an $FC$-group. Hence, from Corollary 8 in Chapter 7, we derive that $Z(\Delta(S)) \neq \emptyset$ and $\Delta(S)\Delta(S)^{-1} = \Delta(S)Z(\Delta(S))^{-1}$.

Let $z \in Z(\Delta(S))$, and let $D_S(z) = \{z_1 = z, z_2, \ldots, z_m\}$. By Proposition 2 (iv), $D_S(z) \subseteq Z(\Delta(S))$. Consider the element $w = z_1 z_2 \ldots z_m$. Since the elements $z_1, \ldots, z_m$ commute and $\{z_1, \ldots, z_m\} = \{z_1^x, \ldots, z_m^x\}$ for any $x \in S$ by Proposition 2 (iv), then $w = w^x$ for any $x \in S$. This means that $w \in Z(S)$. Hence, $w^{-1}, z_2, z_3, \ldots, z_m \in Z(\Delta(S))Z(S)^{-1}$, and so $z^{-1} = z_2 z_3 \ldots z_m w^{-1} \in Z(\Delta(S))Z(S)^{-1}$. Therefore, $Z(\Delta(S))^{-1} \subseteq Z(\Delta(S))Z(S)^{-1}$. Consequently,

$$\Delta(S)\Delta(S)^{-1} = \Delta(S)Z(\Delta(S))^{-1} \subseteq \Delta(S)Z(\Delta(S))Z(S)^{-1} \subseteq \Delta(S)Z(S)^{-1}$$

which implies that $\Delta(S)\Delta(S)^{-1} = \Delta(S)Z(S)^{-1}$.

If $s \in \Delta(S)$, $t \in Z(S)$, then, clearly, $(t^{-1})^u = t^{-1}$ for every $u \in \hat{S}$, and since we know from (ii) and Proposition 2 (iii) that $s^u \in \Delta(S)$, then $D_{\hat{S}}(st^{-1}) = \{(st^{-1})^u \mid u \in \hat{S}\} = \{s^u t^{-1} \mid u \in \hat{S}\} \subseteq \Delta(S)Z(S)^{-1}$. Consequently, $\Delta(S)Z(S)^{-1} \subseteq \Delta(\hat{S}) = \hat{\Delta}$, which completes the proof of (iii).

Now, $\hat{S} = S\Delta(S)^{-1} = S\Delta(S)\Delta(S)^{-1} = S\Delta(S)Z(S)^{-1} = SZ(S)^{-1}$, which proves (iv).

Finally, (iv) applied to $\hat{S}$ yields $\hat{S}\hat{\Delta}^{-1} = \hat{S}Z(\hat{S})^{-1}$. Assume that $xy^{-1} \in Z(\hat{S})$ for some $x \in S$, $y \in Z(S)$. Then $xy^{-1}w = wxy^{-1}$ for any $w \in S$, and so $xwy^{-1} = xy^{-1}w = wxy^{-1}$. It follows that $xw = wx$, which means that $x \in Z(S)$. Therefore, $Z(\hat{S}) = Z(S)Z(S)^{-1}$. Now, we get $\hat{S}\hat{\Delta}^{-1} = \hat{S}Z(\hat{S})^{-1} = \hat{S}Z(S)Z(S)^{-1} = \hat{S}$. Thus, by (iii), $\hat{\Delta}\hat{\Delta}^{-1} \subseteq \Delta(\hat{S}\hat{\Delta}^{-1}) = \Delta(\hat{S}) = \hat{\Delta}$, so $\hat{\Delta}$ is a group and, also by (iii), $\hat{\Delta}$ is an $FC$-group. This completes the proof.

**Remark 9** As we observed in the above proof, $(\hat{\hat{S}}) = \hat{S}\Delta(\hat{S})^{-1} = \hat{S}$. This shows that an iteration of the construction of $\hat{S}$ from $S$ does not lead to a new object. We note that $\hat{\Delta}$ and $\hat{S}$ were originally approached through a direct limit of semigroups $S_n$, $n \geq 1$, where $S_0 = S$ and, for any $n \geq 1$, $S_n = \hat{S}_{n-1}$; see [124].

As the first application of the $FC$-center $\Delta(S)$, we state a description of the center of the algebra $K[S]$ given by Krempa in [123]; see [124], which is analogous to that known for group algebras; see [203], Lemma 4.1.1.

**Theorem 10** Let $S$ be a cancellative semigroup. Then $Z(K[S])$ is the $K$-subspace of $K[S]$ spanned by the elements of the form $\sum_{t \in D_S(s)} t$, where $s$ is any element of $\Delta(S)$.

*Proof.* Let $s \in \Delta(S)$ and let $x \in S$. Then $x \sum_{t \in D_S(s)} t = \sum_{t \in D_S(s)} t^x x$ and since $t \to t^x$ is a permutation of $D_S(s)$ by Proposition 2, then the latter sum is equal to $\sum_{t \in D_S(s)} tx$. This shows that $\sum_{t \in D_S(s)} t \in Z(K[S])$. Assume now that $a = \sum \alpha_s s \in Z(K[S])$ for some $\alpha_s \in K$, $s \in S$. Since $ax = xa$ for any $x \in S$, then from the definition of $\Delta(S)$ it follows that $\mathrm{supp}(a) \subseteq \Delta(S)$. Moreover, if $s \in \mathrm{supp}(a)$, then $\alpha_t = \alpha_s$ for all $t \in D_S(s)$.

Then $a_0 = \alpha_s(\sum_{t \in D_S(s)} t) \in Z(K[S])$ and $|\text{supp}(a - a_0)| < |\text{supp}(a)|$. Thus, an easy induction on $|\text{supp}(a)|$ establishes the assertion.

**Corollary 11**  Let $S$ be a cancellative semigroup such that $\Delta(S) \neq \varnothing$. Then,

(i) $Z(K[S]) \subseteq Z(K[\Delta(S)])$ and $Z(K[\hat{S}]) \subseteq Z(K[\hat{\Delta}])$.
(ii) $Z(K[S]) = Z(K[\hat{S}]) \cap K[S]$ and $Z(K[\hat{S}]) = Z(K[S])Z(S)^{-1}$.

*Proof.*  (i) is a direct consequence of Theorem 10 applied to $S$ and $\hat{S}$, respectively.

(ii) From Proposition 8 we know that $\hat{S} = SZ(S)^{-1}$. Hence, by Lemma 13 in Chapter 7, $K[\hat{S}]$ is the localization of $K[S]$ with respect to the central subset $Z(S)$. The former equality is straightforward. If $a \in Z(K[\hat{S}])$, then there exists $s \in Z(S)$ such that $as \in K[S]$. Then,

$$a = (as)s^{-1} \in (K[S] \cap Z(K[\hat{S}])Z(S))s^{-1} \subseteq Z(K[S])s^{-1} \subseteq Z(K[S])Z(S)^{-1}$$

Thus, $Z(K[\hat{S}]) \subseteq Z(K[S])Z(S)^{-1}$. The converse inclusion is clear.

Clearly, one can have $Z(K[S]) \neq Z(K[\Delta(S)])$, $Z(K[\hat{S}]) \neq Z(K[\hat{\Delta}])$ above. For example, $S$ can be chosen as a group with nontrivial abelian, but noncentral, $FC$-subgroup $\Delta(S)$.

The above result allows us to extend the assertion of Proposition 8.

**Corollary 12**  Let $S$ be a cancellative semigroup such that $\Delta(S) \neq \varnothing$. Then, $\hat{\Delta} = \Delta(S)\Delta(S)^{-1} = \Delta(S)Z(S)^{-1}$.

*Proof.*  We know that $Z(K[\hat{S}]) = Z(K[S])Z(S)^{-1}$. From the description of $Z(K[\hat{S}])$ and $Z(K[S])$ given in Theorem 10, it then follows that, for every $s \in \hat{\Delta}$, we have $\sum_{t \in D_{\hat{S}}(s)} t \in Z(K[S])Z(S)^{-1} \subseteq K[\Delta(S)]Z(S)^{-1}$ and, consequently, $s \in \Delta(S)Z(S)^{-1}$. Therefore, $\hat{\Delta} \subseteq \Delta(S)Z(S)^{-1}$, and the result follows.

We can also derive an analog of a result on the center of group algebras; see [203], Theorem 4.3.8.

**Corollary 13**  Assume that $S$ is a cancellative semigroup. If $e$ is a nonzero central idempotent in $K[S]$, then $\langle \text{supp}(e) \rangle$ is a finite subgroup of $\Delta(S)$.

*Proof.*  If $e = e^2 \in Z(K[S])$, $e \neq 0$, then, by Corollary 11, $e$ is a central idempotent of the group algebra $K[\hat{\Delta}]$ and $e \in K[\Delta]$. It is well known that the subgroup $H$ of $\hat{\Delta}$ generated by $\text{supp}(e)$ must be finite in this case; see [203], Theorem 4.3.8. Then $H = \langle \text{supp}(e) \rangle$ and, since $\text{supp}(e) \subseteq \Delta(S)$, $\text{supp}(e)$ is a finite subgroup of $\Delta(S)$.

One of the most important applications of the $\Delta$-methods in group algebras is that providing a description of prime and semiprime algebras

$K[G]$ in terms of the group $G$, and also $\Delta(G)$. It is natural to connect the same problem for an arbitrary cancellative semigroup $S$ with some groups derived from $S$, so that the group algebra results would yield criteria for $K[S]$ to be prime and semiprime. The best general result of this type, beyond Theorem 19, in Chapter 7, settling the case where $S$ has a group of right fractions, is given below.

**Theorem 14** Let $S$ be a cancellative semigroup such that $\Delta(S) \neq \emptyset$. Consider the following statements:

   (i)   $K[S]$ is prime (semiprime, respectively).
  (ii)   $K[\hat{S}]$ is prime (semiprime).
 (iii)   $Z(K[\hat{S}])$ is prime (semiprime).
 (iv)   $K[\hat{\Delta}]$ is prime (semiprime).
  (v)   $K[S \cap \hat{\Delta}]$ is prime (semiprime).
 (vi)   $K[\Delta(S)]$ is prime (semiprime).
(vii)   $Z(K[S])$ is prime (semiprime).

     Then the following implications hold: (i) $\Leftrightarrow$ (ii) $\Rightarrow$ (iii) $\Leftrightarrow$ (iv) $\Leftrightarrow$ (v) $\Leftrightarrow$ (vi) $\Leftrightarrow$ (vii). Moreover, if $\hat{S}$ is a group, then all statements are equivalent.

*Proof.* In view of Lemma 13, in Chapter 7, Proposition 8 implies that $K[\hat{S}]$ is a central localization of $K[S]$. Similarly, Corollaries 11 and 12 imply that $Z(K[\hat{S}]) \supseteq Z(K[S])$, $K[\hat{\Delta}] \supseteq K[S \cap \hat{\Delta}]$, $K[\hat{\Delta}] \supseteq K[\Delta(S)]$ are all central localizations of their subalgebras in question. Therefore, the equivalences (i) $\Leftrightarrow$ (ii), (iii) $\Leftrightarrow$ (vii), and (iv) $\Leftrightarrow$ (v) $\Leftrightarrow$ (vi) follow from Lemma 14 in Chapter 7.

     The implication (ii) $\Rightarrow$ (iii) is clear. Thus, it is enough to establish the implications (iii) $\Rightarrow$ (iv) and (vi) $\Rightarrow$ (vii).

     Assume that $g \in \hat{\Delta}$ is a nontrivial periodic element. Then, by Proposition 2 (iv), $D_{\hat{S}}(g)$ is a finite subset of $\hat{\Delta}$ consisting of periodic elements. Since $\hat{\Delta}$ is an $FC$-group, then it is well known that the subsemigroup $G$ of $\hat{\Delta}$ generated by $D_{\hat{S}}(g)$ is a finite subgroup of $\hat{\Delta}$; see [203], Lemma 4.1.5. Let $h \in G$. Then $h = g^{x_1} \ldots g^{x_n}$ for some $n \geq 1$, $x_i \in \hat{S}$. Now, if $y \in \hat{S}$, then we know that $h^y$ exists in $\hat{\Delta}$ and that $h^y = (g^{x_1} \ldots g^{x_n})^y = (g^{x_1})^y \ldots (g^{x_n})^y = g^{yx_1} \ldots g^{yx_n} \in G$. Therefore, $G$ is a union of some sets $D_{\hat{S}}(z)$, $z \in G$, and Theorem 10 implies that $e = \sum_{z \in G} z$ lies in $Z(K[\hat{S}])$. Clearly, $e^2 = (\sum_{z \in G} z)(\sum_{z \in G} z) = |G|(\sum_{z \in G} z) = |G|e$. Consequently, either $e^2 = 0$ if $|G| = 0$ in $K$, or $|G|^{-1}e$ is an idempotent if $|G| \neq 0$. Since $e \in Z(K[\hat{S}])$, then $Z(K[\hat{S}])$ is not semiprime, and it is not prime in the latter case. While the order of $g$ divides $|G|$, then the implication (iii) $\Rightarrow$ (iv) follows from Lemma 16 in Chapter 7.

That $Z(K[S]) \subseteq Z(K[\Delta(S)])$ (see Corollary 11) establishes the implication (vi) $\Rightarrow$ (vii).

Finally, if $\hat{S}$ is a group, then, by Lemma 16 in Chapter 7, $K[\hat{S}]$ is prime (semiprime) if and only if $Z(K[\hat{S}])$ is also, and so the remaining assertion follows.

Observe that, by Proposition 8 and Lemma 16 in Chapter 7, the primeness of $K[\Delta]$, $K[S \cap \hat{\Delta}]$, $K[\hat{\Delta}]$ is equivalent to the fact that $\Delta$, $S \cap \hat{\Delta}$, $\hat{\Delta}$, respectively, is a commutative torsion-free semigroup. Similarly, the semiprimeness of any of these algebras is equivalent to the fact that the group $\hat{\Delta}$ has no $p$-torsion, where $p = \text{ch}(K)$. In particular, if $\text{ch}(K) = 0$, then $K[\hat{\Delta}]$ always is semiprime, and so all algebras listed in (iii), ..., (vii) above are semiprime.

Finally, we extract some useful consequences for the case where $S$ embeds into a group.

**Corollary 15** Let $G$ be a group generated by its subsemigroup $S$. Then,

(i)   If $\Delta(S) \neq \varnothing$, then $\hat{\Delta} \subseteq \Delta(G)$ is a normal subgroup of $G$.
(ii)  If $G$ is an $FC$-group, then $\hat{S} = SZ(S)^{-1} = G$ and $\Delta(G) = \Delta(S)Z(S)^{-1}$.

*Proof.* (i) Since $G$ is the group generated by $\hat{S}$, then, by Lemma 4, $D_G(s) = D_{\hat{S}}(s)$ for every $s \in \hat{\Delta}$. Therefore, Proposition 8 implies that $\hat{\Delta}$ is a normal subgroup of $G$ contained in $\Delta(G)$.

(ii) From Corollary 8 in Chapter 7, we know that $Z(S) \neq \varnothing$ and that $SZ(S)^{-1}$ is a group of fractions of $S$. Therefore, $\hat{S} = G$ is a group, and $\Delta(G) = \Delta(S)Z(S)^{-1}$ by Corollary 12.

**Corollary 16** The following conditions are equivalent for a cancellative semigroup $S$:

(i)   $\Delta(S) \neq \varnothing$ and $\hat{\Delta} = \hat{S}$.
(ii)  $S$ is a subsemigroup of an $FC$-group $G$.
(iii) $Z(S) \neq \varnothing$, and $SZ(S)^{-1}$ is an $FC$-group.

Moreover, in this case, $Z(K[S]) = Z(K[\Delta(S)])$.

*Proof.* If (i) holds, then $\hat{\Delta}$ is an $FC$-group by Proposition 8. Thus, $S \subseteq \hat{S} = \hat{\Delta}$ establishes (ii).

If (ii) holds, then, by Corollary 15, $SZ(S)^{-1}$ is a group of fractions of $S$, and $SZ(S)^{-1}$ embeds into $G$. This implies (iii).

Assume (iii). Then $\Delta(S) \supseteq Z(S) \neq \varnothing$, $\hat{S} = SZ(S)^{-1}$ is an $FC$-group, and so $\hat{\Delta} = \Delta(\hat{S}) = \hat{S}$.

Now, from the fact that $Z(K[\Delta(S)]) \subseteq Z(K[\hat{\Delta}]) = Z(K[\hat{S}])$ (see Corollary 12), it follows that $Z(K[\Delta(S)]) \subseteq Z(K[S])$. Thus, the equality $Z(K[\Delta(S)]) = Z(K[S])$ is a consequence of Corollary 11.

It seems that the remaining implication in Theorem 14 may not hold in general. It would be interesting to find counterexamples or to determine cases in which the equivalence of all conditions may be established. Observe that an obstacle arises even if we know that $S$ has a group of right fractions $G$. We do not know any other connections between $\hat{\Delta}$ and $\Delta(G)$ beyond these given in Corollaries 15 and 16 in this case. To improve the information on an arbitrary cancellative semigroup $S$ carried by $\hat{\Delta}$, it may seem reasonable to consider some other $FC$-groups associated with $S$. Namely, let $T$ be an arbitrary maximal right Ore subset in $S$. Then we may deal with $\hat{\Delta}_T = \Delta(ST^{-1}Z(ST^{-1})^{-1})$ instead of $\hat{\Delta} = \Delta(SZ(S)^{-1})$. In particular, this allows us to overcome the above-mentioned problem in the case where $S$ has a group of right fractions because we come to $\hat{\Delta}_S = \Delta(G)$.

## Comments on Chapter 9

We refer to [203] as a general reference on the role of the $FC$-groups in the theory of group algebras. The first attempt to find a similar powerful tool in the case of arbitrary cancellative semigroups was made by Krempa in [123], where a definition of the $FC$-center (that of our Lemma 1) was given. In [123], and its extension [124], $\Delta(S)$ was particularly used for the study of the center and the special elements of the semigroup algebra $R[S]$ with coefficients in an arbitrary associative ring $R$. As noted in Remark 9, Proposition 8 is the key to a simplified approach to $\hat{\Delta}$ and allows us to apply $\Delta(S)$ to the problem of determining prime and semiprime algebras $K[S]$. Recently, Dauns [35], [36] applied a similar idea to a study of the center of more general structures—skew semigroup rings and skew power series rings of cancellative semigroups.

We note that, from the results on the behavior of semiprimeness under tensor products (see [125]), it follows that, for any semigroup $S$ and any fields $K$, $L$ with $\mathrm{ch}(K) = \mathrm{ch}(L)$, the algebra $K[S]$ is semiprime if and only if $L[S]$ is semiprime, so that there exists an intrinsic characterization of the semiprimeness of $K[S]$ in terms of $S$ and $\mathrm{ch}(K)$. Such a characterization, beyond the cases settled in Theorem 19 in Chapter 7 and Theorem 14, is known for some classes of not necessarily cancellative semigroups. In

[64], Theorem 9.17 (see also our Chapter 21), this is given with respect to the class of commutative semigroups, while we refer to [162], [165] for a discussion of the case of inverse semigroups. The situation is more complicated when the primeness of $K[S]$ is considered. Here, from the results of Krempa on tensor products (see [125]), it follows only that $K[S]$ is prime if and only if $L[S]$ is prime, provided that $K, L$ contain the algebraic closure of their prime subfield. However, for commutative semigroups (see [64], Theorem 8.1), for some inverse semigroups (see [165]), and some cancellative semigroups (see Theorem 19 in Chapter 7 and Theorem 14), the primeness of $K[S]$ may be characterized in terms of $S$ and ch($K$) only. Semiprimeness of certain twisted semigroup rings was studied by Quesada in [230]. Interesting examples of semigroups satisfying $S = \Delta(S)$ arise from the class of algebras considered in Chapter 25.

# 10
# Unique-/Two-Unique-Product Semigroups

For reasons similar to those of the group algebra case, see [203], one may define the classes of u.p. and t.u.p. semigroups. Namely, $S$ is said to be a u.p. (unique-product) semigroup if, for any nonempty finite subsets $X$, $Y$ of $S$ with $|X| + |Y| > 2$, there exists an element in the set $XY = \{xy \mid x \in X, y \in Y\}$ that has a unique presentation in the form $xy$, where $x \in X, y \in Y$. Similarly, $S$ is called a t.u.p. (two-unique-product) semigroup if, for any nonempty finite subsets $X$, $Y$ with $|X| + |Y| > 2$, there exist at least two elements in $XY$ that have unique presentations as $xy$, for some $x \in X, y \in Y$. Clearly, every t.u.p. semigroup is a u.p. semigroup. Groups with the above-defined properties are called u.p. and t.u.p. groups, respectively.

**Lemma 1** Let $S$ be a u.p. (t.u.p., respectively) semigroup. Then

(i)  $S$ is cancellative.

(ii) If $S$ has a group of right fractions $G$, then $G$ is a u.p. (t.u.p., respectively) group.

(iii) If $T$ is a subsemigroup of $S$, then $T$ is a u.p. (t.u.p., respectively) semigroup.

*Proof.* (i) Let $s,t,u \in S$, $s \neq t$. Put $X = \{s,t\}$, $Y = \{u\}$. The hypothesis on $S$ implies that $su \neq tu$, and similarly $us \neq ut$. Thus, (i) holds.

(ii) If $X$, $Y$ are finite nonempty subsets of $G$ with $|X| + |Y| > 2$, then, by Lemma 13 of Chapter 7, there exists $s \in S$ such that $X = \bar{X}s^{-1}$ for a finite subset $\bar{X}$ of $S$. Now, again, since $s^{-1}Y$ is a finite subset of $G$, then there exists $t \in S$ such that $s^{-1}Yt \subseteq S$. By the hypothesis, $\bar{X}(s^{-1}Yt)$ has an element (two such elements if $S$ is assumed to be a t.u.p. semigroup) with a unique presentation in the form $xy$, where $x \in \bar{X}$, $y \in s^{-1}Yt$. Then $xyt^{-1} = (xs^{-1})(syt^{-1}) \in XY$ is the unique presentation of $xyt^{-1}$ as an element of the product of $X$, $Y$. Suppose that $x' \in X$, $y' \in Y$ are such that $x'y' = xyt^{-1}$. Then $(xs^{-1})(syt^{-1}) = x'y' = (x_1s^{-1})(sy_1t^{-1})$ for $x_1 = x's \in \bar{X}$, $y_1 = s^{-1}y't \in s^{-1}Yt$, and so $xy = x_1y_1$. The choice of $x$, $y$ implies that $x = x_1$, $y = y_1$. Hence, $x' = x_1s^{-1} = xs^{-1}$, $y' = sy_1t^{-1} = syt^{-1}$, which establishes our claim. Thus, (ii) follows.

(iii) is obvious.

We state another simple auxiliary result.

**Lemma 2** Let $S$ be a cancellative semigroup, and let $I$ be an ideal of $S$. Then $S$ is a u.p. (t.u.p.) semigroup if and only if $I$ is a u.p. (t.u.p., respectively) semigroup.

*Proof.* Consider two nonempty subsets $X$, $Y$ of $S$. If $s \in I$, then $sX$, $Ys \subseteq I$. Let $(sx)(ys) \in (sX)(Ys)$ be the unique presentation of an element of the product of the sets $sX$, $Ys$. Then $xy$ is an element of $XY$ with a unique presentation as $x'y'$, where $x' \in X$, $y' \in Y$. In fact, if $xy = x'y'$ for some $x' \in X$, $y' \in Y$, then $(sx')(y's) = (sx)(ys) \in (sX)(Ys)$ and, consequently, $sx = sx'$, $ys = y's$. The cancellativity of $S$ implies that $x = x'$ and $y = y'$. Thus, $S$ is a u.p. (t.u.p., respectively) semigroup whenever $I$ is also. The converse follows from Lemma 1.

Let $X,Y$ be nonempty subsets of $S$. Then, for brevity, an element $s \in XY$ will be said to be a u.p. element in the product of the sets $X$, $Y$ if it is uniquely presented in the form $s = xy$, where $x \in X$, $y \in Y$.

The main idea of introducing the classes of u.p. and t.u.p. semigroups comes from the theorem given below. Before stating it, we need one more auxiliary result. We say that a semigroup $S$ is idempotent-free if $S$ has no idempotents, except possibly identity and zero.

**Lemma 3** Let $S$ be an idempotent-free semigroup. If $K[S]$ is an algebra with unity, then $S$ is a monoid.

*Proof.* Let $e = \sum_{i=1}^{n} \lambda_i s_i$ be the unity of $K[S]$. We can assume that $n > 1$ because, otherwise, $e = e_1$ and $S$ is a monoid. It is enough to show that $S$ has a nonzero idempotent. We construct inductively a sequence of nonzero elements of the set $T = \{s_1, \ldots, s_n\}$ such that $t_j t_k = t_k$ for all $j$, $k$ with $j > k$. Let $t_1 \in T$ be such that $t_1 S^1$ is maximal among all right ideals, $s_i S^1$, $i = 1, \ldots, n$. Then $t_1$ is not a zero in $S$ because, otherwise, $n = 1$, a contradiction. Assume that $t_1, t_2, \ldots, t_m$ have been chosen for some $m \geq 1$. From the equality $e t_m = t_m$, it follows that there exists $s \in T$ such that $s t_m = t_m$. Put $t_{m+1} = s$. Then, for every $k < m$, the induction hypothesis yields $t_k = t_m t_k = (t_{m+1} t_m) t_k = t_{m+1}(t_m t_k) = t_{m+1} t_k$ as desired. Since $T$ is a finite set, then $t_p = t_r$ for some $r > p$. Therefore, $t_p t_p = t_p$. Clearly, $t_p$ is not a zero in $S$ because $t_p t_1 = t_1 \neq \theta$. The result follows.

We say that a monoid algebra $K[S]$ has trivial units if every unit in $K[S]$ is of the form $\lambda s$, where $0 \neq \lambda \in K$ and $s$ is a unit in $S$.

**Theorem 4** Assume that $S$ is a u.p. semigroup. Then $K[S]$ is a domain. Moreover, if $S$ is a t.u.p. semigroup and $K[S]$ is an algebra with unity, then $S$ is a monoid and $K[S]$ has trivial units.

*Proof.* If $a$, $b \in K[S]$ are such that $ab = 0$, then there is no element in the product $(\text{supp}(a))(\text{supp}(b))$ with a unique presentation as $xy$, for $x \in \text{supp}(a)$, $y \in \text{supp}(b)$. If $S$ is a t.u.p. semigroup and $K[S]$ has a unity, then $S$ is a monoid by Lemma 3. Then the condition $ab = 1$, $a$, $b \in K[S]$ implies that there is at most one u.p. element in the product of $\text{supp}(a)$ and $\text{supp}(b)$. Hence, the assertion is a direct consequence of the definition of u.p. and t.u.p. semigroups.

**Corollary 5** If $S$ is a t.u.p. semigroup, then $\mathcal{J}(K[S]) = 0$.

*Proof.* Since $S^1$ is a t.u.p. semigroup by Lemma 2, then Theorem 4 implies that $\mathcal{J}([K[S^1]]) = 0$. Thus, the result follows because $\mathcal{J}(K[S]) = \mathcal{J}(K[S^1])$ by Lemma 11 in Chapter 4.

In the case where $S$ has a group of fractions, the two considered properties coincide. The following result was essentially obtained by Strojnowski in [256].

**Theorem 6** Let $S$ be a semigroup with a group of right fractions $G$. Then the following conditions are equivalent:

(i)   $S$ is a u.p. semigroup.
(ii)  $G$ is a u.p. group.
(iii) $S$ is a t.u.p. semigroup.
(iv)  $G$ is a t.u.p. group.

*Proof.* In view of Lemma 1, it is enough to show that (ii) implies (iv). Assume that $G$ is a u.p. group. Suppose that there exist two nonempty finite subsets $X, Y$ of $S$ such that $|X| + |Y| > 2$, and there exists exactly one element $st \in XY$, $s \in X$, $t \in Y$, that has a unique presentation in the form $xy$, $x \in X$, $y \in Y$. Let $U = s^{-1}X$, $V = Yt^{-1}$. It is clear that the only element with a unique presentation in the product $UV$ is the identity element $e = ee$. Put $W = V^{-1}U$, $Z = VU^{-1}$. Every element of $WZ$ may be written in the form $(v_1^{-1}u_1)(v_2u_2^{-1})$, where $u_1, u_2 \in U$, $v_1, v_2 \in V$. We consider three cases:

(1) If $u_1 \neq e$ or $v_2 \neq e$, then there exist $u_3 \in U$, $v_3 \in V$ such that $u_1v_2 = u_3v_3$ and $u_1 \neq u_3$. Hence, the element $(v_1^{-1}u_1)(v_2u_2^{-1})$ may also be written as $(v_1^{-1}u_3)(v_3u_2^{-1}) \in WZ$.

(2) If $u_2 \neq e$ or $v_1 \neq e$, then there exist $u_4 \in U$, $v_4 \in V$ such that $u_2v_1 = u_4v_4$ and $u_2 \neq u_4$. Then $v_1^{-1}u_2^{-1} = v_4^{-1}u_4^{-1}$, and so the element $(v_1^{-1}e)(eu_2^{-1})$ may be written as $(v_4^{-1}e)(eu_4^{-1}) \in WZ$.

(3) Assume that $u_1 = v_1 = u_2 = v_2 = e$. Since $|U| + |V| > 2$, then one of these sets contains an element $t \neq e$. If $t \in U$, then the product $(ee)(ee)$ may be written as $(et)(et^{-1}) \in WZ$. If $t \in V$, then $(ee)(ee) = (t^{-1}e)(te) \in WZ$.

We have shown that there is no element in $WZ$ with a unique presentation in the form $xy$, $x \in W$, $y \in Z$. This contradicts the fact that $G$ is a u.p. group. Thus, (ii) $\to$ (iv) follows.

**Remark 7** If $X, Y$ are two nonempty subsets of a cancellative semigroup $S$ such that any element of $YX$ has at least two different presentations in the form $xy$, where $x \in X$, $y \in Y$, then put $U = XY$. If $y_1 \in Y$, $u_2 \in U$, then it is easy to see that $y_1u_2 = y_3u_3$ for some $y_3 \in Y$, $u_3 \in U$ with $y_1 \neq y_3$. Therefore, for any $y_2 \in Y$, $u_1 \in U$, we have $(u_1y_1)(u_2y_2) = (u_1y_3)(u_3y_2) \in (UY)(UY)$, and $u_1y_1 \neq u_1y_3$. This shows that there are no elements with unique presentations as products in $(UY)(UY)$. Therefore, when verifying whether $S$ is a u.p. semigroup, it is enough to show that, for every subset $Z$ of $S$ with $|Z| > 1$, there exists an element of $ZZ$ with a unique presentation in the form $xy$, where $x, y \in Z$.

Let us note that every u.p. group $G$ must be torsion-free; see [203], Section 13.1. In fact, for every finite subgroup $H$ in $G$, each product in $HH$ occurs with multiplicity $|H|$. Hence, $|H| = 1$ and $G$ is torsion-free. An example of a torsion-free group that is not a u.p. group has recently been constructed in [236].

We say that a semigroup $S$ is ordered if there is a linear order $<$ on $S$ such that, for every $s, t, x \in S$, the condition $s < t$ implies that $sx < tx$ and $xs < xt$. Clearly, in this case, $S$ must be cancellative because any of the conditions $xs = xt$, $sx = tx$ implies $s = t$ (note that one of $s < t$, $t < s$, $s = t$ must hold).

**Lemma 8**  Let $S$ be a semigroup. If $S$ is ordered, then $S$ is a t.u.p. semigroup.

*Proof.*  Let $X, Y \subseteq S$ be nonempty subsets such that $|X| + |Y| > 2$. Let $x_0, y_0$ be the elements that are minimal in $X$, $Y$, respectively, and let $x_1$, $y_1$ be maximal in $X$, $Y$ with respect to the ordering of $S$. Then, for every $x \in X$, $y \in Y$, we have $x_0 y_0 \le x_0 y \le xy \le x_1 y \le x_1 y_1$. If $x_0 y_0 = xy$, then $x_0 y_0 = x_0 y = xy$, and the cancellativity of $S$ implies that $x_0 = x$, $y_0 = y$. Therefore, $x_0 y_0$ is a u.p. element in the product of the sets $X$, $Y$. Similarly, $x_1 y_1$ is a u.p. element in the product $XY$. Since $|X| + |Y| > 2$, then the pairs $(x_0, y_0)$, $(x_1, y_1)$ are not equal. Moreover, $x_0 y_0 \ne x_1 y_1$ because, as above, the equality would imply that $x_0 = x = x_1$, $y_0 = y = y_1$. This shows that, whenever $|X| + |Y| > 2$, then $x_0 y_0$, $x_1 y_1$ are distinct u.p. elements in the product of the sets $X$, $Y$.

It is known that every torsion-free nilpotent group is ordered; see [203], Lemma 13.1.6. The proof of this result employs the fact that, if the center of a nilpotent group if torsion-free, then the group itself must be torsion-free; see [203], Lemma 13.1.3. We extend the former to the class of cancellative weakly nilpotent semigroups.

**Proposition 9**  Let $S$ be a cancellative weakly nilpotent semigroup. Then the following conditions are equivalent:

(i)    $S$ is a u.p. semigroup.
(ii)   $S$ is a t.u.p. semigroup.
(iii)  $S$ is an ordered semigroup.
(iv)   For every $s, t \in S$, $n \ge 1$, the condition $s^n = t^n$ implies that $s = t$.
(v)    For every $s, t \in S$, the conditions $s^n = t^n$, $st = ts$ imply that $s = t$.

*Proof.*  From Theorem 3 in Chapter 7, we know that $S$ has a nilpotent group of fractions $G$. It is known that $G$ is ordered if and only if $G$ is torsion-free; see [203], Lemma 13.1.6. Since u.p. groups must be torsion-free, then, from Theorem 6 and Lemma 8, it follows that conditions (i), (ii), (iii) are equivalent and that they hold if and only if $G$ is torsion-free.

Assume that $<$ is a linear order on $G$ that is compatible with the multiplication in $G$. If $s < t$, then $s^n = ss^{n-1} < ts^{n-1}$. Since $ts^{n-1} < tt^{n-1} = t^n$, then $s^n \neq t^n$. Thus, (iv) is a consequence of (iii).

It is clear that (iv) implies (v).

Finally, assume that $G$ is not torsion-free. We know that $Z(G)$ is not torsion-free in this case. Therefore, there exist $s, t \in S, s \neq t$ such that $st^{-1}$ is a torsion element in the center of $G$. This means that $s = ut$ for some $u \in Z(G)$ and that $u^n = 1$ for some $n > 1$. Now, $s^n = (ut)^n = u^n t^n = t^n$ and, clearly, $st = ts$. This establishes the implication (v) $\Rightarrow$ (i) and completes the proof.

We derive a consequence of Proposition 9.

**Corollary 10**  Let $S$ be a cancellative weakly nilpotent semigroup. Then the following conditions are equivalent:

  (i)   $K[S]$ is prime.
  (ii)  $K[S]$ is a domain.
  (iii) Every commutative subsemigroup of $S$ is torsion-free.

*Proof.*  (iii) implies, in view of Proposition 9, that $S$ is a t.u.p. semigroup, so that (ii) follows by Theorem 4. If $K[S]$ is prime, then the group algebra $K[G]$ of the group of fractions $G$ of $S$ is prime by Theorem 19 in Chapter 7. Consequently, Lemma 16, in Chapter 7 implies that $Z(G) \subseteq \Delta(G)$ is torsion-free. Since $G$ is a nilpotent group, then $G$ must be torsion-free. Thus, (i) implies (iii). The result follows.

**Proposition 11**  Let $S$ be a subsemigroup of an $FC$-group. Then the following conditions are equivalent:

  (i)   $S$ is a u.p. semigroup.
  (ii)  $S$ is a t.u.p. semigroup.
  (iii) $S$ is a torsion-free commutative semigroup.

*Proof.*  If (i) holds, then $K[S]$ is a domain by Theorem 4. Therefore, (iii) follows by Corollary 20 in Chapter 7. The assertion follows.

We derive the following immediate consequence for the class of semigroups studied in Chapter 9.

**Corollary 12**  Let $S$ be a cancellative semigroup. Then $\Delta(S)$ is a u.p. semigroup if and only if it is a torsion-free commutative semigroup.

The situation is much more complicated in the case of arbitrary cancellative semigroups, as the following example due to Krempa shows.

**Example 13**  Let $T$ be the free monoid on letters $x_1, x_2, x_3, y_1, y_2, y_3$. Let $\rho$ be the least congruence on $T$ such that

(a) $\qquad\qquad (x_1y_1, x_2y_3) \in \rho \qquad (x_1y_3, x_2y_2) \in \rho$
$\qquad\qquad\quad (x_1y_2, x_3y_1) \in \rho \qquad (x_3y_2, x_2y_1) \in \rho$

We put $S = T/\rho$, and we define $\underline{x}_1, \underline{x}_2, \underline{x}_3, \underline{y}_1, \underline{y}_2, \underline{y}_3$ as the images of the generators of $T$ under the natural homomorphism $T \to S$. Observe that, since the defining relations for $S$ are homogeneous, every equivalence class of $\rho$ consists of elements of the same length in the free generators of $T$. Then there is a natural notion of length of elements in $S$ coming from the length of words in $T$. Moreover, from the form of the relations, it follows easily that

(b)  Two words $w = a_1 \ldots a_n$, $v = b_1 \ldots b_m$, where $n, m \geq 1$ and $a_i$, $b_j \in \{x_1, x_2, x_3, y_1, y_2, y_3\}$, yield equal elements in $S$ if and only if $n = m$, and there are integers $k, i_1, i_2, \ldots, i_k$ such that $1 \leq i_1$, $i_j + 1 < i_{j+1}$ for $j = 1, \ldots, k - 1$, and $i_k < n$ such that any $(a_{i_j} a_{i_j+1}, b_{i_j} b_{i_j+1})$ is one of the defining relations (a) for $S$, and $a_l = b_l$ for all $l \notin \{i_j, i_j + 1 \mid j = 1, \ldots, k\}$.

We first show that $S$ is cancellative. If $(y_i w, y_i v) \in \rho$ for some $i \in \{1, 2, 3\}$, and some $w, v \in T$, then (b) implies that $(w, v) \in \rho$ because none of $y_i x_j$, $y_i y_j$ is involved in (a).

Assume that $(x_i w, x_i v) \in \rho$. If $w = x_j w_1$ for some $j \in \{1, 2, 3\}$, $w_1 \in T$, then, as above, $(w, v) \in \rho$ because $x_i x_j$ is not involved in (a). If $w = y_j w_1$ for some $j \in \{1, 2, 3\}$, then $(x_i y_j w_1, x_i v) \in \rho$. From (b) and the form of the defining relations (a) of $S$, it follows that $(v, y_j v_1) \in \rho$, where $v_1 \in T$, and $(v_1, w_1) \in \rho$. Thus, $(w, v) \in \rho$ in this case, too. This proves that the cancellation on the left with respect to any element of the set $\{\underline{x}_1, \underline{x}_2, \underline{x}_3, \underline{y}_1, \underline{y}_2, \underline{y}_3\}$ may be performed in $S$. It follows that $S$ is left cancellative. A similar argument shows that $S$ is right cancellative.

Now, we show that $S$ is a u.p. semigroup. Let $X, Y$ be nonempty subsets of $S$. Define $X_0, Y_0$ as the subsets of $X, Y$ consisting of the elements of minimal length in $X, Y$, respectively. It is clear that, if $z = xy$, $x \in X_0$,

$y \in Y_0$ is an element with a unique presentation in the product $X_0 Y_0$, then it is also a u.p. element in the product $XY$. Therefore, we may assume that $X = X_0$ and $Y = Y_0$.

Let $n(X)$, $n(Y)$ be the lengths of all elements in $X$, $Y$, respectively. We proceed by induction on $n(X) + n(Y)$. This will be done in two steps.

Step 1: $n(X) + n(Y) \leq 2$, or $n(X) = 0$, or $n(Y) = 0$. If $n(X) = 0$, then the identity of $S$ is the only element of $X$. Therefore, every element of $XY$ is a u.p. element. Similarly, we are finished if $n(Y) = 0$. Hence, it remains to consider the case where $n(X) = n(Y) = 1$, that is, $X, Y \subseteq \{\underline{x_1}, \underline{x_2}, \underline{x_3}, \underline{y_1}, \underline{y_2}, \underline{y_3}\}$. By (a) we may assume that $X \subseteq \{\underline{x_1 x_2 x_3}\}$, $Y \subseteq \{\underline{y_1 y_2 y_3}\}$. A direct examination of all possible cases shows that always there exists a u.p. element in the product $XY$.

Step 2: $n(X) + n(Y) > 2$, $n(X) > 0$, $n(Y) > 0$. Then $n(X) > 1$, or $n(Y) > 1$. We consider the former case only, the proof in the latter case going symmetrically. We have $zs \in X$ for some $s \in S$ and some $z \in \{\underline{x_1}, \underline{x_2}, \underline{x_3}, \underline{y_1}, \underline{y_2}, \underline{y_3}\}$. Let $X' = \{v \mid zv \in X\}$. Then $n(X') < n(X)$ and, by the induction hypothesis, there exists a u.p. element $vu$, $v \in X'$, $u \in Y$ in the product $X'Y$. Suppose that $(zv)u = tu_1$ for some $t \in X$, $u_1 \in Y$. Since $t \in X$, then $t$ has length exceeding 1. From (b) it then follows that $t = zw$ for some $w \in S$ (if $z = \underline{y_i}$ for some $i$, then $\underline{y_i}$ must be the initial letter in $t$; if $z = \underline{x_i}$, then (a) allows us to write $t = \underline{x_i} w$ for some $w \in S$). Moreover, the cancellativity of $S$ implies that $vu = wu_1$ because $zvu = zwu_1$. Note that $w \in X'$ because $t \in X$. Thus, $u, u_1 \in Y$, $v, w \in X'$, and so the fact that $vu$ is a u.p. element in the product $X'Y$ implies that $v = w$, $u = u_1$. Therefore, $t = zv$. This shows that $(zv)u$ is a u.p. element in the product of the sets $X, Y$.

Finally, put $X = \{\underline{x_1}, \underline{x_2}, \underline{x_3}\}$ and $Y = \{\underline{y_1}, \underline{y_2}, \underline{y_3}\}$. Then $x_3 y_3$ is easily seen to be the only u.p. element in the product $XY$. Therefore, $S$ is not a t.u.p. semigroup.

Let us note that, in view of Theorem 6, the semigroup $S$ constructed above cannot have a group of fractions. However, it may be verified that $S$ is embeddable into a group. Moreover, the classical example of a semigroup not embeddable into a group, due to Malcev [142], may be shown to be a t.u.p. semigroup. This shows that, beyond the class of semigroups with groups of fractions, the u.p. and t.u.p. properties may be very difficult to handle.

Our last result in this chapter provides some criteria for determining whether a given semigroup satisfies one of the unique product properties.

We follow the corresponding result for the class of groups; see [203], Lemma 13.1.8.

**Lemma 14** Let $S$ be a cancellative semigroup. Then any of the following conditions implies that $S$ is a u.p. semigroup (t.u.p. semigroup, respectively).

(i) There is a family of congruences $\rho_\alpha$, $\alpha \in A$, on $S$ such that $\bigcap_{\alpha \in A} \rho_\alpha$ is the trivial congruence on $S$ and any $S/\rho_\alpha$ is a u.p. (t.u.p., respectively) semigroup.

(ii) $S$ has no nonidentity periodic elements and every noncyclic finitely generated subsemigroup of $S$ can be homomorphically mapped onto a noncyclic u.p. (t.u.p., respectively) semigroup.

*Proof.* (i) Let $X$, $Y$ be nonempty finite subsets of $S$ with $|X| + |Y| > 2$. We show, proceeding by induction on $|X| + |Y|$, that there is a u.p. element (there are two u.p. elements, respectively) in the product $XY$. Assume, for example, that $|X| \geq 2$, and choose $s_1, s_2 \in X$, $s_1 \neq s_2$. Then there exists $\alpha \in A$ such that $(s_1, s_2) \notin \rho_\alpha$. Let $\bar{X}$, $\bar{Y}$ denote the images of $X$, $Y$ in $\bar{S} = S/\rho_\alpha$, respectively, and let $\bar{s}$ be the image of $s \in S$ in $\bar{S}$. Then $|\bar{X}| \geq 2$, and so $|\bar{X}| + |\bar{Y}| > 2$. By the hypothesis on $\rho_\alpha$, there exist elements $\bar{z}_1 = \bar{x}_1 \bar{y}_1$, $\bar{z}_2 = \bar{x}_2 \bar{y}_2 \in \overline{XY}$ with unique presentations as products in $\overline{XY}$ (in the u.p. case, we put $\bar{z}_1 = \bar{z}_2$, but $\bar{z}_1 \neq \bar{z}_2$ in the t.u.p. case). Let $X_1$, $X_2$, $Y_1$, $Y_2$ denote the subsets of $X$, $Y$ consisting of the elements mapping onto $\bar{x}_1$, $\bar{x}_2$, $\bar{y}_1$, $\bar{y}_2$ in $\bar{S}$, respectively. Observe that, by the choice of $\rho_\alpha$, $|X_i| + |Y_i| < |X| + |Y|$, $i = 1, 2$, because $X_i \neq X$.

Assume that $u_i = v_i w_i$ is a u.p. element in the product $X_i Y_i$, where $v_i \in X_i$, $w_i \in Y_i$. If $u_i = xy$ for some $x \in X$, $y \in Y$, then $\bar{x}\bar{y} = \bar{v}_i \bar{w}_i = \bar{x}_i \bar{y}_i = \bar{z}_i$. Since $\bar{z}_i$ is a u.p. element in the product $\overline{XY}$, then $\bar{x}_i = \bar{x}$, $\bar{y}_i = \bar{y}$. Hence, $x \in X_i$, $y \in Y_i$, and the fact that $v_i w_i$ is a u.p. element in the product $X_i Y_i$ implies that $v_i = x$, $w_i = y$. It then follows that $v_i w_i$ is a u.p. element in the product $XY$.

Now, if $|X_i| + |Y_i| = 2$, then $X_i = \{x_i\}$, $Y_i = \{y_i\}$, so that $x_i y_i$ is a u.p. element in the product $X_i Y_i$. On the other hand, if $|X_i| + |Y_i| > 2$, then, by the induction hypothesis, there also exists a u.p. element in the product $X_i Y_i$. Thus, by the foregoing, there are always u.p. elements $v_i w_i$, $i = 1, 2$, in the product $XY$ such that $\bar{v}_i \bar{w}_i = \bar{z}_i$. Moreover, if $\bar{z}_1 \neq \bar{z}_2$, then $v_1 w_1 \neq v_2 w_2$. This establishes the desired assertion also in the t.u.p. case.

Assume that (ii) holds, and let $X$, $Y$ be nonempty subsets of $S$ with $|X| + |Y| > 2$. If $s \in X$, $t \in Y$, then $ts \in tX \cap Ys$. Since $S$ is cancellative, it is enough to show that there exists a u.p. element (two u.p. elements, respectively) in the product $(tX)(Ys)$. Consequently, we may assume that $X \cap Y \neq \varnothing$ because $tX \cap Ys \neq \varnothing$. We may also assume that $S$ is the semigroup generated by the subset $X \cup Y$ because $\langle X \cup Y \rangle$ inherits hypothesis (ii) on $S$. If $S$ is a cyclic semigroup, then we are done (see Proposition 11). Otherwise, there exists a congruence $\rho$ on $S$ such that $S/\rho$ is a noncyclic u.p. (t.u.p., respectively) semigroup. Clearly, $|\bar{X}| + |\bar{Y}| \geq 2$, where $\bar{X}$, $\bar{Y}$ denote the images of $X$, $Y$ in $S/\rho$, and the fact that $\bar{X} \cap \bar{Y} \neq \varnothing$ and $S/\rho = \langle \bar{X} \cap \bar{Y} \rangle$ is noncyclic implies that $|\bar{X}| + |\bar{Y}| > 2$. Thus, the assertion follows by induction through a reasoning similar to that used in case (i) ($\rho$ plays the role of $\rho_\alpha$ here).

## Comments on Chapter 10

The unique-product properties of semigroups and groups first appeared in the context of the corresponding semigroup algebras in the papers of Rudin and Schneider [240] and Schneider and Weissglass [242]. The basic facts (see our Theorem 4), motivating the definition were obtained in connection with the study of special elements and radicals of semigroup rings. The u.p. groups have been especially exploited for the zero divisor problem for group rings; see [203]. The long-standing problem of whether the classes of torsion-free and u.p. groups coincide was answered negatively by Rips and Segev [236]. We refer to [205] for a presentation of another example of this type due to Promyslow. Krempa [124], [126], Groenewald [79], and Mehrvarz and Wallace [151], continued the study of special elements, especially units, and homomorphisms of semigroup rings of u.p. groups and semigroups over an arbitrary coefficient ring. Some results have also been obtained in the more general setting of rings graded by u.p. groups and semigroups by Jespers et al. [100], and Jespers [96], [99]. For some results on the role of ordered groups in the theory of group algebras, we refer to [203]. We note that Malcev–Neumann's construction allows us to embed semigroup algebras of ordered semigroups into ordered series algebras, which are division algebras in the group algebra case; see [203], Section 13.2, [31], § 0.8.7. For results and bibliography on certain more general structures arising from ordered semigroups, we refer to the papers of Dauns [35], [36]. The problem of characterizing semigroup algebras with unity (see our Lemma 3) is discussed in the survey paper of Ponizovskii [221].

# 11

# Subsemigroups of Polycyclic-by-Finite Groups

A group $G$ is polycyclic-by-finite if it admits a subnormal series $G = G_0 \supset G_1 \supset \cdots \supset G_n = \{1\}$ such that every factor group $G_i/G_{i+1}$ is either cyclic or finite. Groups of this type have been extensively studied—the corresponding group algebras are noetherian, and they are of importance for the theory of some other general classes of algebras; see [203], [204], [149].

While any semigroup algebra of a semigroup $S$ contained in a group $G$ may be regarded as a $G$-graded algebra, the methods of the theory of graded algebras are especially efficient in the case of gradings by a polycyclic-by-finite group, where the general theory is especially well developed. Thus, in this chapter, we will be concerned with the cancellative semigroups that are embeddable into polycyclic-by-finite groups.

Let $R$ be an algebra graded by a group $G$. Recall from Chapter 6 that the grading is nondegenerated if $rR(g^{-1}) \neq 0$ and $R(g^{-1})r \neq 0$ for every $g \in G$ and every $0 \neq r \in R(g)$. Gradations of this type are of interest because of the following simple observation.

**Lemma 1** Let $H$ be a normal subgroup of a group $G$ and, let $S$ be a submonoid of $G$ that generates $G$ as a group. Then,

(1) If $K[S]$ is right noetherian, then, for every $g \in G$, $K[S] \cap gK[H]$ is a noetherian right $K[S \cap H]$-module.

(2) Assume that $[G : H] < \infty$. Then $K[S](g) = K[S] \cap gK[H]$, $g \in G$, defines a nondegenerated $G/H$-grading on $K[S]$. Moreover, there is an embedding of right $K[S \cap H]$-modules $K[S] \rightarrow K[S \cap H]^{|G/H|}$, and the following assertions are true:

   (i) The right Goldie dimension of $K[S]$ is finite whenever the right Goldie dimension of $K[S \cap H]$ is finite.

   (ii) $K[S]$ is right noetherian if and only if $K[S \cap H]$ is right noetherian, and then $K[S]$ is a noetherian right $K[S \cap H]$-module.

*Proof.* Let $X$ be a set of the coset representatives for $H$ in $G$. We observed in Example 2 in Chapter 6 that the rule $K[S](g) = K[S] \cap gK[H]$, $g \in X$, defines a $G/H$-gradation on $K[S]$.

If $M_1 \subseteq M_2 \subseteq \cdots$ is a chain of right $K[S \cap H]$-submodules of $K[S] \cap gK[H]$, then $M_1 K[S] \subseteq M_2 K[S] \subseteq \cdots$ are right ideals of $K[S]$. Since $M_i K[S] \cap (K[S] \cap gK[H]) = M_i K[S \cap H] \cap (K[S] \cap gK[H]) = M_i$ for all $i$, then (1) follows.

2) From Lemma 4 in Chapter 7, we know that there exists $s_g \in S \cap Hg^{-1}$. Clearly, $s_g(S \cap Hg) \subseteq S \cap H$, from which it follows that the gradation is nondegenerated. Further, the left multiplication by $s_g$ defines an embedding of right $K[S \cap H]$-modules $K[S](g) \rightarrow K[S \cap H]$. Consequently, $K[S] \cong \oplus_{g \in X} K[S](g)$ embeds into $K[S \cap H]^{|G/H|}$ as a right $K[S \cap H]$-module. Now, (i) is a direct consequence of the fact that the finiteness of the Goldie dimension is inherited by finite direct products and by submodules. A similar argument shows that $K[S]$ is a right noetherian $K[S \cap H]$-module (and, hence, $K[S]$ is a right noetherian algebra) whenever $K[S \cap H]$ is also. The converse follows from (1).

We will exploit the fundamental result on the radical of group-graded rings given in Corollary 15 in Chapter 6. This is applied to derive the following auxiliary result.

**Lemma 2** Let $G$ be a group generated by its submonoid $S$. Assume that $H$ is a normal subgroup of finite index in $G$ such that $\mathcal{J}(K[S \cap H]) = 0$. Then $\mathcal{J}(K[S])^{|G/H|} = 0$, and $\mathcal{J}(K[S]) = 0$ if $|G/H| \neq 0$ in $K$.

*Proof.* We treat $K[S]$ as a $G/H$-graded algebra subject to the gradation given in Lemma 1. Since the grading is nondegenerated, then the result is a direct consequence of Corollary 15 in Chapter 6.

It is well known that every polycyclic-by-finite group $G$ has a normal subgroup $W$ of finite index that is poly-(infinite cyclic); see [203], Lemma 10.2.5. In other words, $W$ has a subnormal chain $W = W_0 \supset W_1 \supset \cdots \supset W_n = \{1\}$, with all factors $W_i/W_{i+1}$ being infinite cyclic. Since every subgroup $H$ of $G$ is also polycyclic-by-finite, then the same applies to $H$. This fact is used below when the foregoing observations are summarized.

**Theorem 3** Let $S$ be a submonoid of a polycyclic-by-finite group. Assume that $W$ is a poly-(infinite cyclic) normal subgroup of finite index in the group $H$ generated by $S$. Then,

(i)   $\mathcal{J}(K[S])$ is nilpotent, $\mathcal{J}(K[S \cap W]) = 0$, and $\mathcal{J}(K[S]) = 0$ if $|H/W| \neq 0$ in $K$ or if $H$ has no normal $p$-subgroups of order divisible by $p$ if $\mathrm{ch}(K) = p > 0$.

(ii)  If $H$ is torsion-free, then $K[S]$ is a domain and $\mathcal{J}(K[S]) = 0$.

(iii) The following conditions are equivalent.

    (1) $S$ has a group of right fractions.

    (2) $K[S]$ is a right Goldie ring.

    (3) $K[S \cap W]$ is a semiprime right Goldie ring.

Moreover, in this case, $\mathcal{J}(K[S]) = \mathcal{J}(K[H]) \cap K[S]$, and $K[S]$ is an Ore domain if $H$ is a torsion-free group.

*Proof.* (i) From [203], Lemma 13.1.8, we know that $W$ is a u.p. group. Consequently, $S \cap W$ is a t.u.p. semigroup by Theorem 6 in Chapter 10. Now, Corollary 5 in Chapter 10 implies that $\mathcal{J}(K[S \cap W]) = 0$. If $\mathrm{ch}(K) = p > 0$ and $H$ has no normal subgroups of order divisible by $p$, then $K[S]$ is semiprime by Lemma 16 and Theorem 19 both in Chapter 7. Therefore, (i) follows from Lemma 2.

(ii) If $H$ is torsion-free, then, by the result of Farkas and Snider [57] if $\mathrm{ch}(K) = 0$ (see [203], Theorem 13.4.18) and of Cliff [25] if $\mathrm{ch}(K) > 0$, $K[H]$ is a domain. Now, $\mathcal{J}(K[S]) = 0$ follows from (i).

(iii) Assume that $S$ has a group of right fractions. Then Corollary 5 in Chapter 7 implies that $S \cap W$ also has a group of right fractions and $(S \cap W)(S \cap W)^{-1} = W$. Thus, $K[S \cap W]$ is a right order in the noetherian algebra $K[W]$. Since $K[W]$ is a Goldie ring, then $K[S \cap W]$ also must be a right Goldie ring; see [203], Lemma 10.2.13. Hence, (3) follows from (i).

If we assume that (3) holds, then $K[S]$ has a finite Goldie dimension by Lemma 1. Clearly, $K[S]$ has a.c.c. on right annihilator ideals because it is true for the noetherian algebra $K[H] \supseteq K[S]$. Hence (2) follows.

The implication (2) $\Rightarrow$ (1) is established in Proposition 12 in Chapter 7.

Now, the fact that $K[H] = K[SS^{-1}]$ is a noetherian algebra implies, in view of Lemma 15 in Chapter 7, that $\mathcal{J}(K[S]) = \mathcal{J}(K[H]) \cap K[S]$ because $\mathcal{J}(K[S]) = \mathcal{B}(K[S])$ by (i). If $H$ is torsion-free, then $K[S]$ is a right Goldie domain by (ii). This implies that $K[S]$ is an Ore domain (see [31], Proposition 0.8.9), and this completes the proof of (iii).

The information on the radical given in the above theorem can be extended to a large class of cancellative semigroups. Since $\mathcal{J}(K[S]) = \mathcal{J}(K[S^1])$ by Lemma 11 in Chapter 4, then Lemma 2 and the assertions on $\mathcal{J}(K[S])$ in Theorem 3 may be used for not necessarily monoids $S$. Here, $\mathcal{L}(K[S])$ stands for the locally nilpotent radical of $K[S]$.

**Proposition 4** Let $G$ be a group every finitely generated subgroup of which is polycyclic-by-finite. Assume that $S$ is a subsemigroup of $G$ that has no free noncommutative subsemigroups. Then $\mathcal{J}(K[S]) = \mathcal{L}(K[S]) = \mathcal{J}(K[H]) \cap K[S]$, where $H$ is the subgroup of $G$ generated by $S$. In particular, $\mathcal{J}(K[S]) = 0$ if $\mathrm{ch}(K) = 0$ or if $\mathrm{ch}(K) = p > 0$ and $H$ has no $p$-torsion.

*Proof.* Let $a_1, \ldots, a_n \in \mathcal{J}(K[S])$. Then $F = \langle\langle \bigcup_{i=1}^{n} \mathrm{supp}(a_i) \rangle\rangle$ is a polycyclic-by-finite group. Since $S \cap F$ is a group-like subsemigroup in $S$, then Corollary 16 in Chapter 4 implies that $\mathcal{J}(K[S]) \cap K[S \cap F] \subseteq \mathcal{J}(K[S \cap F])$. By Lemma 1 in Chapter 7, $F$ is a group of fractions of $S \cap F$. From Theorem 3, it follows that $\mathcal{J}(K[S \cap F]) = \mathcal{J}(K[F]) \cap K[S]$ is nilpotent, and it is zero if $\mathrm{ch}(K) = 0$ or if $\mathrm{ch}(K) = p > 0$ and $H$ has no $p$-torsion. It follows that $\mathcal{J}(K[S]) = \mathcal{L}(K[S])$, and $\mathcal{J}(K[S]) = 0$ if any of the latter conditions is assumed. Then $\mathcal{J}(K[H]) \cap K[S] \subseteq \mathcal{J}(K[S])$ because the above reasoning with $S = H$ shows that $\mathcal{J}(K[H])$ is a nil ideal of $K[H]$. Let $b \in K[H]$. Then $a_1 b \in K[F']$ for a finitely generated subgroup $F'$ of $H$. As above, we see that $a_1 b$ is a nilpotent element. Since $b$ is arbitrary, then $a_1 \in \mathcal{J}(K[H])$. This shows that $\mathcal{J}(K[S]) \subseteq \mathcal{J}(K[H])$ and completes the proof.

We can now apply the above fact to the special case of cancellative semigroups discussed in the foregoing chapters. The following corollary depends heavily on several results on the radicals of group algebras, especially the description of the prime radical; see [203], Theorem 8.4.16.

**Corollary 5**  Let $S$ be a cancellative semigroup that either has a weakly nilpotent subsemigroup of finite index or is contained in a finite extension of an $FC$-group. Then $\mathcal{J}(K[S]) = \mathcal{B}(K[S]) = \mathcal{J}(K[H]) \cap K[S]$, where $H$ is the group of fractions of $S$. Moreover, $\mathcal{J}(K[S]) = 0$ if $\mathrm{ch}(K) = 0$ or if $\mathrm{ch}(K) = p > 0$ and $H$ has no normal subgroups of order divisible by $p$.

*Proof.*  The existence of a group of fractions $H$ of $S$ was established in Chapters 7 and 8. Moreover, we know that $H$ is a finite extension of a group $F$ that is an $FC$-group or a nilpotent group. Therefore, $H, S$ satisfy the hypotheses of Proposition 4 (see Lemma 7 in Chapter 7). Thus, $\mathcal{J}(K[S]) = \mathcal{J}(K[H]) \cap K[S]$. Since the prime radical $\mathcal{B}$ is hereditary on subalgebras, then it is enough to show that $\mathcal{J}(K[H]) = \mathcal{B}(K[H])$.

Suppose we know that $\mathcal{J}(K[F]) = \mathcal{B}(K[F])$. It is well known that $\mathcal{J}(K[H])^n \subseteq \mathcal{J}(K[F])K[H] = \mathcal{B}(K[F])K[H]$, where $n = [H : F]$ [203], Theorem 7.2.7; see Corollary 15 in Chapter 6. Since $\mathcal{B}(K[F]) \subseteq \mathcal{B}(K[H])$ by [149], Corollary 10.2.10, then $\mathcal{J}(K[H]) = \mathcal{B}(K[H])$ follows.

If $\mathrm{ch}(K) = 0$, then $\mathcal{J}(K[F]) = 0$ by Proposition 4.

Assume that $\mathrm{ch}(K) = p > 0$. Now $\mathcal{J}(K[F]) = \mathcal{B}(K[F])$ is established in [203], Lemma 8.1.8, for any $FC$-group $F$. If $F$ is nilpotent, then let $N$ be the maximal normal $p$-subgroup of $F$. From [203], Theorem 8.4.16, it follows that $\omega(K[N]) \subseteq \mathcal{B}(K[F])$. Since $F/N$ has no nontrivial $p$-subgroups, then, again, [203], Theorem 7.4.6, implies that $\mathcal{J}(K[F/N]) = 0$. This shows that $\mathcal{J}(K[F]) \subseteq \omega(K[N])K[F] \subseteq \mathcal{B}(K[F])$, completing the proof.

Clearly, a polycyclic-by-finite group cannot contain a free noncommutative subgroup. However, this is not generally true with respect to free noncommutative subsemigroups, and so the conditions listed in Theorem 3 (iii) do not hold for arbitrary subsemigroups of polycyclic-by-finite groups. In fact, adopting some results from [121], we will show that a polycyclic-by-finite group all of which subsemigroups have groups of right fractions must be nilpotent-by-finite, that is, a finite extension of a nilpotent group. This is a consequence of Theorem 7 below, the proof of which slightly modifies that of [121], Theorem 11.9.

**Lemma 6**  Let $G$ be a group that does not have free noncommutative subsemigroups, and let $A$ be an abelian normal subgroup of $G$. Fix some elements $g \in A$, $h \in G$, and set $g_k = h^k g h^{-k}$, $k = 1, 2, \ldots$. Then there exists a natural number $z \geq 1$ and integers $i(1), \ldots, i(z)$, each equal to either 0 or $\pm 1$ (and not all zero), such that $g_1^{i(1)} \ldots g_z^{i(z)} = 1$.

*Proof.* If $h^j \in A$ for some $j > 0$, then the fact that $A$ is abelian implies that $g_j = g = g_{2j}$ and $g_j g_{2j}^{-1} = 1$, establishing the assertion. Thus, assume that $h^j \notin A$ for every $j > 0$. By the hypothesis, the semigroup $\langle h, hg \rangle$ is not free noncommutative. Hence, there exist two distinct words, $v(x,y)$, $w(x,y)$, in the free semigroup $\langle x,y \rangle$ such that $v(h,hg) = w(h,hg)$. Let $v(x,y) = x^{k(1)} y^{l(1)} \ldots x^{k(r)} y^{l(r)}$, $w(x,y) = x^{m(1)} y^{n(1)} \ldots x^{m(t)} y^{n(t)}$ for some $r$, $t \geq 1$, and some $k(2), \ldots, k(r), l(1), \ldots, l(r-1), m(2), \ldots, m(t), n(1), \ldots, n(t-1) \geq 1$, $k(1), l(r), m(1), n(t) \geq 0$. Clearly, we may assume that the pairs $(k(1), l(1))$ and $(m(1), n(1))$ are distinct. Then it is easy to see that

(a)    $v(h,hg) = h^{k(1)}(hg)^{l(1)} \ldots h^{k(r)}(hg)^{l(r)}$
    $= (g_{k(1)+1} g_{k(1)+2} \cdots g_{s(1)})(g_{s(1)+k(2)+1} \cdots g_{s(2)}) \ldots h^{s(r)}$

and similarly,

$w(h,hg) = h^{m(1)}(hg)^{n(1)} \ldots h^{m(t)}(hg)^{n(t)}$
    $= (g_{m(1)+1} g_{m(1)+2} \cdots g_{p(1)})(g_{p(1)+m(2)+1} \cdots g_{p(2)}) \ldots h^{p(t)}$

where $s(j) = k(1) + l(1) + \cdots + k(j) + l(j)$ for $j = 1, 2, \ldots, r$, and $p(j) = m(1) + n(1) + \cdots + m(j) + n(j)$ for $j = 1, 2, \ldots, t$. Since $v(h,hg) = w(h,hg)$, $A$ is normal in $G$, and $h^j \notin A$ for every $j > 0$, then $s(r) = p(t)$. Therefore, canceling by $h^{s(r)}$, we derive from (a) that

$(g_{k(1)+1} \cdots g_{s(1)})(g_{s(1)+k(2)+1} \cdots g_{s(2)}) \cdots$
$\qquad\qquad\qquad = (g_{m(1)+1} \cdots g_{p(1)})(g_{p(1)+m(2)+1} \cdots g_{p(2)}) \cdots$

Since $(k(1), l(1)) \neq (m(1), n(1))$, the sequences of indices appearing in the left and right sides of the above equality are increasing, then the assertion follows.

The above is a refined version of [121], Lemma 11.7, where the same assertion is proved under the stronger assumption that $G$ has no exponential growth. The proof of [121], Lemma 11.8, remains valid once we know that the assertion of Lemma 6 holds. Thus, proceeding as in the proof of the theorem of Milnor and Wolff [121], Theorem 11.9 (which shows that finitely generated solvable groups of subexponential growth must be nilpotent-by-finite), we get the following result.

**Theorem 7** Let $G$ be a polycyclic-by-finite group. Then $G$ has no free noncommutative subsemigroups if and only if $G$ is nilpotent-by-finite.

**Example 8**  Let $G$ be the group given by the generators $x, y, z$, subject to the relations $zxz^{-1} = y$, $zyz^{-1} = xy^2$, $xy = yx$. It is easy to see that $G$ is a polycyclic group that is not nilpotent-by-finite. In particular, by Theorem 7, $G$ has a free noncommutative subsemigroup. It may be checked that $\langle zx, z \rangle$ is such a semigroup.

We close this chapter with a discussion of the noetherian property for semigroup algebras of semigroups embeddable in polycyclic-by-finite groups.

**Theorem 9**  Let $S$ be a submonoid of a polycyclic-by-finite group $G$. Then $K[S]$ is right noetherian if and only if $S$ satisfies a.c.c. on right ideals. Moreover, in this case, $S$ is finitely generated.

*Proof.*  We can assume that $S$ generates $G$ as a group, replacing $G$ by $\langle\langle S \rangle\rangle$ if necessary. Assume that $S$ has a.c.c. on right ideals. We treat $K[S]$ as a $G$-graded algebra, with the components $K[S](s) = Ks$, $s \in S$, $K[S](g) = 0$ for $g \in G \setminus S$. Then $K[S]$ is graded right noetherian because every graded right ideal of $K[S]$ must be of the form $K[I]$ for a right ideal $I$ of $S$. From the right-left symmetric analog of Corollary 17 in Chapter 6, it then follows that $K[S]$ is a right noetherian algebra.

On the other hand, it is clear that every semigroup $S$ must satisfy a c.c. on right ideals whenever $K[S]$ is right noetherian.

Assume now that $K[S]$ is right noetherian. Suppose first that $G$ is a poly-(infinite cyclic) group. We proceed by induction on the Hirsch number of $G$, that is, the length of the subnormal chain $G = G_0 \supset G_1 \supset \cdots \supset G_n = \{1\}$ with infinite cyclic quotients. If $n = 0$, then $G = S = \{1\}$, and the assertion is trivial. Suppose that $n > 1$, and put $F = G_1$. Let $G = \bigcup_{j \in \mathbb{Z}} g^j F$ for some $g \in G$, and let $S_+ = \bigcup_{j \geq 0}(S \cap g^j F)$, $S_- = \bigcup_{j < 0}(S \cap g^j F)$. Let $T = \bigcup_{j > 0}(S \cap g^j F)$. Clearly, $T$ is an ideal of $S_+$, and $TS$ is a right ideal in $S$. By the hypothesis, $TS = s_1 S \cup \cdots \cup s_m S$ for some $s_i \in T$, $m \geq 1$. Let $s_i \in g^{j_i} F$. If $t \in S \cap g^l F$, where $l > j_i$ for $i = 1, \ldots, m$, then $t \in T \subseteq TS$, and so $t = s_{i_1} u$ for some $i_1 \in \{1, \ldots, m\}$, $u \in S$. Thus, $u \in S_+$, which shows that $S \cap g^l F \subseteq s_1 S_+ \cup \cdots \cup s_m S_+$. Moreover, if $u \notin \bigcup_{j < k}(S \cap g^j F)$, where $k = \max\{j_i \mid i = 1, \ldots, m\}$, then we can repeat the above procedure to find $i_2$ such that $u \in s_{i_2} S_+$. After a finite number of such steps, we come to $t = s_{i_1} \ldots s_{i_r} z$, where $z \in \bigcup_{j < k}(S \cap g^j F)$, $r \geq 1$. Therefore, $S_+ = \langle s_1, \ldots, s_m, \bigcup_{j < k}(S \cap g^j F)\rangle$. On the other hand, every $K[S] \cap g^j K[F]$, $j = 1, \ldots, k$, is a noetherian right $K[S \cap F]$-module by Lemma 1. In particular, the induction hypothesis implies then that $K[S \cap F]$ is a finitely generated semigroup, and so $S_+$ also is finitely generated. A similar argument shows that $S_-$ is a finitely generated semigroup. Then $S = S_+ \cup S_-$ is finitely generated as desired.

Finally, let $G$ be an arbitrary polycyclic-by-finite group. Then, by Lemma 1, $K[S \cap W]$ is right noetherian for a poly-(infinite cyclic) normal subgroup $W$ of finite index in $G$. The foregoing implies that $S \cap W$ is a finitely generated semigroup. Since Lemma 1 implies also that $K[S]$ is a finitely generated right $K[S \cap W]$-module, then $S = \bigcup_{i=1}^{r} t_i(S \cap W)$ for some $t_i \in S$. It follows that $S$ is finitely generated. This completes the proof.

It is easy to give examples showing that the situation described in Theorem 3 (iii) is more general than that of Theorem 9. For example, let $S = \langle 1, x, xy, xy^2, xy^3, \ldots \rangle$ be the subsemigroup of the free commutative group $G = \langle\langle x, y \rangle\rangle$ of rank 2. Clearly, $S$ is not finitely generated and has no a.c.c. on ideals, while $G$ is the group of fractions of $S$.

Let us also note that, in general, a semigroup algebra of a finitely generated subsemigroup of a polycyclic-by-finite group need not be right noetherian. Algebras of this type may be provided through Example 8 above.

Finally, we consider the noetherian property in the special case of cancellative semigroups studied in Chapter 9. Since a.c.c. on right ideals implies the existence of a group of right fractions, this reduces to the case of subsemigroups of $FC$-groups. For this class, the second assertion of Theorem 9 may be strengthened.

**Theorem 10** Let $S$ be a submonoid of an $FC$-group. Then the following conditions are equivalent:

(i)   $S$ is finitely generated.
(ii)  $K[S]$ is right noetherian.
(iii) $K[S]$ is left noetherian.

*Proof.* If $S$ is finitely generated, then $G = \langle\langle S \rangle\rangle$ is a finitely generated group, so that, from Lemma 7 in Chapter 7, we know that $|G/Z(G)| < \infty$. On the other hand, if $K[S]$ is right noetherian, then $S$ has a.c.c. on right ideals, and Proposition 12 in Chapter 7 implies that $G = SS^{-1}$. Thus, by Lemma 15 in Chapter 7, $K[G]$ is a right noetherian group algebra. It is well known that $G$ must be a finitely generated group in this case, too; see [203], Lemma 10.2.2. Therefore, when proving the equivalence (i) $\Leftrightarrow$ (ii), we can assume that $G$ is a finite extension of its center $Z(G)$.

Suppose first that $S$ is a finitely generated semigroup, say $S = \langle s_1, \ldots, s_n \rangle$ for some $s_i \in S$. Put $m = |G/Z(G)|$, and let $T$ denote the set of all elements in $S \cap Z(G)$ that are products of at most $m$ elements from the set $\{s_1, \ldots, s_n\}$. Consider an element $s = t_1 \ldots t_r \in S \cap Z(G)$, where $r > m$ and $t_j \in \{s_1, \ldots, s_n\}$ for $j = 1, 2, \ldots, r$. Then there exist $k, l$ with $k < l$ such that the elements $t_1 \ldots t_k, t_1 \ldots t_l$ are in the same coset of $Z(G)$ in $G$. Therefore, $z = t_{k+1} \ldots t_l$ lies in $Z(G)$. Now $s = t_1 \ldots t_r = z(t_1 \ldots t_k t_{l+1} \ldots t_r)$ and, clearly, $t_1 \ldots t_k t_{l+1} \ldots t_r \in Z(G)$. Thus, proceeding this way, we eventually come to $s = z_1 z_2 \ldots z_j t$, where $t$ and all $z_i$ are products of at most $m$ elements of the set $\{s_1, \ldots, s_m\}$, and $z_1, \ldots, z_j \in Z(G)$. Then, also, $t \in Z(G)$. Therefore, $s \in \langle T \rangle$ and, since $s \in S \cap Z(G)$ is arbitrary, $S \cap Z(G) = \langle T \rangle$ is a finitely generated monoid. It follows that $K[S \cap Z(G)]$ is a noetherian algebra. Therefore, Lemma 1 implies that $K[S]$ is right noetherian, which establishes the implication (i) $\Rightarrow$ (ii).

Assume now that $K[S]$ is right noetherian. Since $G/Z(G)$ is finite, then $Z(G)$ is a finitely generated group, so that $G$ is polycyclic-by–finite. Thus, $S$ must be finitely generated by Theorem 9.

The equivalence (i) $\Leftrightarrow$ (iii) may be established through a symmetric proof.

The following result is an immediate consequence of Theorem 10 and Lemma 1 applied to the monoid $S^1$.

**Corollary 11** Let $G$ be an $FC$-group generated by its subsemigroup $S$. Assume that $H$ is a normal subgroup of finite index in $G$. Then $S$ is finitely generated if and only if $S \cap H$ is finitely generated.

We close this chapter with the following observation. As we noticed before, if $K[S]$ is right noetherian, and $S$ is cancellative, then $S$ has a group $G$ of right fractions, and $K[G]$ is a noetherian group algebra. It is conjectured that, in this case, $G$ must be a polycyclic-by-finite group. This shows the role of the case of subsemigroups of polycyclic-by-finite groups in the search for a characterization of noetherian semigroup algebras of arbitrary cancellative semigroups. In view of Theorem 9, this also motivates the following question:

Is $S$ finitely generated whenever $K[S]$ is right noetherian? An affirmative answer to this question will be given for several important classes of semigroup algebras in Chapters 12 and 19.

## Comments on Chapter 11

We refer to [203], § 10.2, and [204] for the results known on noetherian group algebras. If $S$ is a subsemigroup of a polycyclic-by-finite group $G$, then the fact that $G$ is residually finite (see [203], Lemma 10.2.11) implies that $S$ is a subdirect product of finite groups. Thus, using Proposition 4 in Chapter 4 and Maschke's theorem, one can deduce that $\mathcal{J}(K[S]) = 0$ if $\text{ch}(K) = 0$. This provides an alternative proof of one of the statements of Theorem 3. Problems concerning Goldie and noetherian rings graded by subsemigroups of polycyclic-by-finite groups were studied by Jespers in [99]. He established some of the assertions of Theorem 3 in this more general setting. In fact, some of the methods used in this chapter come from general techniques in group graded rings; see [168]. In [238], Rosenblatt showed that every finitely generated solvable group with no free noncommutative subsemigroup must be nilpotent-by-finite; see our Theorem 7. Recently, Boffa and Bryant [13] extended this assertion to the class of all algebraic groups using Rosenblatt's result and Tit's alternative. We also note that a link between polycyclic-by-finite and solvable groups results from the fact that every polycyclic-by-finite group has a normal subgroup of finite index $H$, with $H'$ being nilpotent; see [203], Theorem 12.1.5. The above results can be viewed as facts concerning the growth of semigroups; see Chapter 8. Another class of semigroups for which the absence of noncommutative free subsemigroups implies polynomial growth will be discussed in Chapter 24. Theorem 10 will be used in Chapter 20 to derive a characterization of noetherian $PI$ algebras of cancellative semigroups.

# Part III

# Finiteness Conditions

This part is devoted to a study of some classical ring theoretic finiteness conditions for $K[S]$. The general problem considered is to characterize semigroup algebras of a given type in terms of the underlying semigroup $S$ only, or possibly in terms of the field $K$ as well. Thus, in particular, we are looking for extensions of the important results describing artinian, self-injective, and regular group algebras.

We start with the chain conditions. Chapter 12 extends the results on the noetherian property for $K[S]$ presented in Chapter 11. Then we turn to a family of related descending chain conditions (d.c.c.). In Chapter 13, we discuss a specific auxiliary type of algebras more general than algebraic algebras, which is useful when considering the d.c.c. in Chapter 14. The results on semisimple artinian algebras are extended to the class of von Neumann regular algebras in Chapter 15. Self-injectivity and related conditions for $K[S]$ are studied in Chapter 11. Finally, we review finiteness conditions coming from some other types of restrictions, such as freeness or projectivity, on right $K[S]$-modules or ideals. While hereditary and semihereditary algebras can be viewed as an extension of semisimple artinian, respectively, von Neumann regular algebras, the methods involved here are very different from those used before, and so the proofs are not given.

# 12
# Noetherian Semigroup Algebras

It is well known, and easy to prove, that, for a group $G$, the algebra $K[G]$ is right noetherian if and only if it is left noetherian. This is due to the antiautomorphism of $K[G]$ determined by $g \rightarrow g^{-1}$ for $g \in G$. Moreover, in this case, $G$ has the a.c.c. on subgroups; see [203], § 10.2. This seems to be the only general result on noetherian group algebras. On the other hand, a classical result of Higman asserts that the group algebra of a polycyclic-by-finite group is noetherian. A very deep structure, and representation theory of these algebras have recently been developed. For a survey of this topic, we refer to [204]. This motivated the results in Chapter 11 dealing with subsemigroups of polycyclic-by-finite groups. We note that no known examples of noetherian group algebras arise from groups that are not polycyclic-by-finite.

In the case where $S$ is a commutative semigroup, it is known that the a.c.c. on congruences in $S$ implies that $S$ is finitely generated; see [64],

Theorem 5.10. Hence, noetherian commutative semigroup algebras are finitely generated. By the Hilbert basis theorem, the converse holds if $S$ is a monoid.

Our main aim is to extend the above-mentioned results on the finite generacy of $G$ and $S$ to the class of arbitrary noetherian semigroup algebras. Recall that the a.c.c. on right congruences in $S$ is a necessary condition for $K[S]$ to be right noetherian; see Lemma 5 in Chapter 4.

**Proposition 1**   Assume that $K[S]$ is right noetherian. Then,

(i)   $S/\rho_{\mathcal{B}(K[S])}$ embeds into the multiplicative semigroup of a matrix algebra $M_n(D)$ for some $n \geq 1$, and a division $K$-algebra $D$.

(ii)   $S$ has finitely many right ideals of the form $eS$, where $e$ is an idempotent in $S$.

(iii)   Any 0-simple principal factor of $S$ is completely 0-simple.

*Proof.*   (i) It is well known that $K[S]/\mathcal{B}(K[S])$ is a right Goldie ring; hence, it embeds via the quotient ring construction into a ring of the form $M_{n_1}(D_1) \oplus \cdots \oplus M_{n_r}(D_r)$ for some $r \geq 1$, $n_i \geq 1$, and some division algebras $D_i$. From the results of Cohn and Bergman (see [30], Theorems 5.32 and 4C), it follows that there exists a division algebra $D$ such that $D \supseteq D_i$ for $i = 1, \ldots, r$. Therefore, the algebra $K[S]/\mathcal{B}(K[S])$ embeds into $M_n(D)$, where $n = n_1 + \cdots + n_r$. On the other hand, from Lemma 5 in Chapter 4, it follows that $S/\rho_{\mathcal{B}(K[S])}$ embeds into the multiplicative semigroup of $K[S]/\mathcal{B}(K[S])$. Thus, (i) follows.

(ii) Let $\leq$ denote the natural partial order in the set $E(S)$ of idempotents of $S$, that is, $e \leq f$ if $ef = e = fe$ for $e, f \in E(S)$. Clearly, $S$ has no infinite chains (with respect to $<$) of idempotents because, otherwise, $K[S]$ has an infinite chain of orthogonal idempotents, which contradicts the finiteness of the right Goldie dimension of $K[S]$. Let $e_1 S \supset e_2 S$, where $e_1, e_2 \in E(S)$. Then $e_2 = e_1 e_2 = e_2 e_1 e_2$ and, consequently, $e_2 e_1 = (e_2 e_1)^2$, $e_2 e_1 S = e_2 S$. Further, $(e_2 e_1)e_1 = e_2 e_1 = e_1(e_2 e_1)$, which means that $e_2 e_1 \leq e_1$. Clearly, $e_2 e_1 \neq e_1$ because $e_1 S \neq e_2 S$. Any chain $e_1 S \supset e_2 S \supset \cdots e_n S \supset \cdots$, where $e_i \in E(S)$, may thus be written as $e_1 S \supset e_2 e_1 S \supset e_3 S \supset \cdots \supset e_n S \supset \cdots$. Repeating this procedure, we would come to a chain of idempotents $e_1 > e_2 e_1 > e_3 e_2 e_1 > \cdots > e_n e_{n-1} \ldots e_1 > \cdots$. This contradiction shows that $S$ has d.c.c. on right ideals of the form $eS$, $e \in E(S)$.

Fix an element $f \in E(S^1)$. Suppose that there are infinitely many right ideals $e_i S$, $e_i \in E(S)$ for $i = 1, 2, \ldots$, which are maximal among the ideals of the form $eS$, $e \in E(S)$, satisfying $eS \subset fS$. It is clear that $e_1 S \subset e_1 S \cup e_2 S \subset \cdots \subset \bigcup_{i=1}^{n} e_i S$ for every $n > 1$ because $e_n \notin e_i S$ for $n \neq i$.

This contradicts the fact that a.c.c. on right ideals is satisfied in $S$. Therefore, the set of right ideals $eS$, $e \in E(S)$, which are maximal with respect to the property $eS \subset fS$, is finite. This, along with the fact that there are no infinite descending chains of ideals $eS$, $e \in E(S)$, implies, in view of Konig's lemma ([232],§ 1.4.4), that (ii) holds.

(iii) Assume that the principal factor $S_t$ of $S$ determined by an element $t \in S$ is 0-simple. Let $I$ be the ideal of nongenerators of the ideal $StS$. Passing, if necessary, to the semigroup $S/I$, we may assume that $I = \theta$, or $I = \varnothing$, that is, $S_t = StS$ is a 0-simple ideal of $S$. Clearly, we may also assume that $S_t$ is nontrivial. We will show that $StS$ has a nonzero idempotent. Let $\bar{S} = S/\rho_{\boldsymbol{\mathcal{B}}(K[S])}$ and $T = (StS)/\rho_{\boldsymbol{\mathcal{B}}(K[S])}$. By (i), $\bar{S} \subseteq M_n(D)$ for some $n \geq 1$, and a division algebra $D$. Moreover, $K[\bar{S}]$ is right noetherian, and $T$ is a 0-simple ideal of $\bar{S}$. The latter implies that $T$ embeds into one of the completely 0-simple principal factors $I_j/I_{j-1}$ of the multiplicative semigroup of $M_n(D)$ defined in Chapter 1. Let $s_1$ be the image of $t$ under the natural homomorphism $StS \to T$. By the 0-simplicity of $T$, $\theta \neq s_1 = s_2 s_1 t_1$ for some $s_2, t_1 \in T$. Similarly, we construct elements $s_n \in T$, $n > 2$, such that $s_n = s_{n+1} s_n t_n$ for some $t_n \in T$. Then $s_n(\bar{S})^1 \subseteq s_{n+1}(\bar{S})^1$ for every $n > 1$ and, by the a.c.c. on right ideals in $\bar{S}$, there exists $m > 1$ such that $s_m(\bar{S})^1 = s_{m+1}(\bar{S})^1$. Then $s_{m+1} = s_m z$ for some $z \in (\bar{S})^1$, which implies that $s_{m+1} s_m t_m z = s_m z = s_{m+1} \neq \theta$. Since $s_m t_m z \in T$ and $T$ embeds into a completely 0-simple semigroup, then, by Lemma 4 in Chapter 1, $s_m t_m z$ is a nonzero idempotent in $T$. It then follows that any inverse image $x \in S$ of $s_m t_m z$ is a periodic nonzero element because $x - x^2$ is a nilpotent in $K[S]$. Thus, $StS$ has a nonzero idempotent. This establishes our claim.

Now, the fact that $S$ has no infinite chains of idempotents, by (ii), implies that $StS$ has a primitive idempotent. Hence, (iii) follows; see Lemma 1 of Chapter 1.

As a direct consequence of assertions (ii) and (iii) in the above proposition, we derive the following result.

**Corollary 2**   Assume that $K[S]$ is right noetherian. Then $S$ has finitely many 0-simple principal factors.

Observe that the finiteness of the set of the left ideals of the form $Se$, $e = e^2 \in S$, is not a necessary condition for $K[S]$ to be right noetherian.

**Example 3**   Let $S = \{1, s^n, e_n \mid n = 1, 2, \ldots\}$ be the monoid, with multiplication given by the rules

$$se_n = e_n, \; e_n e_m = e_m, \; e_n s = e_{n+1} \qquad \text{for every } m, n \geq 1$$

Then $K[S]$ is right noetherian since it is a finitely generated right module over the polynomial ring $K[\langle s \rangle^1]$, $(1, e_1$ is a generating set). Clearly, $e_n S = e_m S$, and $S e_m = \{e_m\}$ for every $m, n \geq 1$. In particular, $K[S]$ is not left noetherian (use the left-right symmetric analog of Proposition 1 (ii)).

We note that assertions (ii) and (iii) of Proposition 2, as well as case (ii) in Theorem 6 below, were proved in [88] through some purely semigroup theoretic methods under the formally weaker assumption that $S$ has the a.c.c. on right congruences.

Example 3 shows that completely 0-simple principal factors of $S$ may have infinitely many idempotents. This is not the case if $S$ itself is completely 0-simple, as the following result shows. Here we use the notation, and results of Chapter 1.

**Proposition 4** Let $S$ be a subsemigroup of a semigroup of matrix type $T = \mathfrak{M}^0(G, I, M, P)$. If $K[S]$ is right noetherian, then $S \cap T_{(i)}^{(m)} \neq \varnothing, \theta$ for only finitely many subsemigroups $T_{(i)}^{(m)}$, $i \in I$, $m \in M$.

*Proof.* Let $\rho = \rho_G$ be the congruence defined in Lemma 5 in Chapter 1. Then $K[S/\rho]$ is right noetherian, and $S/\rho \subseteq T/\rho = \mathfrak{M}^0(\{1\}, I, M, \bar{P})$ for an $M \times I$ matrix $\bar{P}$ with entries in $\{0, 1\}$. If $(S/\rho) \cap T_{(i)} \neq \varnothing, \theta$, then it is a right ideal of $S/\rho$. Thus, the a.c.c. on right ideals in $S/\rho$ implies that $S/\rho$ intersects nontrivially only finitely many rows $T_{(i)}$, $i \in I$, of $T/\rho$. Therefore, $S/\rho \subseteq \mathfrak{M}^0(\{1\}, t, M, Q)$ for some $t \geq 1$, and an $M \times t$ submatrix $Q$ of $\bar{P}$. Suppose that $(S/\rho) \cap T^{(m_i)} \neq \varnothing, \theta$ for infinitely many $m_i \in M$, $i = 1, 2,$ .... It is clear that, among the corresponding $m_i$th rows of $Q$, $i = 1, 2, ...,$ there are infinitely many identical rows. Thus, refining the sequence $m_1$, $m_2, ...,$ we may assume that for every $i, j \geq 1$ the $m_i$th and the $m_j$th rows of $Q$ are identical. Then, $(1, k, m_i)(1, l, m) = (p_{m_i l}, k, m) = (p_{m_j l}, k, m) = (1, k, m_j)(1, l, m)$ for every $k, l \in \{1, 2, ..., t\}$, $m \in M$, $i, j \geq 1$. This shows that $(1, k, m_i) - (1, k, m_j)$ is in the left annihilator of the algebra $K_0[T/\rho]$ in $K_0[T/\rho]$. Now, by our supposition, there exists $k \in \{1, 2, ..., t\}$ such that $S/\rho$ contains infinitely many elements of the form $(1, k, m_i)$, $i \geq 1$. Therefore, the left annihilator of the algebra $K_0[S/\rho]$ is infinite dimensional over $K$. This contradicts the fact that $K_0[S/\rho]$ is right noetherian and shows $S/\rho$ intersects nontrivially (that is, $\neq \varnothing, \theta$) only finitely many $T^{(m)}$, $m \in M$. Now, the assertion is a direct consequence of the definition of the congruence $\rho$.

If a completely 0-simple semigroup $J = \mathfrak{M}^0(G, I, M, P)$ is an ideal of $S$ and if $\rho$ is a congruence on $J$ used in Proposition 4, then, define a relation $\mu$ on $S$ by $(s, t) \in \mu$ if $s = t$, or $s, t \in J$ and $(s, t) \in \rho$. If $s, t \in J$ are such that $(s,t) \in \rho$, then there exists $e \in J$ such that $s = se$, $t = te$. Hence, for every $x \in S$, $sx = s(ex)$, $tx = t(ex)$, and $ex \in J$. Therefore, $(sx,tx) \in \rho$. This shows that $\mu$ is a right congruence on $S$. Similarly, one shows that $\mu$ is a left congruence. This observation will be used in the proof of the main result of this chapter, when some of the ideas of Lemma 4 are extended to a more general situation.

When proving Theorem 6, we will also exploit the following known fact.

**Lemma 5** Let $R$ be a right noetherian $K$-algebra, and let $J$ be a nil ideal of $R$. If $R/J$ is a finitely generated $K$-algebra, then $R$ is a finitely generated $K$-algebra.

*Proof.* While $J$ must be nilpotent (see [55], 17.24), then an induction on the nilpotency index of $J$ may be used. This allows us to consider only the case where $J^2 = 0$. By the hypothesis $J = \sum_{i=1}^{n} a_i R'$ for some $a_i \in R$, $n \geq 1$(here $R'$ is the ring with an identity adjoined to $R$ via the ring of integers). Let $\{b_1,\ldots,b_m\}$ be the subset of $R$ mapping onto a set of generators of the $K$-algebra $R/J$ under the natural homomorphism $R \to R/J$. We claim that $R$ is generated by the set $\{a_1,\ldots,a_n,b_1,\ldots,b_m\}$. If $a \in R$, then $a = b + j$, where $b$ lies in the subalgebra $B$ generated by $\{b_1,\ldots,b_m\}$, and $j \in J$. Hence, $a = b + \sum a_i c_i$ for some $c_i \in R'$. Now, any $c_i$ may be written as $d_i + j_i$, where $d_i \in B'$, $j_i \in J$, and, hence, $a = b + \sum a_i(d_i + j_i) = b + \sum a_i d_i$ is in the subalgebra of $R$ generated by the set $\{a_1,\ldots,a_n,b_1,\ldots,b_m\}$, which proves our claim.

To prove the main theorem, we need one more preparatory observation. Let $e$ be an idempotent of a ring $R$. If $I \subset J$ are right ideals of $eRe$, then $IRe = IeRe = I$ and, similarly, $JRe = J$. Now, $IR \cap Re = IRe = I$ and $JR \cap Re = J$, which implies that $IR \subset JR$. Therefore, $eRe$ is right noetherian whenever $R$ is right noetherian.

**Theorem 6** Assume that $K[S]$ is right noetherian. Then $S$ is a finitely generated semigroup if either of the following holds:

(i)   $S$ has a.c.c. on principal left ideals.
(ii)  $S$ is a weakly periodic semigroup.
(iii) Every cancellative subsemigroup of any homomorphic image of $S$ has a finite-by-abelian-by-finite group of fractions.

*Proof.* Since $K[S^0]$ is a right noetherian algebra by Corollary 9 in Chapter 4, then we may assume that $S$ has a zero element, adjoining it to $S$ if necessary. Observe first that the noetherian property, as well as any of the hypotheses (i), (ii), (iii), is inherited by homomorphic images of $S$.

Suppose that $S$ is not finitely generated. Since $S$ has a.c.c. on congruences, then there exists a congruence $\rho$ in $S$ that is maximal with respect to the property that $S/\rho$ is not finitely generated. Hence, replacing $S$ by $S/\rho$, we may assume that the following condition is satisfied:

(a)   Any nontrivial homomorphic image of $S$ is finitely generated.

If $\tau$ is a nontrivial congruence on $S$, then, from (a) and Lemma 5, it follows that the kernel of the natural homomorphism $K_0[S] \to K_0[S/\tau]$ cannot be a nilideal. Therefore, we deduce that

(b)   For any nilideal $I$ of $K_0[S]$, the relation $\rho_I$ on $S$ defined by $(s,t) \in \rho_I$ if $s - t \in I$ is a trivial congruence on $S$ (note that the zero of $S$ is identified with that of $K_0[S]$ here).

We will consider two cases.

Case 1: $S$ has a minimal nonzero ideal $R$.

By (b), $R$ is not nilpotent. Hence, Proposition 1 implies that it is completely 0-simple. We identify $R$ with its Rees presentation $\mathfrak{M}^0(G,I,M,P)$. As noted earlier, the relation $\mu$ defined by $(s,t) \in \mu$ if $s = t$, or $s,t \in R$, and if they are in the same row and in the same column of $R$, is a congruence in $S$. Suppose first that $\mu$ is trivial. Then $G$ must be a trivial group. Let $L$ be the left annihilator of the algebra $K_0[R]$, and let $\rho_L$ be the congruence in $S$ defined by $(s,t) \in \rho_L$ if $s = t$, or $s,t \in R$ and $s - t \in L$. Since $L$ is an ideal of $K_0[S]$, $L^2 = 0$, then (b) implies that $\rho_L$ is trivial. Hence, the mapping $\phi : R \to M_I(\{0,1\})$ given by $\phi(s) = s \circ P$ is an embedding. Since $|I|$ must be finite because, by Proposition 1, $S$ has finitely many ideals of the form $eS$, $e = e^2 \in S$, then $R$ is finite. Now, in view of (a), $S/R$ is finitely generated, so that $S$ must be finitely generated, a contradiction.

Thus, assume that $\mu$ is nontrivial. Then (a) implies that $S/\mu$ is finitely generated. Let $\{s_1,\ldots,s_p\} \subseteq S$ be a subset that maps onto a generating set of $S/\mu$ under the natural homomorphism $S \to S/\mu$. If $\theta \neq e = e^2 \in R$, then $eSe = eRe \cong G^0$. Since $K[eSe] \cong eK[S]e$ is right noetherian, then $G$ is a finitely generated group; see [203], Lemma 10.2.2. Take $t_1, \ldots, t_r \in eSe$ such that $eSe = \langle t_1,\ldots,t_r \rangle$. We claim that $S = \langle s_1,\ldots,s_p,t_1,\ldots,t_r \rangle$. While $\mu$ is trivial on $S \setminus R$, then $S \setminus R \subseteq \langle s_1,\ldots,s_p \rangle$. Take $\theta \neq u \in R = \mathfrak{M}^0(G,I,M,P)$. Let $u = (g,j,m)$, where $g \in G$, $j \in I$, $m \in M$. Then there exist $x,y \in R$ such that $u = xey$ because $R$ is a 0-simple semigroup. Moreover, by the definition of $\mu$, there exist $x',y' \in R$ such that $(x,x') \in \mu$, $(y,y') \in \mu$, and $x',y' \in \langle s_1,\ldots,s_p \rangle$. Let $x' = (h,j',n)$, $y' = (f,l,m')$ with $h,f \in G$, $j',l \in I$,

$m', n \in M$. From the fact that $\theta \neq u = xey$, it follows that $j' = j$, $m' = m$, because $(x, x') \in \mu$, $(y, y') \in \mu$. If $e = (k, i, q)$ for some $k \in G$, $i \in I$, $q \in M$, then the fact that $\theta \neq u = xey$ implies also that $p_{ni}, p_{ql} \neq \theta$. Put $z = (p_{ni}^{-1} h^{-1} g f^{-1} p_{ql}^{-1}, i, q)$. Then $x'zy' = (g, j, m) = u$, and $z \in eSe$ because $eSe = \{(v, i, q) \mid v \in G^0\}$. Now, $x', y', z \in \langle s_1, \ldots, s_p, t_1, \ldots, t_r \rangle$ and, since $u \in R$ is arbitrary, then $R \subseteq \langle s_1, \ldots, s_p, t_1, \ldots, t_r \rangle$. In view of the inclusion $S \setminus R \subseteq \langle s_1, \ldots, s_p \rangle$, this again contradicts the supposition that $S$ is not finitely generated and completes the proof in case 1.

Before considering the second case, we show that, when dealing with weakly periodic semigroups, one meets the hypothesis of case 1. In fact, since $S$ has finitely many 0-simple principal factors by Corollary 2, then, from Theorem 3 in Chapter 3, it follows that $S$ has a nonzero ideal $I$ that is either nil or 0-simple. The former case is impossible in view of (b). Thus, $I$ is 0-simple and, hence, it is a minimal nonzero ideal of $S$. This proves the theorem under hypothesis (ii).

Case 2: $S$ has no minimal nonzero ideal.

By (b), the congruence $\rho_{\mathcal{B}(K[S])}$ is trivial. Now, Proposition 1 allows us to consider $S$ as a subsemigroup of a matrix algebra $M_n(D)$ over a division algebra $D$, that contains the zero matrix, see Remark 14 in Chapter 2. Let $T$ be the ideal of $S$ consisting of elements of the least nonzero rank and the zero matrix. Then $T$ embeds into a completely 0-simple semigroup $I_r / I_{r-1}$ defined in Chapter 1 for the monoid $M_n(D)$. By Proposition 1, we know that $S$ has finitely many $\mathcal{J}$-classes containing idempotents. Let $w \in T$ be such that $S^1 w S^1$ is minimal among the ideals $S^1 x S^1$, $x \neq \theta$, determining principal factors containing idempotents. Since $S$ has no nonzero minimal ideals, then $S^1 w S^1$ is not such an ideal, and so the set $J$ of nongenerators of $S^1 w S^1$ is a nonzero ideal of $S$. By the choice of $S^1 w S^1$, $J$ has no nonzero idempotents. We will show that $S = \langle S \setminus J^2 \rangle$.

First, we claim that

(c)  If $s = tu \neq \theta$, $t, u \in J$, then $sJ \subset tJ$, $Js \subset Ju$

Suppose that $sJ = tJ$. Then $s = tu \in sJ$ and, since $J$ lies in the completely 0-simple semigroup $I_r / I_{r-1}$ and has no nonzero idempotents, this contradicts Lemma 4 of Chapter 1. Thus, $sJ \neq tJ$ and, clearly, $sJ \subseteq tJ$, so that $sJ \subset tJ$ holds. The second assertion in (c) may be established similarly.

Since $K[S]$ is right noetherian, then $J^2 = \bigcup_{i=1}^{m} z_i J$ for some $m \geq 1$ and $z_i \in J$. Clearly, we may assume that all $z_i J$ are maximal among right ideals of the form $xJ$, $x \in J$ (note that $S$ has a.c.c. on right ideals). From (c), it then follows that $z_i \in J \setminus J^2$ for $i = 1, \ldots, m$. Put $Z = \{z_1, \ldots, z_m\}$. Let

$s \in J^2$. Then $s = t_1u_1$, where $t_1 \in Z$, $u_1 \in J$. If $u_1 \in J^2$, then, similarly, $u_1 = t_2u_2$, where $t_2 \in Z$, $u_2 \in J$. This procedure may be continued, so that either $s = t_1t_2 \ldots t_ku_k$ for some $k \geq 1$, $t_i \in Z$, $u_k \in J \setminus J^2$ or, for all $n \geq 1$, $s = t_1t_2 \ldots t_nu_n$, where $t_i \in Z$, $u_i \in J^2$. If the former holds, then $s \in \langle t_1,\ldots,t_k,u_k\rangle \subseteq \langle J \setminus J^2\rangle \subseteq \langle S \setminus J^2\rangle$, as desired. Thus, assume that the latter holds. Then (c) implies that $J^1s \subset J^1u_1 \subset J^1u_2 \subset \cdots$. Since $J$ is an ideal of $S$, then, by Lemma 4 in Chapter 2, $S^1s \subset S^1u_1 \subset S^1u_2 \subset \cdots$ is an ascending chain of principal left ideals of $S$. This contradicts (i) and shows that $S \subseteq \langle S \setminus J^2\rangle$ if (i) holds.

Assume that (iii) holds. Since $u_n = t_{n+1}u_{n+1}$ for every $n \geq 1$, then it is clear that all $u_n$ lie in the same column of $I_r/I_{r-1}$. From Theorem 6 in Chapter 1 and the noetherian property of $K[S]$ it follows that $J$ intersects nontrivially only finitely many rows of $I_r/I_{r-1}$. This and the fact that $t_n$ are taken from the finite set $Z$ imply that there exists a sequence of integers $i_1 < i_2 < \cdots$ such that $t_{i_1} = t_{i_2} = \cdots$ and the elements $u_{i_1}, u_{i_2}, \ldots$ lie in the same row and in the same column of $I_r/I_{r-1}$. Put $z = t_{i_1}$. We claim that there exists $x \in T$ such that $u_{i_1}x$, and so all $u_{i_j}x$ for $j \geq 1$, lie in a subgroup $X$ of $I_r/I_{r-1}$. If this is not the case, then $u_{i_1}T$ is a nil semigroup; hence, it must be nilpotent as a subsemigroup of $M_n(D)$; see [55], 17.24. This contradicts (b) because $T$ is a nonzero ideal of $S$, and it establishes our claim.

Let $v_j = t_{i_j+1} \ldots t_{i_{j+1}}$. We know that $u_{i_j}x = v_ju_{i_{j+1}}x$ for $j \geq 1$. Since $u_{i_j}x$ are non-nilpotents and they are in the same row of $I_r/I_{r-1}$ as the elements $v_j$, then $(u_{i_j}x)v_1 \neq \theta$. Let $w_j = u_{i_j}xv_1$, $j \geq 1$. Then,

(d)  $$w_j = v_jw_{j+1} \qquad \text{for } j \geq 1$$

and all $w_j$, $v_j$ lie in the same row and the same column of $I_r/I_{r-1}$ (note that all $v_j$ lie in the column in which $z = t_{i_1} = t_{i_2} = \cdots$ lies). This means that there is a subgroup $Y$ of $I_r/I_{r-1}$ such that $w_j,v_j \in Y$, $j \geq 1$. Let $W = Y \cap S$. Then $W$ is a cancellative subsemigroup of $S$. From Lemma 21 in Chapter 7, it follows that $K[WW^{-1}]$ is right noetherian, so that $WW^{-1}$ is a finitely generated group; see [203], Lemma 10.2.2. Since every group $H$ with finite commutator subgroup $H'$ is an $FC$-group, then, from Lemma 7 in Chapter 7, it follows, in view of our condition (iii), that $WW^{-1}$ is abelian-by-finite (see Chapter 19, particularly Corollary 13, for an extended description of this case). Let $A$ be a normal abelian subgroup of finite index in $WW^{-1}$. Then there exists a subsequence $w_{j_k}$, $k = 1, 2, \ldots$, of $w_j$ such that $w_{j_1}A = w_{j_2}A = \cdots$. In this case, $y_k = v_{j_k} \ldots v_{j_{k+1}-1} \in A$ for $k \geq 1$ by (d). From Lemma 4 in Chapter 7, we know that there exists $u \in W$ such that

$w_{j_k}u \in A$, $k \geq 1$. Then $w_{j_k}u = y_k(w_{j_{k+1}}u) = (w_{j_{k+1}}u)y_k$ because $A$ is abelian. Therefore, (c) implies that $w_{j_k}uJ \subset w_{j_{k+1}}uJ$, $k \geq 1$. This contradicts the fact that $S$ has a.c.c. on right ideals. Hence, $S \subseteq \langle S \setminus J^2 \rangle$ in this case, too.

Now, since $S/J^2$ is finitely generated by (b), then the fact that $S \subseteq \langle S \setminus J^2 \rangle$ implies that $S$ itself is finitely generated. This contradicts our original supposition and completes the proof of the theorem.

While condition (iii) in the above theorem may seem artificial at this point, it will appear crucial for the class of algebras considered in Part IV.

As an immediate consequence of Theorem 6, we get an extension of the result on commutative semigroup algebras; see [64], Theorem 20.7.

**Corollary 7** Assume that $K[S]$ is (right and left) noetherian. Then $S$ is finitely generated.

Our next consequence of Theorem 6 will be applied in Chapter 14, where periodic semigroups will be mainly involved.

**Corollary 8** Assume that $S$ is a weakly periodic semigroup. If $K[S]$ is right noetherian, and all subgroups of $S$ are locally finite, then $S$ is finite.

*Proof.* From Proposition 1 and from Theorem 3 in Chapter 3, we know that $S$ has a chain of ideals $S = S_n \supset S_{n-1} \supset \cdots \supset S_1$ such that $S_1$ and all factors $S_i/S_{i-1}$, $i > 1$, are completely 0-simple or nil. In the latter case, $S_1$, or $S_i/S_{i-1}$, must be nilpotent as a nil subsemigroup of the right noetherian ring $K_0[S]$, or $K_0[S/S_{i-1}]$, respectively [55], 17.22, or Proposition 13 in Chapter 2. In the former case, it is locally finite by Theorem 7 in Chapter 2. Thus, by Proposition 2 in Chapter 2, it follows that $S$ is a locally finite semigroup. Since it is finitely generated by Theorem 6, it is finite.

In the case of inverse semigroups, a criterion for $K[S]$ to be right noetherian may be given in terms of the maximal subgroups of $S$.

**Corollary 9** Assume that $S$ is an inverse semigroup. Then the following conditions are equivalent:

(i) $K[S]$ is right noetherian.
(ii) $S$ has finitely many idempotents, and all group algebras $K[G]$, $G$—a subgroup of $S$, are noetherian.
(iii) $K[S]$ if (right and left) noetherian.

*Proof.* If (i) holds, then, by Proposition 1, $S$ has finitely many right ideals of the form $eS$, $e = e^2 \in S$. Since $S$ is inverse, then the idempotents of $S$ commute. Thus, $eS = fS$ for $e = e^2, f = f^2 \in S$, implies that $e = f$. Therefore,

$S$ has finitely many idempotents. If $G$ is a subgroup of $S$ with an identity $e$, then $K[eSe]$ is right noetherian by the remark preceding Theorem 6. From Lemma 12 in Chapter 4, it follows that $K[G]$ is a homomorphic image of $K[eSe]$, and so it is right noetherian.

Assume that (ii) holds. Clearly, $S$ has finitely many principal factors, all of which are completely 0-simple by Lemma 1 in Chapter 1. Since the sandwich matrices in the Rees presentations of the principal factors of $S$ are the identity matrices (see [26], Theorem 3.9), then from Corollary 27 in Chapter 5, it follows that

$$K[S] \cong M_{n_1}(K[G_1]) \oplus \cdots \oplus M_{n_r}(K[G_r])$$

for some $r$, $n_i \geq 1$, and some maximal subgroups $G_i$ of $S$. Thus, $K[S]$ is noetherian because all $K[G_i]$ are noetherian.

The implication (iii) $\Rightarrow$ (i) is trivial.

It is not known whether, in general, $S$ must be finitely generated whenever $K[S]$ is right noetherian, even in the case where $S$ is a cancellative semigroup. On the other hand, the proof of Theorem 6 (going through a reduction to some cancellative subsemigroups arising from $S$) indicates that once we are able to prove this, the same might be true for arbitrary semigroups. Note also that, as observed in Chapter 11, the only known noetherian group rings are those of polycyclic-by-finite groups. If $S$ is a subsemigroup of such a group, then the problem is settled in Theorem 9 in Chapter 11.

## Comments on Chapter 12

Some of the results of this chapter can be obtained under the weaker hypothesis that $S$ has a.c.c. on right congruences. This is the case with respect to assertions (i) and (ii) of Proposition 1 and Theorem 6 (ii) proved by Hotzel in [88]. Further results on semigroups with maximal conditions on congruences and ideals are given in the papers of Hotzel [88], [89], and Kozhukhov [117]. Theorem 6 (i) and (ii) was obtained by Okniński in [188]. The description of noetherian semigroup algebras of inverse semigroups (see Corollary 9) comes from [188]; see also [165]. In [250], Shoji was concerned with the problem of describing noetherian semigroup algebras of weakly periodic semigroups. However, his Lemma 5 and Theorem 6 are in error. Richter [235] studied the noetherian property in a more general setting of small categories, giving some applications to semigroup algebras. Noetherian commutative semigroup rings $R[S]$ of a semigroup $S$, which is not necessarily a monoid, over an arbitrary ring $R$ are discussed by Gilmer in [65]. This extends the results presented in his monograph [64]. Some related noetherian structures are studied in the monograph of McConnell and Robson [149], which is our main reference for the theory of noetherian rings. Certain aspects of the noetherian group and semigroup graded rings are discussed in the recent papers of Quinn [231], Chin and Quinn [23], and Jespers [99].

# 13

# Spectral Properties

We start with considering some auxiliary "local" properties of elements of $K[S]$. These will be properties of subsemigroups of $S$ generated by the supports of elements in $K[S]$, connected with the global finiteness conditions examined for $K[S]$ in subsequent chapters. In the case of a group algebra $K[G]$, results of this type are often obtained through the projection mapping $K[G] \to K[H]$, where $H$ is a subgroup of $G$; see [203]. Here we use an extension of this method discussed in Chapter 4. We start with elements $a \in K[S]$, which are von Neumann regular, that is, satisfying $aba = a$ for some $b \in K[S]$.

**Lemma 1** Let $f(x) \in K[x]$ be a polynomial that is not a monomial. If $s \in S$ is such that $f(s)$ is a von Neumann regular element in $K[S]$, then $s$ is a periodic element.

*Proof.*  Let $a = \sum_{i=1}^{n} \lambda_i s_i \in K[S]$ be an element such that

(a)                                      $f(s)af(s) = f(s)$

Consider the subset $T = \{x \in S \mid uxt \in \langle s \rangle$ for some $u, t \in \langle s \rangle\}$. From Lemma 20 in Chapter 4, it follows that $\pi_T(ab) = \pi_T(a)b$, and $\pi_T(ba) = b\pi_T(a)$ for every $b \in K[\langle s \rangle]$, where $\pi_T$ is the projection of $K[S]$ onto $T$ discussed in Chapter 4. In particular, from (a) we derive that

$$f(s) = \pi_T(f(s)af(s)) = f(s)\pi_T(af(s)) = f(s)\pi_T(a)f(s)$$

Since $\pi_T(a)$ has a finite support, then there exists $m \geq 1$ such that the element $z = s^m \pi_T(a)s^m$ lies in $K[\langle s \rangle]$, and

(b)                                      $s^{2m}f(s) = f(s)zf(s)$

Suppose that $s$ is not a periodic element. Then $s$ is not an algebraic element in the $K$-algebra $K[S]$, $f(s) \neq 0$, and (b) may be considered as an equality in the polynomial algebra $K[\langle s \rangle^1]$. By the unique decomposition in $K[\langle s \rangle^1]$, it follows that $f(s) = \lambda s^k$ for some $k \geq 0$ and $\lambda \in K$. This contradicts the hypothesis on $f(x)$ and proves the assertion.

In the case where $K = \mathbb{Q}$, basing the proof on an idea of Formanek [59], more may be shown on regular elements of $K[S]$ of some special type.

**Lemma 2**  Let $a = \sum_{i=1}^{n} \lambda_i s_i \in \mathbb{Q}[S]$ be such that $1 - a$ is a von Neumann regular element in $\mathbb{Q}[S^1]$, where $\lambda_i > 0$ and $\sum_{i=1}^{n} \lambda_i < 1$. Then $\langle s_1, \ldots, s_n \rangle$ is a finite subsemigroup of $S$.

*Proof.*  For any $c = \sum_{i=1}^{m} \mu_i t_i \in \mathbb{Q}[S^1]$, define $\|c\| = \sum_{i=1}^{m} |\mu_i|$. It is well known that $\| \ \|$ is a submultiplicative norm on the algebra $\mathbb{Q}[S^1]$; see [203], § 2.1. Let $R$ be the completion of the $\mathbb{Q}$-algebra $\mathbb{Q}[S^1]$ with respect to $\| \ \|$. Since $\|a\| < 1$ by the hypothesis, then $1 - a$ is an invertible element in $R$ because $\sum_{i=0}^{\infty} a^i(1-a) = 1$, and $\sum_{i=0}^{\infty} a^i \in R$. From the hypothesis, it follows that $(1-a)b(1-a) = 1-a$ for some $b \in \mathbb{Q}[S^1]$. Then $b(1-a) = (1-a)b = 1$, so that $1 - a$ is invertible in $\mathbb{Q}[S^1]$. Let $b_m = 1 + a + a^2 + \cdots + a^m$ for any $m \geq 1$. Then,

$$b - b_m = b(1-a)(b - b_m) = b[1 - (1 - a^{m+1})] = ba^{m+1}$$

Therefore

(c)                    $\|b - b_m\| \leq \|b\| \|a\|^{m+1} \xrightarrow[m \to \infty]{} 0$ because $\|a\| < 1$.

Let $t \in \langle s_1, \ldots, s_n \rangle$. Then $t = s_{i_1} s_{i_2} \ldots s_{i_r}$ for some $r \geq 1$, $i_j \in \{1, \ldots, n\}$. While, for all $i$, we have $\lambda_i > 0$, then $t \in \mathrm{supp}(a^r)$. Moreover, $t \in \mathrm{supp}(b_m)$ for every $m \geq r$, and the coefficient of $t$ in the element $a^r$ is equal to or greater than $\lambda^r$, where $\lambda = \min\{\lambda_i \mid i = 1, 2, \ldots, n\}$. If $t \notin \mathrm{supp}(b)$, then it follows that $b - b_m \geq \lambda^r$ for every $m \geq r$, which contradicts assertion (c). Thus, $t \in \mathrm{supp}(b)$, which shows that $\langle s_1, \ldots, s_n \rangle \subseteq \mathrm{supp}(b)$ is a finite semigroup.

Let $R$ be a $K$-algebra with unity. By the spectrum of an element $a \in R$, we mean the set $\sigma_R(a) = \{\lambda \in K \mid \lambda - a$ is not invertible in $R\}$. We refer to [16] for the basic properties of the spectrum. $R$ is called a spectrally nondegenerated $K$-algebra if $\sigma_R(a) \cup \{0\} \neq K$ for every $a \in R$. We give two basic examples of algebras of this type.

**Example 3** (i) If $R$ is an algebraic $K$-algebra with unity and $K$ is an infinite field, then $R$ is a spectrally nondegenerated $K$-algebra. In fact, $\sigma_R(a)$ consists of zeros of the minimal polynomial of $a \in R$ over $K$ (see [16]) and, hence, it is a finite set.

(ii) Let $R$ be a semilocal $K$-algebra, that is, $R/\mathcal{J}(R)$ is an artinian algebra. It is well known that $\sigma_R(a) = \sigma_{R/\mathcal{J}(R)}(a + \mathcal{J}(R))$ for every $a \in R$; see [16]. By an easy linear algebra argument, one shows that, for every element $b$ of a matrix algebra over a division $K$-algebra $D$, we have $|\sigma_{M_n(D)}(b)| \leq n$, and the equality holds for some $b \in M_n(D)$. From the Wedderburn–Artin theorem, it then follows that $|\sigma_R(a)| \leq n_R$, where $n_R$ is the length of the $R$-module $R/\mathcal{J}(R)$. Therefore, $R$ is a spectrally nondegenerated $K$-algebra provided that the cardinality of $K$ exceeds $n_R + 1$.

The algebras given as examples above are, in fact, "spectrally finite" (that is, $\sigma_R(a)$ is finite for every $a \in R$), or even "spectrally bounded" (that is, $\sigma_R(a), a \in R$, has cardinality bounded by an integer). It may be shown that spectrally bounded algebras over infinite fields are exactly semilocal algebras [173]. In the proof, we will exploit the obvious fact that the cardinality of the spectrum of a given element does not grow under homomorphisms.

**Theorem 4** Let $R$ be a spectrally bounded algebra with unity over an infinite field $K$. Then $R$ is semilocal, and the length of the $R$-module $R/\mathcal{J}(R)$ is equal to $N = \max\{|\sigma_R(a)| \mid a \in R\}$.

*Proof.* By the remark used in Example 3 (ii), we can assume that $\mathcal{J}(R) = 0$. Suppose first that $R$ is left primitive. Then there exists a faithful and irreducible left $R$-module $M$. Suppose that $m_1, m_2, \ldots, m_{N+1}$ are linearly independent elements of $M$ over the division ring $D = \mathrm{End}_R M$. From the density theorem, it follows that, for every $\lambda_1, \lambda_2, \ldots, \lambda_{N+1} \in K$, there exists $r \in R$ such that $rm_i = \lambda_i m_i, i = 1, \ldots, N + 1$. Thus, $(r - \lambda_i)m_i = 0$, and so $\lambda_1, \ldots, \lambda_{N+1} \in \sigma_R(r)$. This contradicts the hypothesis and shows that the dimension of $M$ over $D$ does not exceed $N$. It is well known that $R \cong M_n(D)$ for some $n \leq N$ in this case; see [86].

Now, let $R$ be an arbitrary semisimple algebra. Assume that $P_1$, ..., $P_k$, $k \geq 1$, are left primitive ideals of $R$. From the first part of the proof, it follows that all $P_j$ are maximal ideals of $R$. Hence, the Chinese Remainder Theorem implies that

$$R/(P_1 \cap \cdots \cap P_k) \cong (R/P_1) \oplus \cdots \oplus (R/P_k)$$

However, $R/(P_1 \cap \cdots \cap P_k)$ inherits the assumption on the cardinality of the spectra of elements. Therefore, $k \leq N$, and, consequently,

$$R \cong R/ \left( \bigcap_{j=1}^{t} P_j \right) \cong (R/P_1) \oplus \cdots \oplus (R/P_t)$$

for some $t \geq 1$. By the first part of the proof, it follows that $R \cong \oplus_{j=1}^{t} M_{n_j}(D_j)$ for some division $K$-algebras $D_j$ and some $n_j$. From Example 3, we know that $\sum_{j=1}^{t} n_j$ is the upper bound on the cardinality of spectrum in $R$. This completes the proof.

On the other hand, we note that it was shown in [175] that the class of "spectrally finite" algebras does not behave well under natural algebraic constructions.

Theorem 4 may be completed in the following way, allowing the cardinality restriction on $K$ to be overcome.

**Lemma 5** Let $R$ be a semilocal algebra with unity over a field $K$. Then there exists a finite field extension $L$ of $K$ such that $A \otimes_K L$ is a spectrally nondegenerated and semilocal $L$-algebra.

*Proof.* If $K$ is infinite, then, by Example 3, it is enough to put $L = K$. Thus, assume that $K$ is finite, and let $F$ be a finite field extension of $K$. Since $F$ is a separable extension of $K$, it is well known that $\mathcal{J}(R \otimes_K F) \cap R = \mathcal{J}(A) \otimes_K F$ (see [203], Theorem 7.2.13). Then $R/\mathcal{J}(R) \subseteq (R \otimes_K F)/\mathcal{J}(R \otimes_K F)$. Moreover, $(R \otimes_K F)/\mathcal{J}(R \otimes_K F)$ is a finitely generated module over the artinian algebra $R/\mathcal{J}(R)$; hence, it is artinian. Further, $n_{R \otimes_K F} \leq n_R[F : K]$, where $n_R$, $n_{R \otimes_K F}$ denote the lengths of the $R$-module $R/\mathcal{J}(R)$ and the $R \otimes_K F$-module $(R \otimes_K F)/\mathcal{J}(R \otimes_K F)$, respectively. Let $L$ be a finite field extension of $K$ satisfying $|L| = |K|^{[L:K]} > n_R[L : K] + 1$. Then $n_{R \otimes_K L} < |L| - 1$, and the assertion follows through Example 3.

We will need one more result on spectrally nondegenerated tensor products.

**Proposition 6** Let $R$, $T$ be $K$-algebras with unities. If $R \otimes_K T$ is a spectrally nondegenerated $K$-algebra, then $R$ or $T$ is an algebraic $K$-algebra.

*Proof.* Suppose that there are elements $a \in R$, $b \in T$ that are not algebraic over $K$. By the hypothesis, there exists $\lambda \in K$ such that the element $u = \lambda + (a \otimes 1 + 1 \otimes b) = (\lambda + a) \otimes 1 + 1 \otimes b$ is invertible in $R \otimes_K T$. Then the element $c = \lambda + a$ is not algebraic over $K$. Let $R_1$ denote the centralizer of $c$ in $R$. Since $R_1 \otimes_K T$ coincides with the centralizer of $c \otimes 1$ in $R \otimes_K T$, and $u \in R_1 \otimes_K T$, then it is easy to see that $u^{-1} \in R_1 \otimes_K T$. The same argument shows that $u$, $u^{-1} \in R_2 \otimes_K T$, where $R_2$ is the centralizer of $R_1$ in $R_1$, that is, $R_2$ is the center of $R_1$. Similarly, one shows that $u$, $u^{-1} \in R_2 \otimes_K T_2$, where $T_2$ is the center of the centralizer of $b$ in $T$. Thus, replacing $R$, $T$ by $R_2$, $T_2$, we may assume that $R$, $T$ are commutative algebras.

Let $X = \{f(c) \mid 0 \neq f(x) \in K[x]\}$. Then $X$ is a multiplicatively closed set, and $0 \notin X$. If $I$ is an ideal of $R$ that is maximal with respect to the property $I \cap X = \varnothing$, then it is a prime ideal of $R$. Thus, $R/I$ is a domain and, from the choice of $I$, it follows that the image $c'$ of $c$ under the natural homomorphism $R \to R/I$ is not algebraic over $K$. Since $c' \otimes 1 + 1 \otimes b$ is invertible in $(R/I) \otimes_K T$, then we may assume that $R$ is a domain. Similarly, replacing $T$ by a prime homomorphic image of $T$, we may assume that $T$ is a domain. Let $K(c)$ be the subfield of the field of fractions $R^*$ of $R$ generated by $c, K$. Then $u$, $u^{-1} \in R^* \otimes_{K(c)} (K(c) \otimes_K T)$ and $u \in K(c) \otimes_K T$ easily imply that $u^{-1} \in K(c) \otimes_K T$. Analogously, we come to $u$, $u^{-1} \in K(c) \otimes_K K(b)$. Now, $K(c) \otimes_K K(b) \cong Z^{-1}K[c,b]$, where $Z = (K[c] \setminus \{0\})(K[b] \setminus \{0\})$, and the element $u$ may be identified with $c + b$. Then $(c + b)^{-1} = f(c,b)g(c)^{-1}h(b)^{-1}$ for some polynomials $f$, $g$, $h$, and we may assume that this is an irreducible representation of $(c + b)^{-1}$ in the field $K(c,b)$. Then $c + b = g(c)h(b)$, which is impossible. This contradiction completes the proof of the proposition.

**Proposition 7** Assume that $S$ is a monoid. Then,

(i)   If $K[S]$ is a spectrally nondegenerated $K$-algebra, then $S$ is a periodic semigroup.
(ii)  If the transcendence degree of $K$ over its prime subfield is infinite and $K[S]$ is a spectrally finite $K$-algebra, then $K[S]$ is an algebraic $K$-algebra.

*Proof.* (i) Let $s \in S$. Then, for some $0 \neq \lambda \in K$, $\lambda - s$ is an invertible element in $K[S]$. From Lemma 1, it follows that $s$ is periodic.

(ii) Let $\deg_{\mathrm{tr}}[K : F] = \infty$, where $F$ is the prime subfield of $K$. If $a = \sum_{i=1}^{n} \lambda_i s_i \in K[S]$, then put $L = F(\lambda_1, \ldots, \lambda_n)$. Since $K[S]$ is a spectrally finite $L$-algebra, then, from the fact that $K[S] \cong L[S] \otimes_L K$, Proposition 6, and the assumption on $K$, it follows that $L[S]$ is an algebraic $L$-algebra. Consequently, $a$ is an algebraic over $L$ element. Hence, $a$ is algebraic over $K$, which establishes (ii).

It is not known whether $S$ must be locally finite whenever $K[S]$ is algebraic over $K$, even in the case where $S$ is a group. This is the case if $\mathrm{ch}(K) = 0$, as the following direct consequence of Lemma 2 and Example 3 shows.

**Theorem 8** Assume that $K$ is a field of characteristic zero, and $S$ is a monoid. Then $K[S]$ is a spectrally finite $K$-algebra if and only if $S$ is a locally finite semigroup.

## Comments on Chapter 13

The first significant results employing the spectrum in ring theory were obtained by Amitsur in [1]. He proved that, if $K$ is an infinite field and $R$ a $K$-algebra satisfying $|K| > \dim_K R$, then an element $a \in R$ is algebraic over $K$ if and only if $|K| = |K \setminus \sigma_R(a)|$. This showed, in particular, that the notions of "algebraic" and "spectrally finite" coincide within the class of finitely generated algebras over a nondenumerable field. Spectrally finite algebras, their connections with semilocal algebras, and questions motivated by some important open problems on algebraic algebras were studied by Okniński in [173] and [175]. Formanek's spectral idea was first used in the context of semilocal group algebras in characteristic zero by Lawrence and Woods in [136]; see also [174]. Proposition 6 is due to Krempa and Okniński [127]. It extends the corresponding result of Lawrence [134] on semilocal tensor products. Some more results on the spectral properties of semigroup algebras, including Proposition 7, come from [174].

# 14

# Descending Chain Conditions

In this chapter, we discuss ring theoretic finiteness conditions for $K[S]$ coming from certain descending chain conditions (d.c.c.). These will include artinian, semiprimary, perfect, semilocal, local, and chain semigroup algebras. We start with semilocal algebras, which are basic for the other classes considered here. Recall that an algebra $A$ is semilocal if $A/\mathcal{J}(A)$ is artinian.

Throughout, the following general result will be used.

**Lemma 1** Let $A$, $B$ be $K$-algebras. Then,

(i)  If $B$ is an ideal of $A$, then $A$ is semilocal if and only if the algebras $B$, $A/B$ are semilocal.

(ii)  If $A \otimes_K B$ is semilocal and $A$, $B$ are algebras with unities, then,

   (1)  $A$, $B$ are semilocal algebras.
   (2)  If $B$ is a separable field extension of $K$, $\mathcal{J}(A \otimes_K B) = \mathcal{J}(A) \otimes_K B$.

*Proof.* (i) Since $\mathcal{J}(B) = \mathcal{J}(A) \cap B$, then, passing to $A/\mathcal{J}(A)$, we can assume that $\mathcal{J}(B) = 0$, $B \neq 0$. If $A$ is semilocal, then $B/\mathcal{J}(B) \cong B$ is an ideal in $A/\mathcal{J}(A)$, and so it is artinian. If $B$ is semilocal, then $B$ has an identity, so that $A \cong B \oplus A/B$. Hence, (i) follows.

(ii) We know that $\mathcal{J}(A)$ is the intersection of all maximal left ideals of $A$. Consider all $L \otimes B$ where $L$ is an intersection of finitely many maximal left ideals of $A$. By the hypothesis on $A \otimes B$, there exists $L \otimes B$ that is minimal modulo $\mathcal{J}(A \otimes B)$, that is,

(a)   $L \otimes B \subseteq \mathcal{J}(A \otimes B) + (L \cap I) \otimes B$ for every maximal left ideal $I$ in $A$

If $L \neq \mathcal{J}(A)$, choose $I$ such that $L \cap I \neq L$. Then $L + I = A$ and the unity $e$ of $A$ may be written as $e = l + i$, $l \in L$, $i \in I$. If $V$ is an irreducible left $B$-module and $v \in V$, then choose $b \in B$ such that $bv = v$. In the $A \otimes B$-module $(A/I) \otimes V$, we have

$$(l \otimes b)(\bar{e} \otimes v) = (e \otimes b)(\bar{e} \otimes v) = \bar{e} \otimes v$$

where $\bar{e}$ is the image of $e$ in $A/I$. Since $(L \cap I) \otimes B$ annihilates $\bar{e} \otimes v$, there exists by (a) an element $r \in \mathcal{J}(A \otimes B)$ with $r(\bar{e} \otimes v) = \bar{e} \otimes v$. But $r$ is quasiregular, so that $\bar{e} \otimes v = 0$. While $\bar{e} \neq 0$, then $v = 0$. It follows that $B$ has no nonzero irreducible modules, so that $\mathcal{J}(B) = B$, a contradiction. This shows that $L = \mathcal{J}(A)$. Now, $A/\mathcal{J}(A)$ is a submodule of the direct sum of the corresponding finite number of irreducible $A$-modules. Then $A/\mathcal{J}(A)$ is a semisimple $A/\mathcal{J}(A)$-module. Thus, (1) follows:

(2) It is known that $\mathcal{J}(A \otimes B) = I \otimes B$ for some ideal $I \subseteq \mathcal{J}(A)$ of $A$; see [203], Section 7.3. Since $A \otimes B$ is semilocal, then $(A/I) \otimes B \cong (A \otimes B)/\mathcal{J}(A \otimes B)$ is artinian. Then $A/I$ also is artinian because, for different left ideals $I_\alpha$ in $A/I$, the corresponding ideals $I_\alpha \otimes B$ of $A \otimes B$ are different. If $I \neq \mathcal{J}(A)$, then $A/I$ has a nonzero nilpotent ideal. This contradicts the semisimplicity of $(A/I) \otimes B$ and establishes (2).

We start with deriving some necessary conditions on $S$.

**Theorem 2**   Assume that $K[S]$ is semilocal. Then,

(i)   $S$ is a periodic semigroup.
(ii)   $S$ is locally finite if $\mathrm{ch}(K) = 0$.
(iii)   $K[G]$ is semilocal for every subgroup $G$ of $S$.

*Proof.* By Example 3 and Lemma 5, both in Chapter 13, $L[S^1]$ is a spectrally finite and spectrally nondegenerated $L$-algebra for a field extension $L$ of $K$. Thus, (i) and (ii) follow from Proposition 7, and Theorem 8, both in Chapter 13. (iii) If $e$ is the identity of $G$, then $\mathcal{J}(eK[S]e) = e\mathcal{J}(K[S])e$ (see

[86], Theorem 1.3.3), and so $(eK[S]e)/\mathcal{J}(eK[S]e) \cong e(K[S]/\mathcal{J}(K[S]))e$, which implies that $eK[S]e$ also is semilocal. Now $K[G]$ is a homomorphic image of $eK[s]e$ by (i) and Lemma 12 in Chapter 4. Hence, (iii) follows from Lemma 1.

We note that it is not known whether, in the case of a positive characteristic, $S$ must be locally finite, even if it is a group; see [203], Section 9.

We first deal with the congruence $\rho_{\mathcal{J}(K[S])}$ for semilocal algebras $K[S]$.

**Lemma 3** Assume that $\mathbb{Q}[S]$ is an algebraic $\mathbb{Q}$-algebra. Let $a_1, \ldots, a_k \in \mathbb{Z}[S]$ be $\mathbb{Q}$-linearly dependent modulo $\mathcal{J}(\mathbb{Q}[S]) \cap \mathbb{Z}[S]$. If $p$ is a prime number, and $\bar{a}_i$ is the image of $a_i$ under the natural homomorphism $\mathbb{Z}[S] \rightarrow \mathbb{F}_p[S]$, then $\bar{a}_1, \ldots, \bar{a}_k$ are $\mathbb{F}_p$-linearly dependent modulo $\mathcal{J}(\mathbb{F}_p[S])$.

*Proof.* Since $\mathcal{J}(\mathbb{Q}[S])$ is a $\mathbb{Q}$-subspace of $\mathbb{Q}[S]$, then we may assume that $b = \sum_{i=1}^{k} n_i a_i \in \mathcal{J}(\mathbb{Q}[S])$ for some integers $n_1, \ldots, n_k$, not all divisible by $p$. Now $\mathcal{J}(\mathbb{Q}[S])$ is a nil ideal because $\mathbb{Q}[S]$ is an algebraic $\mathbb{Q}$-algebra. Therefore, $b \in \mathcal{J}(\mathbb{Q}[S]) \cap \mathbb{Z}[S] \subseteq \mathcal{J}(\mathbb{Z}[S])$. Thus, $\bar{b} = \sum_{i=1}^{k} \bar{n}_i \bar{a}_i \in \mathcal{J}(\mathbb{F}_p[S])$, where $\bar{n}_i = n_i (\mathrm{mod}\, p)$, and there exists $i$ such that $\bar{n}_i \neq 0$. The assertion follows.

**Lemma 4** assume that $K[S]$ is semilocal. Then,

(i)   $\rho_{\mathcal{J}(K[S])}$ coincides with $\rho_{\mathcal{J}(F[S])}$, where $F$ denotes the prime subfield of $K$.

(ii)  If $K = \mathbb{Q}$, we have

    (a)   If $(s,t) \in \rho_{\mathcal{J}(K[S])}$, then $(s,t) \in \rho_{\mathcal{J}(\mathbb{F}_p[S])}$ for every $s, t \in S$, and every prime number $p$.

    (b)   There exist primes $p_1, \ldots, p_n$ such that

$$\rho_{\mathcal{J}(K[S])} = \bigcap_{i=1}^{n} \rho_{\mathcal{J}(\mathbb{F}_{p_i}[S])}$$

*Proof.* (i) follows from the fact that $\mathcal{J}(K[S]) = K\mathcal{J}(F[S])$ for semilocal algebras $K[S]$; see Lemma 1.

(ii) Since $S$ is locally finite by Theorem 2, then (a) follows as in Lemma 3.

For brevity, we will write $\rho_p$ for the congruence $\rho_{\mathcal{J}(\mathbb{F}_p[S])}$ if $p$ is a prime number. Assume that $s - t \notin \mathcal{J}(\mathbb{Q}[S])$ for some $s, t \in S$. Since any nil right ideal of $\mathbb{Q}[S]$ lies in $\mathcal{J}(\mathbb{Q}[S])$, there exists $z = \sum n_i s_i \in \mathbb{Z}[S]$, $n_i \in \mathbb{Z}$, $s_i \in S$, such that the element $u = (s-t)z$ is not nilpotent. Let $T = \langle \mathrm{supp}(u) \rangle$. Then

$T$ is a finite semigroup. It is easy to see that, for any prime $p > (2\sum_i |n_i|)^{|T|}$, we get $\bar{u}^{|T|} \neq 0$ for the image $\bar{u}$ of $u$ under the natural homomorphism $\mathbb{Z}[S] \to \mathbb{F}_p[S]$. Then $\bar{u} = (s-t)\bar{z}$ is not nilpotent since nilpotents in $\mathbb{F}_p[T]$ have nilpotency indices not exceeding $|T|$. Therefore, $s - t \notin \mathcal{J}(\mathbb{F}_p[S])$ because the latter is a nilideal. This shows that $(s,t) \notin \rho_p$, and so the equality

$(*)$                                                $\rho_{\mathcal{J}(\mathbb{Q}[S])} = \cap \rho_p$

where the intersection runs over the set of all primes, follows from (a).

Since the algebra $\mathbb{Q}[S/\rho_p]/\mathcal{J}(\mathbb{Q}[S/\rho_p])$ is semisimple, then we get the following commuting diagram given by the natural homomorphisms:

$$
\begin{array}{ccc}
\mathbb{Q}[S] & \xrightarrow{\ \phi_p\ } & \mathbb{Q}[S/\rho_p] \\
\downarrow{\scriptstyle \pi} & & \downarrow{\scriptstyle \pi_p} \\
\mathbb{Q}[S]/\mathcal{J}(\mathbb{Q}[S]) & \xrightarrow[\ \psi_p\ ]{} & \mathbb{Q}[S/\rho_p]/\mathcal{J}(\mathbb{Q}[S/\rho_p])
\end{array}
$$

Since $\mathbb{Q}[S]$ is semilocal, then there are finitely many different kernels of the homomorphisms of the form $\psi_p \pi = \pi_p \phi_p$, where $p$ is a prime. If $x - y \in \mathcal{J}(\mathbb{Q}[S/\rho_p])$ for some $x, y \in S/\rho_p$, then, by (a), $x - y \in \mathcal{J}(\mathbb{F}_p[S/\rho_p])$, which implies that $x = y$. Hence, the restriction of $\pi_p$ to $S/\rho_p$ is a semigroup isomorphism. Suppose that the congruences $\rho_p$, $\rho_q$ are distinct for some primes $p$, $q$. Then, interchanging $p$, $q$ if necessary, we may assume that there exist $s$, $t \in S$ such that $s - t \in \ker(\phi_p)$, $s - t \notin \ker(\phi_q)$. The above shows that $s - t \in \ker(\pi_p \phi_p) = \ker(\psi_p \pi)$ and $s - t \notin \ker(\pi_q \phi_q) = \ker(\psi_q \pi)$. Thus, the kernels of the homomorphisms $\psi_p \pi$, $\psi_q \pi$ are different. Since $\mathbb{Q}[S]$ is semilocal, this shows that there are finitely many distinct congruences of the form $\rho_p$, where $p$ is a prime number. Hence, (b) follows from $(*)$.

**Lemma 5** Assume that $K[S]/\mathcal{J}(K[S])$ is an artinian and algebraic $K$-algebra. If $\mathrm{ch}(K) > 0$, then $S/\rho_{\mathcal{J}(K[S])}$ is a locally finite semigroup.

*Proof.* Let $F$ be the prime subfield of $K$. By Lemma 1, $K\{F[S]/\mathcal{J}(F[S])\} = K[S]/\mathcal{J}(K[S])$ and, hence, $F[S]/\mathcal{J}(F[S])$ is an algebraic $F$-algebra. Since any division subalgebra of $F[S]/\mathcal{J}(F[S])$ must be a field (see [86], Theorem 3.1.1) and $F[S]/\mathcal{J}(F[S])$ is semisimple artinian by Lemma 1, then the Wedderburn–Artin theorem implies that it is a finite direct product of matrix algebras over fields. Since $S$ is periodic by Theorem 2, then, from Corollary 12 in Chapter 3, it follows that $S/\rho_{\mathcal{J}(K[S])}$ is locally finite.

We summarize what is known on semilocal group algebras.

**Theorem 6**  Let $G$ be a group. Assume that $\operatorname{ch}(K) = 0$, or $G$ is locally finite. Then $K[G]$ is semilocal if and only if

(i)  $G$ is finite if $\operatorname{ch}(K) = 0$.
(ii)  $G$ has a normal $p$-subgroup $N$ of finite index if $\operatorname{ch}(K) = p > 0$.

In the latter case, $\omega(K[N]) = \mathcal{J}(K[N]) \subseteq \mathcal{J}(K[G])$, and so $G/\rho_{\mathcal{J}(K[G])}$ is a finite group and $K[G]/\mathcal{J}(K[G])$ is a finite dimensional $K$-algebra.

*Proof.*  Assume that $K[G]$ is semilocal. If $\operatorname{ch}(K) = 0$, then $G$ is locally finite by Theorem 2. It is known that $\mathcal{J}(K[G]) = 0$ in this case; see[203], Lemma 7.4.2. Hence, $K[G]$ is artinian and so, by Connell's theorem, $G$ is a finite group; see [203], Theorem 10.1.1. The converse is obvious.

Let $\operatorname{ch}(K) = p > 0$. Since $K[G]$ is a locally finite algebra, then (ii) is precisely Theorem 10.1.15 in [203]. Moreover, the equality $\omega(K[N]) = \mathcal{J}(K[N])$ follows from [203], Lemma 8.1.17, while $\mathcal{J}(K[N]) \subseteq \mathcal{J}(K[G])$ is established in [203], Theorem 7.2.7. Since $K[G]/\mathcal{J}(K[G])$ is the image of the algebra $K[G/N] \cong K[G]/(\omega(K[N])K[G])$, then the remaining assertion is clear.

The second part of Theorem 6 may be extended as follows.

**Proposition 7**  Let $G$ be a group such that $K[G]/\mathcal{J}(K[G])$ is an artinian algebraic $K$-algebra. If $\operatorname{ch}(K) = p > 0$, then $G$ has a normal $p$-subgroup $N$ of finite index such that $\omega(K[N]) \subseteq \mathcal{J}(K[G])$.

*Proof.*  By Lemma 5 $\bar{G} = G/\rho_{\mathcal{J}(K[G])}$ is a locally finite group. Theorem 6 implies that $\bar{G}$ has a normal $p$-subgroup of finite index $\bar{N}$ such that $\omega(K[\bar{N}]) \subseteq \mathcal{J}(K[\bar{G}])$. From the choice of $\bar{G}$, it then follows that $\bar{N}$ is a trivial group. Therefore, the inverse image $N$ of $\bar{N}$ in $G$ satisfies $\omega(K[N]) \subseteq \mathcal{J}(K[G])$. It must be a $p$-group by [203], Lemma 10.1.13.

**Remark 8**  In view of Proposition 6 in Chapter 13, the hypotheses of Proposition 7 are satisfied for $F[G]$ if $K[G]$ is semilocal and $K$ is not algebraic over its prime subfield $F$.

We now start considering semilocal semigroup algebras of arbitrary semigroups.

**Proposition 9**  Assume that $K[S]$ is semilocal. If $K[G]/\mathcal{J}(K[G])$ is a finite dimensional $K$-algebra for every subgroup $G$ of $S$, then $K[S]/\mathcal{J}(K[S])$ is a finite dimensional $K$-algebra.

*Proof.* We proceed by induction on the length $n_{K[S]}$ of the $K[S]$-module $K[S]/\mathcal{J}(K[S])$. Clearly, $n_{K[S]} > 0$ because we have the augmentation homomorphism $K[S] \to K$. If $n_{K[S]} = 1$, then it follows that $K[S]/\mathcal{J}(K[S]) \cong K$, and we are done. Assume that $N > 1$ is such that the assertion holds for all semilocal semigroup algebras $K[T]$ with $n_{K[T]} < N$. Consider two cases. Case 1: $S$ contains an identity.

Denote by $G$ the group of units of $S$. Then $I = S \setminus G$ is an ideal of $S$ because $S$ is a periodic semigroup by Theorem 2, and $K[G] \cong K[S]/K[I]$; see Lemma 12 in Chapter 4. Let $M$ be a maximal ideal of $K[S]$. If $K[I] \subseteq M$, then $K[S]/M$ is a simple algebra that is an image of the group algebra $K[G]$. Then, by the hypothesis, $K[S]/M$ is finite dimensional. If $K[I]$ is not contained in $M$, then $M + K[I] = K[S]$, and so $K[S]/M \cong K[I]/(M \cap K[I])$. Moreover, $K[I]/\mathcal{J}(K[I]) \neq K[S]/\mathcal{J}(K[S])$ because otherwise $1 = i + r$ for some $i \in I, r \in \mathcal{J}(K[S])$, so that $i$ is a unit in $K[S]$, which contradicts the fact that $\operatorname{supp}(i) \subseteq I$ consists of nonunits in $S$. Therefore, $n_{K[I]} < n_{K[S]}$. Thus, by the induction hypothesis, $K[S]/M$ is finite dimensional in this case, too. Since $K[S]$ is semilocal, it follows that $K[S]/\mathcal{J}(K[S])$ is finite dimensional. Case 2: $S$ has no identity.

If $M$ is a maximal ideal of $K[S]$, then we claim that there exists an idempotent $e \in S$ such that $e \notin M$. If this is not the case, then, by Theorem 2, $S$ would be nil modulo $M$. Since $K[S]/M$ is a simple artinian algebra, then [55], 17.19, would imply that $S^m \subseteq M$ for some $m \geq 1$. Then $K[S]^m = K[S^m] \subseteq M$, which is impossible. This proves the claim. It is clear that $n_{K[eSe]} \leq n_{K[S]}$. Now, from case 1 it follows that $K[eSe]$ is finite dimensional modulo $\mathcal{J}(K[eSe])$. It is well known that if $K[S]/M \cong M_r(D)$ for some $r \geq 1$ and a division algebra $D$, then $K[eSe]/(M \cap K[eSe]) \cong M_t(D)$ for some $t \leq r$. Consequently, $K[S]/M$ is finite dimensional and, since $M$ is an arbitrary maximal ideal of $K[S]$, then, as above, $K[S]/\mathcal{J}(K[S])$ is finite dimensional.

We derive an important consequence improving the assertion of Lemma 5 and Proposition 7.

**Theorem 10** Assume that $K[S]$ is semilocal. If $K[G]/\mathcal{J}(K[G])$ is a finite dimensional $K$-algebra for every subgroup $G$ of $S$, then $S/\rho_{\mathcal{J}(K[S])}$ is a finite semigroup.

*Proof.* From Proposition 9 it follows that $K[S]/\mathcal{J}(K[S])$ is finite dimensional. If $\operatorname{ch}(K) = p > 0$, then $S/\rho_{\mathcal{J}(K[S])} = S/\rho_{\mathcal{J}(F_p[S])}$ by Lemma 4, the latter being embeddable into the algebra $F_p[S]/\mathcal{J}(F_p[S])$; see Lemma 5 in

Chapter 4. In view of Lemma 1, this is a finite dimensional $F_p$-algebra; and so it is a finite ring, and the assertion follows in this case.

Assume now that $ch(K) = 0$. By Lemma 4,

$$S/\rho_{\mathcal{J}(K[S])} = S/\left(\bigcap_{i=1}^{n}\rho_{\mathcal{J}(F_{p_i}[S])}\right) \subseteq \prod_{i=1}^{n}(S/\rho_{\mathcal{J}(F_{p_i}[S])})$$

for some $n \geq 1$, and some primes $p_1, \ldots, p_n$. Since by Theorem 2 and Lemma 3, any $F_{p_i}[S]/\mathcal{J}(F_{p_i}[S])$ is a finite ring, then $S/\rho_{\mathcal{J}(F_{p_i}[S])}$ is a finite semigroup. Consequently, $S/\rho_{\mathcal{J}(K[S])}$ is finite in this case, too.

**Corollary 11**   Assume that $K[S]$ is semilocal. Then $S/\rho_{\mathcal{J}(K[S])}$ is finite if either of the following holds:

(i)   $ch(K) = 0$.
(ii)   $S$ is locally finite.
(iii)   $K$ is not algebraic over its prime subfield.
(iv)   $S$ has no infinite subgroups.

*Proof.*   In view of Proposition 6 in Chapter 13 and Theorem 2, condition (iii) implies that $F[S]$ is an algebraic $F$-algebra, where $F$ is the prime subfield of $K$. Moreover, $K[G]$ and $F[G]$ are semilocal for every subgroup $G$ of $S$. Thus, in any case, we may use Theorem 6 and Theorem 10, or Proposition 7, Lemma 4 and Theorem 10, to derive the desired assertion.

We continue with a definition of a finiteness condition on the set $E(S)$ of idempotents of $S$, which will turn out to be strongly related to the considered properties of the algebra $K[S]$. Let $p$ be a prime number, or $p = 0$, and let $Z \subseteq E(S)$ be a nonempty set. We say that $Z$ is a left $p$-subset of $E(S)$ if the following condition is satisfied for all $e, f \in Z$:

(c)   For every $s \in S$, the element $ese$ lies in the group of units $U(eSe)$ of the monoid $eSe$ if and only if $efse \in U(eSe)$ and, in this case, $eseN = efseN$ for a normal $p$-subgroup $N$ of $U(eSe)$.

Here, by a $p$-supgroup for $p = 0$, we mean the trivial subgroup. Right $p$-subsets are defined symmetrically.

Our first observation establishes a connection between $p$-subsets and the properties of $S/\rho_{\mathcal{J}(K[S])}$.

**Lemma 12**   Assume that $Z \subseteq E(S)$ is a nonempty set contained in an equivalence class of the congruence $\rho_{\mathcal{J}(K[S])}$. Then $Z$ is a left $p$-subset of $E(S)$ for $p = ch(K)$.

*Proof.* Let $e, f \in Z$, $s \in S$. Then $e - f \in \mathcal{J}(K[S])$ implies that $e(e - f) = e - ef \in \mathcal{J}(K[S])$, and so $ese - efse \in \mathcal{J}(K[S]) \cap K[eSe]$. It is well known that the latter coincides with $\mathcal{J}(K[eSe])$; see [86], Theorem 1.3.3. Therefore, if one of the elements $ese$, $efse$ is invertible in $eSe$, then the other must be also. Since $U(eSe)$ is a group-like subset in $eSe$, then, from Corollary 16 of Chapter 4, we know that $\mathcal{J}(K[eSe]) \cap K[U(eSe)] \subseteq \mathcal{J}(K[U(eSe)])$. Thus, if $ese$, $efse \in U(eSe)$, then $eseN = efseN$ for the maximal normal $p$-subgroup $N$ of $U(eSe)$ by [203], Lemma 10.1.13.

We come to the first main result of this chapter, which was essentially obtained by Okniński in [181].

**Theorem 13**  Let $\mathrm{ch}(K) = p$, and assume that either of the following holds:

(1) $p = 0$.
(2) $S$ is locally finite.

Then the following conditions are equivalent:

(i) $K[S]$ is semilocal.
(ii) $S$ is locally finite, any subgroup of $S$ has a normal $p$-subgroup of finite index, and $E(S)$ is a union of finitely many left (right respectively) $p$-subsets.
(iii) $S$ has a chain of ideals $S = S_n \supseteq S_{n-1} \supseteq \cdots \supseteq S_1$ such that any one of $S_1$, $S_i/S_{i-1}$, $i > 1$, is a locally nilpotent or completely 0-simple semigroup. Moreover, if $\mathfrak{M}^0(G, I, M, P)$ is a Rees matrix presentation of some completely 0-simple factor $S_i/S_{i-1}$ or $S_1$, then $G$ is locally finite and has a normal $p$-subgroup of finite index, and there are finitely many $p$-equivalency classes of rows (respectively columns) of $P$.

*Proof.*  From Theorem 2 and from Proposition 2 and Lemma 3 both in Chapter 2, it follows, in view of the hypothesis, that we can assume that $S$ is locally finite. Assume that $K[S]$ is semilocal. If $G$ is a subgroup of $S$, $K[G]$ is semilocal by Theorem 2. Then Theorem 6 implies that $G$ has a normal $p$-subgroup of finite index. From Theorem 10, we know that $S/\rho_{\mathcal{J}(K[S])}$ is a finite semigroup. Then $E(S/\rho_{\mathcal{J}(K[S])})$ is also finite and, since $E(S)$ maps into $E(S/\rho_{\mathcal{J}(K[S])})$ under the natural homomorphism $S \to S/\rho_{\mathcal{J}(K[S])}$, then the remaining assertion of (ii) follows through Lemma 12.

Assume that (ii) holds. Let $Z$ be a left $p$-subset of $E(S)$, and let $e$, $f \in Z$. Then $e = eee$, $efee \in U(eSe)$. Consequently, $U(eSe) \subseteq SfS$. Similarly, $U(fSf) \subseteq SeS$, which implies that $SeS = SfS$. Thus, that $E(S)$ may be covered by finitely many left $p$-subsets implies that $S$ has finitely many

$\mathcal{J}$-classes with idempotents. From Corollary 4 in Chapter 3, it follows that there exists a chain of ideals $S = S_n \supseteq S_{n-1} \supseteq \cdots \supseteq S_1$, $S_0 = \varnothing$ such that any $S_i/S_{i-1}$ is nil or completely 0-simple. In the former case, $S_i/S_{i-1}$ is locally nilpotent because it is locally finite by the hypothesis. Thus, assume that $S_i/S_{i-1}$ is completely 0-simple with a Rees presentation $\mathfrak{M}^0(G,I,M,P)$. Since $G$ is isomorphic to a subgroup of $S$, then it remains to show that there are finitely many $p$-equivalency classes of rows of $P$. It is clear that condition (ii) is inherited by the semigroup $\mathfrak{M}^0(G,I,M,P)$. Assume that $e, f \in \mathfrak{M}^0(G,I,M,P)$ are nonzero idempotents lying in a left $p$-subset $Y$ of $E(\mathfrak{M}^0(G,I,M,P))$. Let $e = (g,j,m)$, $f = (h,k,n)$ for some $j,k \in I$, $m,n \in M$, $g,h \in G$. We know that $p_{tj} \neq \theta$ for some $t \in M$. Let $l \in I$, and $s = (1,l,t)$. By the definition of a $p$-subset, we know that $e = eee$, $efe = efee$ lie in the same coset of the maximal normal $p$-subgroup $N$ of the group $G_e = e\mathfrak{M}^0(G,I,M,P)e \setminus \{\theta\}$ isomorphic to $G$. In particular, $efe \neq \theta$, and so $p_{mk}, p_{nj} \neq \theta$. Consider the elements $ese = (gp_{ml}p_{tj}g,j,m)$, $efse = (gp_{mk}hp_{nl}p_{tj}g,j,m)$. Since $p_{tj} \neq \theta$, $p_{mk} \neq \theta$, then the fact that $e, f \in Y$ implies that either $p_{ml} = \theta = p_{nl}$ or $p_{mj}gp_{ml}p_{tj}gN = p_{mj}gp_{mk}hp_{nl}p_{tj}gN$ (note that the left multiplication by $p_{mj}$ induces a group isomorphism $G_e = \{(g,j,m) \mid g \in G\} \to G$). In the latter case, $p_{ml}N = p_{mk}hp_{nl}N$. Since $l \in I$ is arbitrary, this shows that the $m$th row of $P$ is a multiple of the $n$th row of $P$ modulo $N$. While $G/N$ is finite by the assumption, then it follows that there are finitely many $p$-equivalency classes of rows of $P$ corresponding to the columns of $\mathfrak{M}^0(G,I,M,P)$ containing idempotents from $Y$. Since any column of $\mathfrak{M}^0(G,I,M,P)$ contains an idempotent, then the fact that $E(\mathfrak{M}^0(G,I,M,P))$ may be covered by finitely many left $p$-subsets implies that there are finitely many $p$-equivalency classes of rows of $P$.

Finally, assume that (iii) holds. In view of Lemma 7 in Chapter 4 and Lemma 1, to prove that $K[S]$ is semilocal, it is enough to show that all algebras $K[S_1]$, $K_0[S_i/S_{i-1}]$ are semilocal. If $S_1$ or $S_i/S_{i-1}$ is locally nilpotent, then it is clear that the corresponding contracted semigroup algebra is a locally nilpotent $K$-algebra, and so we are done. Thus, assume that $T = S_1$ or $T = S_i/S_{i-1}$ is completely 0-simple with a Rees presentation $\mathfrak{M}^0(G,I,M,P)$. If $N$ is the maximal normal $p$-subgroup of $G$, then, by the hypothesis, $G/N$ is a finite group. Corollary 18 of Chapter 5 and Theorem 6 imply that the kernel of the natural homomorphism $K_0[T] \to K_0[\mathfrak{M}^0(G/N,I,M,P_N)]$ (see Chapter 5) lies in $\mathcal{J}(K_0[T])$. Hence, it is enough to show that $K_0[\mathfrak{M}^0(G/N,I,M,P_N)]$ is semilocal. By the hypothesis, there are finitely many 0-equivalency classes of rows of $P_N$, which means that $P_N$ has finitely many distinct rows. Applying Lemma 5 in Chapter 5 and its right-left symmetric analog, we see that $K_0[\mathfrak{M}^0(G/N,I,M,P_N)]$ has a nilpotent ideal, with the corresponding quotient being finite dimensional. Hence (i) follows.

The proof of the equivalence of the corresponding conditions involving right $p$-subsets and $p$-equivalency classes of columns of the sandwich matrices is similar.

**Remark 14**   When proving the implication (iii) $\Rightarrow$ (i) above, we have shown that $\mathcal{J}(K_0[S])$ is a nilpotent ideal for every completely 0-simple semigroup $S = \mathfrak{M}^0(G,I,M,P)$ over a finite group $G$.

**Remark 15**   From Corollary 11 and the proof of Theorem 13, it is clear that, if $K$ is not algebraic over its prime subfield and $\mathrm{ch}(K) = p > 0$, then $K[S]$ is semilocal if and only if the following condition holds:

(iii') $S$ has a chain of ideals $S = S_n \supseteq S_{n-1} \supseteq \cdots \supseteq S_1$ such that all $S_i/S_{i-1}$ and $S_1$ are nil or completely 0-simple. If $\mathfrak{M}^0(G,I,M,P)$ is a Rees presentation of some completely 0-simple factor $S_i/S_{i-1}$ or $S_1$, then $G$ has a normal $p$-subgroup $N$ of finite index such that $\omega(K[N]) \subseteq \mathcal{J}(K[G])$, and there are finitely many $p$-equivalency classes of rows of $P$. If some $T = S_i/S_{i-1}$ or $S_1$ is nil, then $K_0[T]$ is a Jacobson radical algebra. (The necessity of the last condition follows because, if $T$ is nil and $T/\rho_{\mathcal{J}(K[T])}$ is finite, then it must be nilpotent, so that $T/\rho_{\mathcal{J}(K[T])}$ is a trivial semigroup. This shows that $\omega(K[T]) \subseteq \mathcal{J}(K[T])$.)

Unfortunately, it is not known whether, in the above case, $T$ and $G$ must be locally finite.

**Example 16**   Let $p$ be a prime number, and let $S = \mathfrak{M}(G,I,I,P)$ be a completely simple semigroup over a locally finite $p$-group $G$ such that $I$ is an infinite set and $P$ has infinitely many distinct rows. It is clear that all rows of $P$ are $p$-equivalent, and so $K[S]$ is semilocal (even local; see Theorem 18) for every field $K$ with $\mathrm{ch}(K) = p$. On the other hand, if $q \neq p$ is a prime or zero, then distinct rows of $P$ are not $q$-equivalent, and so $L[S]$ is not semilocal for fields $L$ with $\mathrm{ch}(L) = q$.

We mention two particular cases, in which the description obtained in Theorem 13 simplifies considerably; see [105] and [181].

**Theorem 17**   Let $S$ be a commutative semigroup, and let $K$ be a field of characteristic $p$ (possibly 0). Then the following conditions are equivalent:

(i)   $K[S]$ is semilocal.

(ii)   $S$ is periodic, $E[S]$ is finite, and every subgroup of $S$ has a $p$-subgroup of finite index.

(iii)  $S/\rho$ is a finite semigroup where $\rho$ is the least separative congruence $\xi$ on $S$ if $p = 0$ and the least $p$-separative congruence $\xi_p$ on $S$ if $p > 0$.

*Proof.* Observe that, if two idempotents $e, f \in S$ are in the same left $p$-subset of $E(S)$, then $e = ef$ and $f = ef$, so that $e = f$. Thus, (ii) is equivalent to the respective condition in Theorem 13. Since $I(\rho)$ is a nil ideal of $K[S]$ (see [64], § 9), then $K[S]$ is semilocal if and only if $K[S/\rho]$ is also. Thus, (iii) implies (i). Further, if (ii) holds, then, by Proposition 7 in Chapter 6, $S/\rho$ is a finite union of groups with no $p$-torsion if $\rho = \xi_p$. Thus, $S/\rho$ must be a finite semigroup in this case. Hence, (ii) implies (iii), which completes the proof.

Let us note that the above proof may be simplified by using the description of the radical of $K[S]$ given in Theorem 2 of Chapter 21.

We continue with the particular case of local semigroup algebras, that is, the algebras in which the Jacobson radical is a maximal ideal. In this case $K[S]/\mathcal{J}(K[S]) \cong K$, because $K[S]$ may be always mapped onto $K$ via the augmentation homomorphism.

**Theorem 18**   Assume that ch$(K) = 0$ or that $S$ is a locally finite semigroup. Then the following conditions are equivalent:

(i)   $K[S]$ is a local algebra.
(ii)  $S$ is locally finite, and $eSe$ is a $p$-group, $p = $ ch$(K)$ (possibly 0), for every idempotent $e \in S$.
(iii) $S$ has a completely simple ideal $T \cong \mathfrak{M}(G, I, M, P)$, where $G$ is a locally finite $p$-group, $p = $ ch$(K)$ (possibly 0), and $S/T$ is a locally nilpotent semigroup.

*Proof.*   Assume that $K[S]$ is local. From the proof of Theorem 13, it follows that $E(S)$ is a left $p$-subset of $E(S)$, and so $S$ has exactly one principal factor containing idempotents. Since every group algebra $K[G]$, $G$—a subgroup of $S$, is local (see Theorem 2), then (ii) and (iii) follow through Theorem 13 and the fact that $G$ must be a $p$-group in this case; see [203], Lemma 10.1.13.

The local finiteness of $S$ required for the implication (iii) $\Rightarrow$ (ii) is a direct consequence of Proposition 2 and Lemma 3 both in Chapter 2.

If (iii) holds, then, to establish (i), it is sufficient to show that the elements of the form $e - f$, $e - s$, where $e, f \in E(S)$, $s \in S$ and $s^k = e$, are nilpotents. Then, the augmentation ideal $\omega(K[S])$ has a basis consisting of nilpotents and, since $S$ is locally finite, then $\omega(K[S]) \subseteq \mathcal{J}(K[S])$ by [86], Theorem 2.3.11. For $e$, $s$, as above, we have $(e - s)^k \in \omega(K[eSe])$, so it is nilpotent by the hypothesis. For $e, f \in E(S)$, we put $x = efe, y = fef$. It is easy to check that $(e - f)^{2n-1} = (e - x)^n - (f - y)^n$ for every $n \geq 1$. Now,

$x \in eSe$, and so, as above, $e - x$ is nilpotent. Similarly, $f - y$ is a nilpotent, and hence, $e - f$ is a nilpotent, too.

**Remark 19** If $K[S]$ is local, then $\mathcal{J}(K[S]) = \omega(K[S])$, and so $S/\rho_{\mathcal{J}(K[S])}$ always is a trivial semigroup. Therefore, proceeding as in the proof of Theorem 18, we can show that, for $\mathrm{ch}(K) = p > 0$, $K[S]$ is local if and only if the following holds:

(iii') $S$ has a completely simple ideal $T = \mathfrak{M}(G,I,M,P)$ over a $p$-group $G$ satisfying $\omega(K[G]) = \mathcal{J}(K[G])$ and such that $S/T$ is a nil semigroup with $K_0[S/T] = \mathcal{J}(K_0[S/T])$; see Remark 15.

Recall that an algebra $R$ is called right-chained if the right ideals of $R$ that are subspaces form a chain with respect to inclusion. Clearly, $R$ must be a local algebra in this case because it has at most one regular maximal right ideal [86]. The following description of chained semigroup algebras is due to Kozhukhov [116].

**Theorem 20** Let $S$ be a nontrivial semigroup. Then $K[S]$ is a right-chained algebra if and only if either of the following conditions is satisfied:

(i)  $\mathrm{ch}(K) = p > 0$, and $S$ is a cyclic $p$-group or a quasicyclic $p$-group.
(ii) $S$ is a two-element semigroup satisfying the identity $xy = y$.

*Proof.* Assume that $K[S]$ is right-chained. Since, for every $s \in S$, $K[sS^1]$ is a right ideal of $K[S]$, which is not contained in the augmentation ideal $\omega(K[S])$, then $sS^1 = S$. Therefore, from Remark 19 it follows that $S \cong \mathfrak{M}(G,1,M,P)$ is a completely simple semigroup over a $q$-group $G$, where $q = \mathrm{ch}(K)$ (possibly 0). If $|M| > 1$ and $|G| > 1$, then, by Lemma 3 in Chapter 5, the left annihilator of $K[S]$ has dimension exceeding 1. Thus, $K[S]$ has independent right ideals that are $K$-subspaces, which is impossible. The same argument shows that $|M| < 3$. Therefore, either $|M| = 2$ and $|G| = 1$, whence (ii) holds or $|M| = 1$, and so $S \cong G$. In this case, $S$ is a cyclic or quasicyclic $q$–group because the lattice of subgroups of $S$ must be a chain; see [64], Theorem 19.3.

On the other hand, if (i) holds, then $K[S]$ is a chained algebra by [64], Theorem 19.4. If (ii) holds, then it is easy to see that, for $s, t \in S$, $s \neq t$, $(s - t)K[S^1] = (s - t)K = \omega(K[S])$ is the only proper right algebra ideal of $K[S]$, and so the result follows.

Our next aim is to give a complete description of semigroup algebras that are perfect. Recall that an algebra $R$ is right perfect if it is semilocal, and $\mathcal{J}(R)$ is right $T$-nilpotent, that is, for every sequence $r_1, r_2, \ldots \in \mathcal{J}(R)$, there exists $n \geq 1$ such that $r_n r_{n-1} \ldots r_1 = 0$. We will use the fact that every right perfect algebra $R$ satisfies d.c.c. on principal left ideals; see [55]. Moreover, as in Lemma 1, one shows that, for every ideal $I$ of $R$, the algebra $R$ is right perfect if and only if $R/I, I$ are right perfect. The equivalence of (i) and (ii) below was proved by Okniński in [177], and condition (iii) was then derived by Finkelstein and Kozhukhov in [58].

**Theorem 21**   Assume that $ch(K) = p$ (not necessarily nonzero). The following conditions are equivalent:

(i)   $K[S]$ is right perfect.
(ii)  $S$ is periodic, has no infinite subgroups, and satisfies d.c.c. on principal left ideals; and $E(S)$ may be covered by finitely many left (respectively right) $p$-subsets.
(iii) $S$ has a chain of ideals $S = S_n \supseteq S_{n-1} \supseteq \cdots \supseteq S_1$, with all factors $S_i/S_{i-1}$ and $S_1$ being right $T$-nilpotent or completely 0-simple. Moreover, if $\mathfrak{M}^0(G, I, M, P)$ is a Rees matrix presentation of some completely 0-simple factor $S_i/S_{i-1}$ or $S_1$, then $G$ is finite and there are finitely many distinct $p$-equivalency classes of rows (respectively columns) of $P$.

*Proof.*   Assume that $K[S]$ is right perfect. By Theorem 2, $S$ is periodic and, if $G$ is a subgroup of $S$, then $K[G]$ is semilocal. Since $\mathcal{J}(K[G])$ is a homomorphic image of $\mathcal{J}(K[eSe]) = \mathcal{J}(K[S]) \cap K[eSe]$ by Lemma 12 in Chapter 4, then $K[G]$ is right perfect. Therefore, $G$ must be a finite group; see [203], Theorem 10.1.3. Since $K[S]$ has d.c.c. on principal left ideals, then $S$ has d.c.c. on principal left ideals. While $K[S]$ is semilocal, the remaining assertion of (ii) follows through Theorem 13.

Assume that (ii) holds. As in the proof of Theorem 13, $S$ has a desired chain, with all $S_i/S_{i-1}$ and $S_1$ being nil or completely 0-simple. Since $S$ satisfies d.c.c. on principal left ideals, then, in the former case, $S_i/S_{i-1}$ or $S_1$ is right $T$-nilpotent by Proposition 5 and Lemma 10 both in Chapter 2. Moreover, Theorem 7 in Chapter 2 implies that $S$ is locally finite. Therefore, (iii) is a consequence of Theorem 13.

Assume that (iii) holds. Then, in view of Lemma 7 in Chapter 4, the fact that the contracted semigroup algebra of a right $T$-nilpotent semigroup is right $T$-nilpotent by Corollary 12 in Chapter 2, along with the remarks preceding the theorem, it is enough to show that $K_0[T]$ is right perfect for every completely 0-simple factor $T = S_i/S_{i-1}$ or $T = S_1$. Since this algebra is semilocal by Theorem 13, and $\mathcal{J}(K_0[T])$ is nilpotent by Remark 14, then (i) follows.

We state explicitly the following observations used in the above proof.

**Corollary 22**

(i)   If $K[S]$ is right perfect, then $S$ is locally finite.

(ii)  If $S$ is a completely 0-simple semigroup with no infinite subgroup, and $K[S]$ is semilocal, then it is semiprimary (that is, semilocal with a nilpotent Jacobson radical).

(iii) If $S$ has no infinite subgroups and has d.c.c. on principal left ideals, and if $K[S]$ is semilocal, then $K[S]$ is right perfect.

Let us note that, since there are known examples of semigroups that are right but not left $T$-nilpotent, then, by Corollary 12 in Chapter 2, there exist semigroup algebras that are right but not left perfect.

The next class of algebras defined through a d.c.c. is that consisting of artinian algebras. The following result is due to Zelmanov [276].

**Theorem 23**   Assume that $K[S]$ is right artinian. Then $S$ is a finite semigroup. The converse holds if $S$ is a monoid.

*Proof.*   It is clear that $K[S^1]$ is right artinian whenever $K[S]$ is also. Thus, by Hopkin's theorem ([55], 18.13), $K[S^1]$ is right noetherian. Since $S$ is periodic by Theorem 2 and $S$ has no infinite subgroups by Theorem 21, then the finiteness of $S$ follows from Corollary 8 in Chapter 12.

The converse is clear, because any right ideal of $K[S]$ is a $K$-subspace whenever $K[S]$ has a (right) unity.

If $S$ is finite but is not a monoid (or, more generally, $K[S]$ has no right unity), then $K[S]$ can have infinite descending chains of right ideals. For example, this is the case for every $S$ with zero multiplication and an infinite field $K$ because subgroups of the additive group of $K_0[S]$ are ideals in $K_0[S]$.

An alternative proof of the above theorem exploits an induction on the length of a chain $S = S_n \supseteq S_{n-1} \supseteq \cdots \supseteq S_1$ arising from Theorem 13. This is based on the fact that, if $I$ is an ideal of $S$ such that $S/I$ is finite and if $I$ is completely 0-simple or has zero multiplication, then $I$ must be finite provided that $K[S/I]$ is right artinian. The original proof of Theorem 23 exploited the above fact but also used some other more complicated techniques.

Finally, we note that Hotzel proved that any semigroup satisfying d.c.c. on right congruences and not containing infinite subgroups must be finite [89]. This provides the most semigroup theoretical proof of Theorem 23.

The next class of interest is that consisting of semisimple artinian semigroup algebras. This leads to the following generalization of Maschke's theorem obtained independently by Munn [154] and Ponizovskii [212].

**Theorem 24**  The following conditions are equivalent:

(i)   $K[S]$ is semisimple artinian.

(ii)  $S$ has a chain of ideals $S = S_n \supseteq S_{n-1} \supseteq \cdots \supseteq S_1$ such that every $S_i/S_{i-1}$ and $S_1$ is a completely 0-simple semigroup with a Rees presentation $\mathfrak{M}^0(G, m, m, P)$ for some $m \geq 1$, and an invertible in $M_m(K[G])$ matrix $P$, where $G$ is a finite group with the order not divisible by $p$ if $ch(K) = p > 0$.

(iii) $S$ is a finite strongly $p$-semisimple semigroup, where $ch(K) = p$, with no subgroups of order divisible by $p$ if $p > 0$.

*Proof.*  Assume that (i) holds. From Theorem 23 it follows that $S$ is finite, and so there exists a desired chain of ideals of $S$ with nilpotent or completely 0-simple factors; see Corollary 4 in Chapter 3. It is well known that every $K_0[S_i/S_{i-1}]$, $K[S_1]$ is semisimple artinian as an ideal of a homomorphic image of $K[S]$. Therefore, $K_0[S_i/S_{i-1}]$ and $K[S_1]$ all have an identity, and so $S_i/S_{i-1}$, $S_1$ must be completely 0-simple. Moreover, from Corollary 26 in Chapter 5, it follows that each of these semigroups is isomorphic to some $\mathfrak{M}^0(G, m, m, P)$, with $P$ invertible in $M_m(K[G])$, and the corresponding contracted semigroup algebra is isomorphic to $M_m(K[G])$. Consequently, $\mathcal{J}(K[G]) = 0$ and, by Maschke's theorem, $p$ does not divide the order of $G$ if $p > 0$. Thus, (ii) holds.

It is clear that (ii) implies (iii); see Corollary 26 in Chapter 5.

The implication (iii) $\Rightarrow$ (i) is a consequence of Maschke's theorem and Corollary 27 in Chapter 5.

In the case where $S$ is a finite inverse semigroup, the invertibility of the sandwich matrices in (ii) above is automatically satisfied; see Corollary 26 in Chapter 5. Thus, $K[S]$ is semisimple exactly when $ch(K) = 0$ or the maximal subgroups of $S$ have orders relatively prime to $ch(K)$.

If $S$ is a finite semigroup, then a description of the radical of the algebra $K[S]$ may be given in terms of the 0-simple principal factors of $S$. Since essentially the same proof works for locally finite semigroups, we state the general result here. Let $S$ be a locally finite semigroup, and let $S_\alpha$, $\alpha \in \mathcal{A}$, be all 0-simple principal factors of $S$. Then, by Lemma 1 in Chapter 1, any $S_\alpha$ may be treated as semigroups of matrix type $S_\alpha = \mathfrak{M}^0(G_\alpha, I_\alpha, M_\alpha, P_\alpha)$ over a group $G_\alpha$. Denote by $\phi_\alpha$ the natural homomorphism $K[S] \to K_0[S/I_\alpha]$, where $I_\alpha$ is the ideal of nongenerators of the principal ideal $J_\alpha$ of $S$ defining $S_\alpha$, that is, $\phi_\alpha(J_\alpha) = S_\alpha$. As in Chapter 5, for any ideal $I$ of $K[G_\alpha]$, we put $\mathfrak{B}_\alpha(I) = \{x \in K_0[S_\alpha] \mid P_\alpha \circ x \circ P_\alpha$ lies over $I\}$. With this notation, we prove the following result.

**Theorem 25**  Let $S$ be a locally finite semigroup. Then,

$$\mathcal{J}(K[S]) = \{a \in K[S] \mid \phi_\alpha(as) \in \mathfrak{B}_\alpha(\mathcal{J}(K[G_\alpha]))$$

$$\text{for every } \alpha \in \mathcal{A} \text{ and } s \in J_\alpha\}.$$

*Proof.* If $a \in \mathcal{J}(K[S])$, and $s \in J_\alpha$, then $\phi_\alpha(as) \in \phi_\alpha(\mathcal{J}(K[S])) \subseteq \mathcal{J}(K_0[S/I_\alpha])$ and $\phi_\alpha(as) \in \phi_\alpha(K[J_\alpha]) = K_0[S_\alpha]$. Since $K_0[S_\alpha]$ is an ideal of $K_0[S/I_\alpha]$, then $\phi_\alpha(as) \in \mathcal{J}(K_0[S_\alpha])$. From Corollary 18 in Chapter 5, we know that $\mathcal{J}(K_0[S_\alpha]) = \mathfrak{B}_\alpha(\mathcal{J}(K[G_\alpha]))$.

Now, assume that $a \notin \mathcal{J}(K[S])$. Since $K[S]$ is an algebraic $K$-algebra by the hypothesis, then there exists an element $x \in K[S]$ such that $e = xa$ is a nonzero idempotent; see [86], p. 165. Let $I$ be an ideal of $S$ that is maximal with respect to the property $e \notin K[I]$, or $I = \varnothing$ if such an ideal does not exist. Then the image $e'$ of $e$ under the natural homomorphism $K[S] \to K_0[S/I]$ is a nonzero idempotent, and the choice of $I$ implies that $\text{supp}_0(e')$ lies in a minimal nonzero ideal $J/I$ of $S/I$, where $J$ is an ideal of $S$. It follows that $J/I \cong S_\beta$ for some $\beta \in \mathcal{A}$, because $S_\beta$ cannot be a semigroup with zero multiplication. Clearly, $J = J_\beta \cup I$.

Suppose that $a \in Z = \{b \in K[S] \mid \phi_\alpha(bs) \in \mathfrak{B}_\alpha(\mathcal{J}(K[G_\alpha]))$ for every $\alpha \in \mathcal{A}, s \in J_\alpha\}$. It is easy to see that the latter set is a left ideal of $K[S]$. Hence, $e = xa \in Z$. Let $e = e_1 + e_2$, where $\text{supp}(e_1) \subseteq J_\beta$, $\text{supp}(e_2) \subseteq I \setminus J_\beta$. Then,

$$\phi_\beta(e) = \phi_\beta(ee_1) + \phi_\beta(ee_2) \in \mathfrak{B}_\beta(\mathcal{J}(K[G_\beta])) + K_0[I/I_\beta]$$
$$= \mathcal{J}(K_0[S_\beta]) + K_0[I/I_\beta]$$

It follows that the image $e'$ of $\phi_\beta(e)$ under the natural homomorphism $K_0[S/I_\beta] \to K_0[S/I]$ lies in $\mathcal{J}(K_0[S/I])$. This contradicts the fact that $e'$ is a nonzero idempotent and shows that $Z \subseteq \mathcal{J}(K[S])$, completing the proof of the theorem.

We note that the radicals of group algebras of locally finite groups, to which the general problem is reduced in Theorem 25, have been extensively studied [202], [203]. In particular, it is well known that $\mathcal{J}(K[G]) = 0$ for every locally finite group $G$, provided that either $\text{ch}(K) = 0$, or that $\text{ch}(K) = p > 0$, and $G$ has no elements of order $p$.

Our last aim in this chapter is to give another description of the radical of finite dimensional semigroup algebras. Let $R$ be a finite dimensional $K$-

algebra. We fix a basis $\mathcal{E} = \{r_1, \ldots, r_n\}$ of $R$. For every $i, j \in \{1, 2, \ldots, n\}$, there exist scalars $\gamma_{ij}^k$, $k = 1, 2, \ldots, n$, such that $r_i r_j = \sum_{k=1}^n \gamma_{ij}^k r_k$. Put $\lambda_{ij} = \sum_{k,l=1}^n \gamma_{ij}^k \gamma_{lk}^l$. Then the $n \times n$ matrix $\mathbf{L}_{\mathcal{E}} = (\lambda_{ij})$ is called the left structure matrix of the algebra $R$ with respect to $\mathcal{E}$. Similarly, the matrix $\mathbf{R}_{\mathcal{E}} = (\rho_{ij})$ defined by $\rho_{ij} = \sum_{k,l=1}^n \gamma_{ji}^k \gamma_{kl}^l$, $i, j = 1, 2, \ldots, n$, is called the right structure matrix of $R$ with respect to $\mathcal{E}$.

We will be concerned with the special case where $R = K[S]$ for a finite semigroup $S$, and $\mathcal{E} = \{r_1, \ldots, r_n\} = S$. We then write $\mathbf{L} = \mathbf{L}_{\mathcal{E}}$, $\mathbf{R} = \mathbf{R}_{\mathcal{E}}$. For any $s \in S$, consider the sets $\mathcal{L}(s) = \{t \in S \mid ts = t\}$, $\mathcal{R}(s) = \{t \in S \mid st = t\}$. Then $\mathcal{L}(s)$, $\mathcal{R}(s)$, if nonempty, are left and right ideals of $S$, respectively. Let $\lambda(s)$, $\rho(s)$ denote the cardinalities of the sets $\mathcal{L}(s)$ and $\mathcal{R}(s)$, respectively.

**Lemma 26** For every $i, j = 1, 2, \ldots, n$, we have $\lambda_{ij} = \lambda(r_i r_j)$, $\rho_{ij} = \rho(r_i r_j)$.

*Proof.* Let $r_i r_j = r_k$. Then $\gamma_{ij}^k = 1$, $\gamma_{ij}^l = 0$ for $l \neq k$, and so $\lambda_{ij} = \sum_{l,m=1}^n \gamma_{ij}^l \gamma_{ml}^m = \sum_{m=1}^n \gamma_{mk}^m$. Since $\gamma_{mk}^m \neq 0$ if and only if $r_m r_k = r_m$, then $\{r_m \mid \gamma_{mk}^m \neq 0\} = \mathcal{L}(r_k)$, and hence $\lambda_{ij} = |\mathcal{L}(r_k)| = |\mathcal{L}(r_i r_j)|$. The assertion on $\rho_{ij}$ is established by a symmetric argument.

**Lemma 27** For every $s, t \in S$, we have $\lambda(st) = \lambda(ts)$, $\rho(st) = \rho(ts)$, and $\mathbf{L}$, $\mathbf{R}$ are symmetric matrices.

*Proof.* Define a mapping $\phi_{s,t}$ by the formula $\phi_{s,t}(x) = xs$ for $x \in \mathcal{L}(st)$. Then $\phi_{s,t}$ maps $\mathcal{L}(st)$ into $\mathcal{L}(ts)$ because $xst = x$ implies $xsts = xs$. Similarly, $\phi_{t,s}$ maps $\mathcal{L}(ts)$ into $\mathcal{L}(st)$. Since $\phi_{t,s}(\phi_{s,t}(x)) = xst = x$, and $\phi_{s,t}(\phi_{t,s}(y)) = y$ for every $y \in \mathcal{L}(ts)$, then $\phi_{s,t}$ is a bijection. This proves that $\lambda(st) = \lambda(ts)$. Hence, $\mathbf{L}$ is a symmetric matrix by Lemma 26. Then assertions on the $\rho_{ij}$ and the matrix $\mathbf{R}$ follow similarly.

Observe that any reordering of the elements of $S$ produces the corresponding permutation of the rows and columns of the matrices $\mathbf{L}$, $\mathbf{R}$.

**Example 28** Let $S$ be a finite group with identity $e$. Then $\mathcal{L}(e) = S$, and $\mathcal{L}(s) = \varnothing$ for $s \in S$, $s \neq e$. Thus, $\mathbf{L} = |S|P$, where $P$ is an $n \times n$ permutation matrix. Clearly, $\mathbf{R} = \mathbf{L}$.

Let $a \in K[S]$, and let $\varphi_a : K[S] \to K[S]$ be defined by $\varphi_a(x) = xa$. By $\tau(a)$ we denote the trace of $\varphi_a$ as a homomorphism of $K$-spaces. The following, essentially well-known result provides a description of the radical in terms of the functional $\tau : K[S] \to K$.

**Proposition 29** Let $\{r_1, \ldots, r_n\}$ be a basis of a $K$-algebra $R$. Assume that $\text{ch}(K) = 0$, or $\text{ch}(K) > n$. If $a \in R$, then $a \in \mathcal{J}(R)$ if and only if $\tau(r_i a) = 0$ for all $i = 1, 2, \ldots, n$.

*Proof.* Assume first that $\tau(r_i a) = 0$ for $i = 1, 2, \ldots, n$. Then, for every $r \in R$, it follows, by the linearity of $\tau$, that $\tau(ra) = 0$. Let $b \in R^1 a$. Then $b = ua + \alpha a$ for some $\alpha \in K$, $u \in R$, and, for any $m \geq 2$, we have

$$\tau(b^m) = \tau(b^{m-1}b) = \tau(b^{m-1}ua + \alpha b^{m-1}a) = \tau(b^{m-1}ua) + \tau(\alpha b^{m-1}a) = 0$$

Hence, $\text{tr}(\varphi_b^m) = \text{tr}(\varphi_{b^m}) = \tau(b^m) = 0$. It is well known that $\varphi_b$ must be nilpotent homomorphisms in this case. In fact, if $\beta_1, \ldots, \beta_l$ are all distinct eigenvalues of $\varphi_b$ in the algebraic closure of $K$, with the multiplicities $k_1$, $\ldots, k_l$, respectively, then $\sum_{i=1}^{l} k_i \beta_i^m = \text{tr}(\varphi_{b^m}) = 0$ for all $m \geq 2$. Since $k_i \neq 0$ in $K$ by the hypothesis on $K$, then, using the Vandermonde determinant, we deduce that $\beta_1 = \beta_2 = \cdots = \beta_l = 0$. Thus, $l = 1$, $0$ is the only eigenvalue of $\varphi_b$, and so $\varphi_b$ is nilpotent. Now, $0 = \varphi_b^t(b) = b^{t+1}$ for some $t \geq 1$. Hence, the left ideal $R^1 a$ consists of nilpotents, so that $a \in \mathcal{J}(R)$.

Conversely, if $a \in \mathcal{J}(R)$, then $r_i a \in \mathcal{J}(R)$ for $i = 1, 2, \ldots, n$. Hence, the nilpotency of $r_i a$ implies that $\tau(r_i a) = 0$.

**Lemma 30**   Let $S$ be a finite semigroup. Then $\tau(s) = \lambda(s)$ for every element $s \in S$ treated as an element of the algebra $K[S]$.

*Proof.* Since $\varphi_s(r_i) = r_i s$ for any $r_i \in S = \{r_1, \ldots, r_n\}$, then $\text{tr}(\varphi_s) = |\{i \,|\, r_i s = r_i\}| = |\mathcal{L}(s)| = \lambda(s)$.

We are now ready to state the following result of Drazin [51], providing another criterion for a semigroup algebra $K[S]$ to be semisimple artinian.

**Theorem 31**   Let $S = \{r_1, \ldots, r_n\}$ be a finite semigroup. Assume that $\text{ch}(K) = 0$ or $\text{ch}(K) > |S|$. Let $a = \sum \alpha_i r_i \in K[S]$, $\alpha_i \in K$. Then $a \in \mathcal{J}(K[S])$ if and only if $\mathbf{La} = 0$, where $\mathbf{a}$ is the column vector with components $\alpha_1$, $\ldots, \alpha_n$. In particular, $K[S]$ is semisimple if and only if $\mathbf{L}$ is nonsingular as a matrix over $K$.

*Proof.* For any $i = 1, \ldots, n$, we have, in view of Lemma 30, $\tau(r_i a) = \tau(r_i \sum_{j=1}^{n} \alpha_j r_j) = \sum_{j=1}^{n} \alpha_j \tau(r_i r_j) = \sum_{j=1}^{n} \alpha_j \lambda(r_i r_j)$. Since $\lambda(r_i r_j) = \lambda_{ij}$ by Lemma 26, then the assertion follows from Proposition 29.

Observe that, since $\mathbf{R}$ is a symmetric matrix by Lemma 27, the dual version of Theorem 31 provides an alternative criterion involving $\mathbf{R}$:

$$a = \sum \alpha_i r_i \in \mathcal{J}(K[S]) \Longleftrightarrow \mathbf{Ra} = 0$$

We note that the hypothesis on the characteristic of $K$ is essential. From the proof of Theorem 31, it follows that $\mathcal{J}(K[S])$ is semisimple whenever $\mathbf{L}$ is nonsingular over $K$. Moreover, in this case, $\mathbf{R}$ also is nonsingular over $K$.

**Example 32**   Let $S = \mathfrak{M}^0(\{1\}, 2, 2, P)$ be the semigroup of matrix type over the trivial group $\{1\}$ with the identity $2 \times 2$ matrix $P$. Then $\mathcal{J}(K[S]) = 0$ by Theorem 24. In fact, $K[S] \cong K_0[S] \oplus K \cong M_2(K) \oplus K$ by Corollary 8 in Chapter 4, and Corollary 26 in Chapter 5. On the other hand, it may easily be shown that, under a suitable ordering of the elements of $S$,

$$
\mathbf{L} = \mathbf{R} = \begin{pmatrix} 1 & 1 & 1 & 1 & 1 \\ 1 & 3 & 1 & 1 & 1 \\ 1 & 1 & 3 & 1 & 1 \\ 1 & 1 & 1 & 1 & 3 \\ 1 & 1 & 1 & 3 & 1 \end{pmatrix}
$$

Consequently, $\mathbf{L}$ is a nonsingular matrix over $K$ if $\mathrm{ch}(K) \neq 2$, but $\mathbf{L}$ is singular if $\mathrm{ch}(K) = 2$.

Theorem 31 may be used to derive a description of the radical of $K[S]$ for a locally finite semigroup $S$ as the following result shows.

**Proposition 33**   Assume that $S$ is a locally finite semigroup, and let $a \in K[S]$. Then $\mathcal{J}(K[S]) = \mathcal{L}(K[S])$, and $a \in \mathcal{J}(K[S])$ if and only if $a \in \mathcal{J}(K[T])$ for every finite subsemigroup $T$ of $S$ with $T \supseteq \mathrm{supp}(a)$.

*Proof.*   It is well known that the radical of a locally finite algebra is locally nilpotent. Thus, $\mathcal{J}(K[S]) = \mathcal{L}(K[S])$ and, for any subsemigroup $T$ of $S$, we have $\mathcal{J}(K[S]) \cap K[T] \subseteq \mathcal{J}(K[T])$ because the former is a nil ideal of $K[T]$. On the other hand, if $a \in \mathcal{J}(K[T])$ for every finite subsemigroup $T$ containing $\mathrm{supp}(a)$, then $ab \in \mathcal{J}(K[\langle \mathrm{supp}(a), \mathrm{supp}(ab) \rangle])$ for any $b \in K[S]$. Then, by the foregoing, $ab$ is a nilpotent element. It follows that $a \in aK[S]^1 \subseteq \mathcal{J}(K[S])$.

## Comments on Chapter 14

Semilocal group rings $K[G]$ in characteristic zero were first characterized by Lawrence and Woods in [136]. Existence of a $p$-subgroup of finite index in $G$ in the case where $K$ is a field of characteristic $p > 0$, which is not algebraic over its prime subfield (see Remark 8) was established by Okniński [174], extending a former result of Lawrence [134] and Valette (see [203], Theorem 4.1.16). We note that, from the remarks in [203], p. 416, it may be deduced that there exist $p$-groups $G$ for which $K[G]$ is not semilocal for every field $K$ with $\mathrm{ch}(K) = p$. The first assertion of Lemma 1 was originally obtained by Rosenberg and Zelinsky in [237]. Further results on d.c.c. of tensor products with applications to group rings are given by Krempa and

Okniński in [127]. The presented material on general semilocal semigroup algebras is taken from the paper of Okniński [181]. We note that assertion (ii) in Theorem 2 comes essentially from [136], while (i) was proved in [176]. Condition (iii) of Theorem 17, describing commutative semilocal semigroup algebras, was obtained by Jespers and Wauters in [105]. Wauters [266] then extended this to some other d.c.c. Both papers used techniques coming from the theory of graded rings, via semilattice decompositions of the underlying semigroup. The commutative artinian case for nonmonoids $S$ was settled by Gilmer in [65]. The description of semisimple artinian semigroup algebras due to Munn and Ponizovskii was extended to arbitrary coefficient rings by Ponizovskii in [219].

The first step toward a characterization of perfect semigroup algebras was made by Domanov [49]. An intrinsic description, Theorem 21 (ii), was given by Okniński in [177]. The condition on the sandwich matrices involved—see (iii) in Theorem 21—was then derived by Finkelstein and Kozhukhov in [58]. Unaware of these results, Shoji [250] settled the special cases of finitely generated semigroups and inverse semigroups. Munn [165] characterized semigroup algebras of inverse semigroups satisfying d.c.c. on principal left ideals, which is an essentially weaker condition for nonunitary rings than being right perfect.

The description of $\mathcal{J}(K[S])$ for finite $S$ (see Theorem 31) was extended to the case of arbitrary finite dimensional algebras by Drazin in [52]. The assertion on the radical for locally finite $S$, Theorem 25, covers the finite dimensional case considered by Hall in [82]; see also [133]. It will be further extended in Part IV to $PI$-algebras $K[S]$. Ovsyannikov [197] studied the case where the semigroup ring $R[S]$ with arbitrary coefficients is Jacobson radical or, in [198], where the augmentation ideal $\omega(R[S])$ is nil or locally nilpotent. If $R = K$, then the latter results may be derived from our material on local algebras.

Descending chain conditions for general semigroups, and their relations with other finiteness conditions, were studied by Hotzel in [89]. This work was continued by Kozhukhov [117], [120]. Using Hotzel's results Shoji [250], gave an alternative proof of Zelmanov's theorem on artinian semigroup algebras; see also [120].

The problem of characterizing semigroup algebras of finite representation type was stated, and some positive results were obtained by Ponizovskii [220].

# 15

# Regular Algebras

In this chapter, we discuss the von Neumann regularity condition for the class of semigroup algebras. Recall that an algebra $R$ is called regular if, for every $a \in R$, there exists $x \in R$ such that $axa = a$. While every semisimple artinian algebra is regular, the results of this chapter may be viewed as generalizations of the description of semisimple artinian semigroup algebras given in Chapter 14. For the general theory of regular algebras, we refer to [69]. We note that $R$ is regular if and only if every finitely generated right ideal of $R$ is of the form $eR$ for some $e = e^2 \in R$. In particular, $R$ satisfies the right (and similarly left) semihereditariness condition, which is reviewed in Chapter 17.

The case of regular group algebras is completely settled; see [203], Theorem 3.1.5.

**Theorem 1** Let $G$ be a group. Then $K[G]$ is regular if and only if $G$ is a locally finite group with no elements of order $p$ if $\mathrm{ch}(K) = p > 0$.

In the case of an arbitrary semigroup algebra $K[S]$, we start with some necessary conditions on $S$.

**Proposition 2** Assume that $K[S]$ is a regular algebra. Then,

  (i)   $S$ is periodic and $S$ is locally finite if $\mathrm{ch}(K) = 0$.
 (ii)   $S$ is a regular semigroup.
(iii)   Any subgroup of $S$ is locally finite and has no elements of order $p$
        if $\mathrm{ch}(K) = p > 0$.

*Proof.* The class of regular algebras is closed under ideal extensions (see [69]), so that $K[S^1]$ is regular. Since, for any $\lambda \in K$, and $s \in S$, the element $\lambda - s$ is regular in $K[S^1]$, then, from Lemmas 1 and 2 in Chapter 13, it follows that (i) holds. Moreover, $sas = s$ for some $a \in K[S]$. It is clear that $sts = s$ for some $t \in \mathrm{supp}(a)$ and, thus, (ii) follows. If $G$ is a subgroup of $S$ with the unity $e$, then the algebra $K[eSe] = eK[S]e$ is regular (see [69], Lemma 1.6). Now, $K[G]$ is a homomorphic image of $K[eSe]$ by Lemma 12 in Chapter 4, and so it is also regular. Thus, (iii) follows from Theorem 1.

**Corollary 3** If $K[S]$ is regular, then $S$ is a completely semisimple semigroup.

*Proof.* This is a consequence of Proposition 2 and the fact that any 0-simple periodic semigroup must be completely 0-simple; see Lemma 1 in Chapter 1.

It is clear that, if $K[S]$ is regular, then any principal factor algebra $K_0[S_t]$, $t \in S$, is regular as an ideal of a homomorphic image of $K[S]$. Thus, in view of Corollary 3, it is important to find a description of completely 0-simple semigroups, the semigroup algebras of which are regular. Such a condition is given in Theorem 24 in Chapter 14 for the case of finite completely 0-simple semigroups since a finite dimensional algebra is regular if and only if it is semisimple. It appears that a slight generalization of the invertibility of the sandwich matrix provides a desired condition.

Let $S = \mathfrak{M}^0(G,I,M,P)$ be a completely 0-simple semigroup. We say that $P$ is locally left invertible over $K[G]$ if, to any nonempty finite subset $J$ of $I$, there corresponds a nonempty finite subset $N$ of $M$ such that the submatrix $P_{NJ}$ of $P$ (see Chapter 1) is left invertible, i.e., there exists a $J \times N$ matrix $Q$ over $K[G]$ such that $QP_{NJ}$ is the $J \times J$ identity matrix. Similarly, $P$ is locally right invertible over $K[G]$ if, to any nonempty finite subset $N$ of $M$, there corresponds a nonempty finite subset $J$ of $I$ such that the matrix $P_{NJ}$ is right invertible over $K[G]$. $P$ is locally invertible if it is locally left and right invertible.

We start by showing that the above property coincides with the usual invertibility in the case of finite matrices.

**Lemma 4**  Let $S = \mathfrak{M}^0(G,I,M,P)$ be a completely 0-simple semigroup. Assume that one of the sets $I$, $M$ is finite. Then $P$ is locally invertible over $K[G]$ if and only if $|I| = |M|$ and $P$ is invertible over $K[G]$.

*Proof.*  Assume that $|I| = |M|$ and that $P$ is invertible over $K[G]$. Let $J \subseteq I$. Since $QP$ is the $|I| \times |I|$ identity matrix for some $Q \in M_{|I|}(K[G])$, then the submatrix $Q_J$ of $Q$, consisting of all $j$th rows of $Q, j \in J$, is such that $Q_J P_{MJ}$ is the identity in $M_{|J|}(K[G])$. Thus, $P$ is locally left invertible. Similarly, $P$ is locally right invertible.

Now, assume that $P$ is locally invertible over $K[G]$. We consider the case where $|I| < \infty$ only. Using Lemma 24 in Chapter 5 and the local right invertibility of $P$, one shows that $|M| \le |J|$, and $P_{MJ}Q$ is the identity matrix for some $J \subseteq I$ and a $J \times M$ matrix $Q$. A symmetric argument then implies that $|M| \ge |I|$, so that $I = J$ and $RP$ is the identity matrix for an $|I| \times |I|$ matrix $R$. Consequently, $P$ is invertible.

To establish the first main result of this chapter, we need the following fact.

**Lemma 5**  Let $r, s \ge 1$, and let $a$ be an $r \times s$ matrix over a regular algebra $R$. Then there exists an $s \times r$ matrix $b$ over $R$ such that $aba = a$.

*Proof.*  Assume that $r \ge s$. Let $\bar{a} = (\overline{a_{ij}}) \in M_r(R)$ be defined by

$$\overline{a_{ij}} = \begin{cases} a_{ij} & \text{for } j \le s \\ 0 & \text{for } j > s \end{cases}$$

where $a = (a_{ij})$. The regularity of $M_r(R)$ (see [69], Theorem 1.7) implies that $\bar{a}c\bar{a} = \bar{a}$ for some $c \in M_r(R)$. Then $\bar{a}c = \bar{a}\bar{c}$, where

$$\overline{c_{ij}} = \begin{cases} c_{ij} & \text{for } i \le s \\ 0 & \text{for } i > s \end{cases}$$

and so $\bar{a}\bar{c}\bar{a} = \bar{a}$. Hence, $aba = a$, where $b$ is the $s \times r$ matrix defined by $b_{ij} = \overline{c_{ij}} = c_{ij}$ for $i \le s, j \le r$. A similar argument works in the case where $r \le s$.

The following result is due to Munn [157].

**Theorem 6**  Let $S = \mathfrak{M}^0(G,I,M,P)$ be a completely 0-simple semigroup. Then $K_0[S]$ is regular if and only if the following conditions hold:

(i)   $G$ is locally finite.

(ii)   $G$ has no elements of order $p$ if $\mathrm{ch}(K) = p > 0$.

(iii)   $P$ is locally invertible over $K[G]$.

*Proof.*   Since $G$ is isomorphic to a subgroup of $S$, then from Proposition 2 it follows that (i) and (ii) hold if $K_0[S]$ is regular. Thus, in view of Theorem 1, we may further assume that $K[G]$ is regular. Let $\bar{S} = \mathfrak{M}^0(\{1\},I,M,\bar{P})$ be the natural homomorphic image of $S$ defined in Chapter 1. Assume first that one of the sets $I$, $M$ is finite. If $K_0[S]$ is regular, then $K_0[\bar{S}]$ is regular and, from Proposition 23 and Corollary 26 both in Chapter 5, it follows that $|I| = |M| < \infty$.

If $P$ is not invertible in $M_{|I|}(K[G])$, then it must be a zero divisor since this is a locally finite algebra. Then, by Lemma 3 in Chapter 5, $K_0[S]$ has a nonzero left annihilator, which contradicts the fact that it is regular. Thus, $P$ must be invertible in $M_{|I|}(K[G])$, and so it is locally invertible by Lemma 4. On the other hand, if (iii) holds, then Lemma 4 shows that $|I| = |M|$ and $P$ is invertible in $M_{|I|}(K[G])$. From Corollary 26 in Chapter 5, it follows that $K_0[S] \cong M_{|I|}(K[G])$, and so $K_0[S]$ is a regular algebra; see [69], Theorem 1.7.

Assume now that $I$, $M$ are infinite. Let $J \subset I$ be a finite nonempty subset. If $J = \{j_1,\dots,j_n\}$, then put $a_k = (1,j_k,m_k)$ for some distinct elements $m_k \in M$, $k = 1, 2, \dots, n$. If $K_0[S]$ is regular and $a = \sum_{k=1}^{n} a_k$, then there exists $b \in K_0[S]$ such that $aba = a$. In other words, $a \circ P \circ b \circ P \circ a = a$ ($\circ$ stands for the usual product of matrices where $K_0[S]$ is identified with the corresponding Munn algebra). Consequently, $(a \circ P \circ b \circ P)_{JJ} \circ a_{JM} = a_{JM}$. By the definition of the matrix $a$, it follows that $(a \circ P \circ b \circ P)_{JJ}$ is the identity matrix. This shows that $P_{NJ}$ is left invertible over $K[G]$ where $N$ is a finite subset of $M$ such that $\mathrm{supp}(b)$ lies in the union of columns $S^{(m)}$ of $S$ for $m \in N$. Therefore, $P$ is locally left invertible. Analogously, one shows that $P$ is locally right invertible over $K[G]$, and so (iii) follows.

Finally, assume that (iii) holds. Let $a \in K_0[S]$. If $a \in S_{(J)}^{(N)}$ for some finite subsets $J \subset I$, $N \subset M$, then we know that there exist finite subsets $J' \subset I$, $N' \subset M$, such that $P_{NJ'}$ is right invertible and $P_{N'J}$ is left invertible over $K[G]$. Consequently, $a \circ P_{NJ'} \circ b = a$, $c \circ P_{N'J} \circ a = a$ for some $b \in S_{(J')}^{(N)}$, $c \in S_{(J)}^{(N')}$. Moreover, from Lemma 5 it follows that there exists a matrix $d$ over $K[G]$ such that $a \circ d \circ a = a$. Now, $a = (a \circ P_{NJ'} \circ b) \circ d \circ (c \circ P_{N'J} \circ a) = a \circ P_{NJ'} \circ b \circ d \circ c \circ P_{N'J} \circ a = axa$, where $x = b \circ d \circ c$ may be treated as an element of $K_0[S]$. Thus, $a$ is a regular element of $K_0[S]$, and so $K_0[S]$ is a regular algebra.

**Example 7** Assume that $I$ is a totally ordered set, and let $G$ be a locally finite group. Let $P$ by any $I \times I$ matrix over $G^0$ such that

$$p_{ii} \neq \theta \qquad \text{for } i \in I$$
$$p_{ij} = \theta \qquad \text{for } i,j \in I \text{ such that } i > j$$

Put $S = \mathfrak{M}^0(G,I,I,P)$. Then, for any finite set $J \subseteq I$, $P_{JJ}$ is a triangular matrix over $K[G]$ with nonzero entries on the main diagonal. It follows that $P_{JJ}$ is invertible over $K[G]$. Thus, $K_0[S]$ is regular if and only if $\mathrm{ch}(K) = 0$ or $G$ has no elements of order $p$ if $\mathrm{ch}(K) = p > 0$. Observe that $P$ itself may not be invertible over $K[G]$.

**Corollary 8** If $S$ is an inverse completely 0-simple semigroup, then $K_0[S]$ is regular if and only if the maximal subgroups of $S$ are locally finite with no elements of order $p$ if $\mathrm{ch}(K) = p > 0$.

*Proof.* This is a direct consequence of Example 7 because the sandwich matrix involved is the identity matrix; see [26], Theorem 3.9.

Once a description of regular semigroup algebras of completely 0-simple semigroups has been obtained, one may ask, in view of Corollary 3, for a characterization of regular semigroup algebras of arbitrary semigroups. Since being regular is a local property of an algebra, the following question is motivated by Proposition 2: Is $K[S]$ regular whenever $S$ is locally finite and all principal factor algebras $K_0[S_t]$, $t \in S$, are regular?

The answer is not known in general. We will consider some important special cases. The following result was obtained by Okniński in [178].

**Theorem 9** Assume that $S$ has d.c.c. on principal ideals. Then $K[S]$ is regular if and only if every principal factor algebra $K_0[S_t]$, $t \in S$, is regular. Moreover, in this case, $S$ is locally finite.

*Proof.* Assume that every $K_0[S_t]$, $t \in S$, is a regular algebra. Let $a \in K[S]$. Put $a_1 = a$ and $Y_1 = S \{\mathrm{supp}(a)\} S$ if $a \neq 0$. As in the proof of Theorem 7 in Chapter 2, there exists an ideal $X_1 \subseteq Y_1$ of $S$, or $X_1 = \varnothing$ if $Y_1$ is simple, such that $Y_1/X_1$ is a principal factor of $S$. By the hypothesis, there exists $x_1 \in K[S]$ such that $a_2 = a_1 - a_1 x_1 a_1 \in K[X_1]$ (we let $K[\varnothing] = 0$ here). If $a_2 \neq 0$, then put $Y_2 = S \{\mathrm{supp}(a_2)\} S$. Proceeding this way, we construct a sequence of ideals $Y_1 \supseteq X_1 \supseteq Y_2 \supseteq X_2 \supseteq \cdots$ and elements $x_j \in K[S]$ such that $a_j = a_{j-1} - a_{j-1} x_{j-1} a_{j-1} \in K[X_{j-1}]$ for any $j$. Since all $Y_j$ are finitely generated ideals of $S$, then, by the hypothesis on $S$ and Lemma 6 in

Chapter 2, this procedure must terminate. This means that $a_r = 0$ for some $r \geq 1$. It is easy to see that, for any $j$, $a_j = a_1 + a_1 z_j a_1$ for some $z_j \in K[S]$. Therefore, $a_r = 0$ implies that $a = a_1$ is a regular element. Hence, $K[S]$ is regular.

The converse holds for any semigroup $S$ as noted before.

Finally, if $K[S]$ is regular, then $S$ is completely semisimple by Corollary 3, and Theorem 7 in Chapter 2 implies that $S$ must be locally finite.

Another special case is partially settled in the following result; see [178], [268].

**Theorem 10**  Let $S$ be an inverse semigroup. Then,

(i)  $K[S]$ is regular if and only if $S$ is locally finite, if $\mathrm{ch}(K) = 0$.
(ii)  Assume that $S$ is locally finite. Then $K[S]$ is regular if and only if the subgroups of $S$ have no elements of order $p$, if $\mathrm{ch}(K) = p > 0$.

*Proof.*  In both cases, the necessity follows from Proposition 2. Assume that $S$ is locally finite. If $a \in K[S]$, then the semigroup $T$ generated by $\mathrm{supp}(a)$ and the inverses of the elements of $\mathrm{supp}(a)$ is a finite semigroup. From [26], Lemma 1.18, we know that $(st)^{-1} = t^{-1}s^{-1}$ for all $s, t \in S$. It then follows that $T$ is an inverse semigroup. Hence, $K[T]$ is a semisimple artinian algebra by Theorem 24 in Chapter 14, provided that the subgroups of $T$ have no elements of order $p$ if $\mathrm{ch}(K) = p > 0$. Hence, $a$ is regular in $K[T]$, and also in $K[S]$.

It is conjectured that $S$ must be locally finite whenever $K[S]$ is regular. This conjecture is supported, beyond the assertions of Theorems 9 and 10 and Proposition 2, by the following result establishing a complete characterization of the regularity condition in the case of some special inverse semigroups.

**Proposition 11**  Assume that $S$ is a union of groups. Then the following conditions are equivalent:

(i)  $K[S]$ is regular.
(ii)  $S$ is a locally finite inverse semigroup and, if $\mathrm{ch}(K) = p > 0$, then the subgroups of $S$ have no elements of order $p$.

*Proof.*  Assume that $K[S]$ is regular. Then, for any $t \in S$, the principal factor algebra $K[S_t]$ is regular. From the hypothesis it follows that $S_t$ is a union of groups. Therefore, $S_t \cong \mathfrak{M}^0(G, I, M, P)$ is a completely simple semigroup, possibly with zero adjoined, that is, $p_{mi} \in G$ for any $m \in M, i \in I$.

Let $\bar{S}_t = \mathfrak{M}^0(\{1\}, I, M, \bar{P})$ be the homomorphic image of $S_t$ determined by the trivial homomorphism $G \to \{1\}$; see Chapter 1. If $|I| > 1$ or $|M| > 1$, then $K_0[\bar{S}_t]$ has a nonzero nilpotent ideal by Lemma 3 in Chapter 5 or its right-left symmetric analog, which contradicts the regularity of this algebra. Therefore, $S_t$ or $S_t \setminus \{\theta\}$ is a group. We denote it by $G_t$. If $txt = t$, $xtx = x$ for some $x \in S$, then $G_t = G_x$, and so $x$ must be the inverse of $t$ in the group $G_t$. Thus, $x$ is uniquely determined, and $S$ must be an inverse semigroup; see [26], § 1.9. If $u \in S$ and $uy = yu = f$ is an idempotent for some $y \in G_u$, then, putting $e = tx = xt$, we have $xtuy = ef = fe = yutx$. Consequently, $G_{ut} = G_{fe} = G_{ef} = G_e G_f = G_f G_e$. It then follows easily that, for any $t_1$, ..., $t_n \in S$, we get $G_{t_1 \ldots t_n} = G_{t_1} \ldots G_{t_n} = G_{\sigma(t_1)} \ldots G_{\sigma(t_n)}$ for all permutations $\sigma \in \mathcal{S}_n$. Hence, the subsemigroup $\langle t_1, \ldots, t_n \rangle$ lies in the semigroup $T = \bigcup G_{t_{i_1}} \ldots G_{t_{i_j}}$, where the union is taken over all $\{i_1, \ldots, i_j\} \subseteq \{t_1, \ldots, t_n\}$, $j = 1, \ldots, n$. Thus, $T$ is a finite union of groups. Since, by Proposition 2, any subgroup of $S$ is locally finite, then, by Theorem 7 in Chapter 2, $T$ is locally finite. Therefore, $\langle t_1, \ldots, t_n \rangle \subseteq T$ is a finite semigroup, and so $S$ is locally finite. The remaining assertion of (ii) follows through Proposition 2. Then implication (ii)$\Rightarrow$(i) was established in Theorem 10.

Let us note that an inverse semigroup $S$ is a union of groups if and only if the idempotents of $S$ are central; see [90], Theorems IV.1.7 and IV.2.1.

The following corollary was first obtained by Gilmer and Teply in [68]; see also [64], Theorem 17.4.

**Corollary 12**  Let $S$ be a commutative semigroup. Then $K[S]$ is regular if and only if $S$ is a union of torsion groups with no elements of order $p$ if $\mathrm{ch}(K) = p > 0$.

*Proof.*  Since any regular commutative semigroup is a union of groups, then the result follows from Proposition 11.

We note that another special case of semigroup algebras that are regular and (right) self-injective is settled in Chapter 16.

As observed before, it is not known whether $K[S]$—regular, $\mathrm{ch}(K) = p > 0$, implies that $S$ is locally finite even if $S$ is a finitely generated inverse semigroup. We close this chapter by showing that, in this case, the problem may be reduced to the case of strongly $p$-semisimple semigroups.

**Proposition 13**  Let $S$ be a finitely generated completely semisimple semigroup such that any subgroup of $S$ is locally finite. If $S$ is not locally finite (hence finite), then there exists an ideal $M$ of $S$, such that $S/M$ is not locally finite and any principal factor of $S/M$ is finite.

*Proof.* Let $M$ be the intersection of all ideals $J$ of $S$ such that $S/J$ is a finite semigroup. The reasoning of Theorem 7 in Chapter 2 shows that $M \neq S$. Let $s \in S \setminus M$. Then there exists an ideal $T$ of $S$ such that $s \notin T$ and $S/T$ is finite. Since the principal factor of $s$ in $S/M$ coincides with the principal factor of $s$ in $S/T$, then it is finite.

Suppose that $S/M$ is locally finite. Then it is finite and, from Lemma 1 in Chapter 2, it follows that $M$ is a finitely generated semigroup. Thus, again as in Theorem 7 in Chapter 2, either $M$ is completely 0-simple and finite or $M$ contains an ideal $I \neq M$ of $S$ with $M/I$ finite. The latter is impossible by the definition of $M$. Hence, $M$ is finite and $S$ must be locally finite.

**Corollary 14** Let $S$ be a finitely generated semigroup such that $K[S]$ is a regular algebra. If $S$ is not finite, then there exists an ideal $M$ of $S$ such that $S/M$ is a strongly $p$-semisimple infinite semigroup for $p = \mathrm{ch}(K)$.

*Proof.* In view of Proposition 2, $M$ can be chosen as in Proposition 13. Since a finite dimensional regular algebra has an identity, the assertion follows from the fact that the principal factor algebras of $S$ are regular.

## Comments on Chapter 15

The study of the regularity condition for group algebras attracted the attention of several authors before the complete solution given in Theorem 1 was obtained. We refer to [203] for the bibliography and the proof. The first attempt to characterize regular semigroup algebras was made by Weissglass in [268], where some necessary conditions on $S$ were obtained and the case of locally finite inverse semigroups was discussed. The periodicity and local finiteness of $S$ in the characteristic zero case (see Proposition 2) were established by Okniński in [178]. Janeski and Weissglass [95] studied regularity of rings graded by a semilattice, which applies particularly to semigroup algebras of semigroups being semilattices of groups; see Proposition 11. Regular self-injective algebras $K[S]$ were described in [179]; see our Chapter 16. We note that May [145] recently showed that locally finite regular algebras need not be unions of semisimple finite dimensional algebras. On the other hand, this is the case within the class of algebras satisfying polynomial identities, as proved by Chekanu in [19]. Finally, Varricchio showed in [264] that a finitely generated semigroup that is a union of locally finite groups must be finite. This result can be compared to our Proposition 11.

# 16

# Self-Injectivity

Our next aim is to examine self-injective semigroup algebras. Recall that a ring $R$ is right self-injective if $R$ is an injective right $R$-module. This is an important property, strongly related to von Neumann regularity, and coinciding with the quasi-Frobenius property when restricted to finite dimensional unitary algebras; see [55], Chapters 19 and 24. We list some well-known results that will be used in the sequel. Here $l_R(X)$, $r_R(X)$, or $l(X)$, $r(X)$ if unambiguous, stand for the left and right annihilators of a subset $X \subseteq R$ in $R$.

**Lemma 1** Assume that $R$ is a right self-injective $K$-algebra. Then,

(i) (Baer condition) for any right ideal $I$ of $R$ and any homomorphism of right $R$-modules $f : I \to R$, there exists $r \in R$ such that $f(x) = rx$ for all $x \in I$.

(ii) $R$ has a left identity.

(iii) $R/\mathcal{J}(R)$ is a regular algebra and, for any set of orthogonal idempotents $e_1, \ldots, e_n$ of $R/\mathcal{J}(R)$, $n \geq 1$, there exists a set of orthogonal idempotents $f_1, \ldots, f_n$ of $R$ such that $f_i$ maps onto $e_i$ under the natural homomorphism $R \to R/\mathcal{J}(R)$, $i = 1, \ldots, n$.

(iv) $l(r(I)) \subseteq I + l(R)$ for any finitely generated left ideal $I$ of $R$.

(v) If $a, b \in l(R)$, $b \neq 0$, then $a \in Rb$.

(vi) If $e = e^2$, $f = f^2 \in R$ are such that $eR \cap fR = 0$, then there exists $g = g^2 \in R$ such that $eR + fR = gR$.

These are standard results in the theory of self-injective algebras. In the case where $R$ has an identity, the proofs may be found in [54], [55], 18.22, 19.27, and [232]. The nonunitary case may be obtained as an easy modification of the original proofs; see [55], 19.27, and its proof, [232], Proposition 4.5.1.

The following group algebra result of Renault is the starting point for the considerations in this chapter; see [203], Theorem 3.2.8.

**Theorem 2** Let $G$ be a group. Then $K[G]$ is right self-injective if and only if $G$ is finite.

The situation for semigroup algebras is much more complicated. First of all, it is fairly easy to construct finite semigroups with non-self-injective semigroup algebras. This may be done within the class of commutative semigroups. Clearly, by Lemma 1 (ii), any nonunitary commutative algebra $K[S]$ is of this type. Examples of finite commutative monoids with non-self-injective semigroup algebras may be obtained through Proposition 22 below. Thus, the "easy" implication of Theorem 2 does not hold for monoids. However, it is conjectured that $S$ must be finite whenever $K[S]$ is right self-injective. The following sequence of lemmas will lead to a positive answer in the case of semisimple semigroups and in the case of (right and left) self-injective semigroup algebras. Observe that, by Corollary 9 in Chapter 4, if one of the algebras $K[S]$, $K_0[S]$ is right self-injective, then so is the other. This will be repeatedly used without further comment.

**Lemma 3** Let $J$ be an ideal of a right self-injective $K$-algebra $R$. If $J^2 = 0$ and $R/J$ is finite dimensional, then $R$ is finite dimensional.

*Proof.* Suppose that $R$ is not finite dimensional. If $a \in J$, then $R^1 a$ is finite dimensional and $R^1 a = Ra$ by Lemma 1 (ii). Thus, there exists a chain of left ideals $Ra_1 \subset Ra_1 + Ra_2 \subset \cdots \subset Ra_1 + \cdots + Ra_n \subset \cdots \subset J$, $a_n \in J$. Then $r(Ra_1) \supseteq r(Ra_1 + Ra_2) \supseteq \cdots \supseteq r(Ra_1 + \cdots + Ra_n) \supseteq \cdots \supseteq J$ and, since $R/J$ is finite dimensional, then there exists $k \geq 1$ such that

$r(Ra_1 + \cdots + Ra_k) = r(Ra_1 + \cdots + Ra_m)$ for any $m \geq k$. From Lemma 1 (iv) it follows that

$$Ra_m \subseteq l(r(Ra_1 + \cdots + Ra_m)) = l(r(Ra_1 + \cdots + Ra_k)) \subseteq Ra_1 + \cdots + Ra_k + l(R)$$

Then, there exist $c_m \in Ra_1 + \cdots + Ra_k$, $m > k$, such that $a_m - c_m \in l(R)$. But $a_m \neq c_m$ because $a_m \notin Ra_1 + \cdots + Ra_k$ by the choice of the $a_i$. Hence, by Lemma 1 (v), $a_{k+2} - c_{k+2} = c(a_{k+1} - c_{k+1})$ for some $c \in R$. This contradicts the fact that $a_{k+2} \notin Ra_1 + \cdots + Ra_{k+1}$ and proves the assertion.

**Lemma 4**  Let $X$ be an infinite dimensional subspace of a $K$-linear space $V$. If $T$ is a basis of $V$, then, for any $n \geq 1$, there exists $x \in X$ such that $x \notin \sum_{i=1}^m Kt_i$ for all $t_i \in T$.

*Proof.*  If $x \in X$, $x = \sum_{i=1}^m \lambda_i t_i$ for some nonzero $\lambda_i \in K$, $t_i \in T$, then put $\mathrm{sp}(x) = \{t_1, \ldots, t_m\}$. Suppose that the assertion does not hold. Then there exists $n \geq 1$ such that $|\mathrm{sp}(x)| \leq n$ for all $x \in X$. Let $U \subseteq T$ be a maximal subset (possibly empty) for which there exists an infinite linearly independent subset $Y \subseteq X$ such that $U \subseteq \mathrm{sp}(y)$ for every $y \in Y$. Then, $|U| \leq n$ and, for all $t \in T \setminus U$, the set $\{x \in Y \mid t \in \mathrm{sp}(x)\}$ is finite. This allows us to choose an infinite subset $Z$ of $Y$ such that, for any $v, u \in Z$, we have $\mathrm{sp}(v) \cap \mathrm{sp}(u) = U$. Since the elements of $Z$ are independent, then $Z' = \{z \in Z \mid z \notin \mathrm{span}_K(U)\}$ is an infinite set. From the choice of $Z'$, it follows that $|\mathrm{sp}(z_1 + \cdots + z_m)| \geq m$ for any distinct elements $z_1, \ldots, z_m \in Z'$, $m \geq 1$. Since $z_1 + \cdots + z_m \in X$ and $m \geq 1$ is arbitrary, this contradicts our supposition and proves the lemma.

**Lemma 5**  Assume that $K_0[S]$ is right self-injective. Then there is no infinite sequence of elements $a_1, a_2, \ldots$ of $K_0[S]$ such that the principal right algebra ideals generated by the $a_i$ are independent and $\dim(K_0[S]a_i) = \infty$ for all $i = 1, 2, \ldots$.

*Proof.*  Suppose that a sequence of elements $a_1, a_2, \ldots \in K_0[S]$ satisfying the conditions of the lemma is given. Since $K_0[S]a_i$ is infinite dimensional, then, by Lemma 4, there exists $b_i \in K_0[S]a_i$ such that $|\mathrm{supp}(b_i)| > i |\mathrm{supp}(a_i)|$. Because the sum $\sum_{i=1}^\infty a_i(K_0[S])^1$ is direct, we may define a homomorphism of right $K_0[S]$-modules $\phi : \sum_{i=1}^\infty a_i(K_0[S])^1 \to K_0[S]$ by putting $\phi(a_i) = b_i$ for $i \geq 1$. The self-injectivity of $K_0[S]$ implies that there exists $z \in K_0[S]$ such that $\phi(x) = zx$ for all $x \in \sum_{i=1}^\infty a_i(K_0[S])^1$. In particular, $za_i = \phi(a_i) = b_i$ for every $i \geq 1$. Then,

$$|\mathrm{supp}(z)| |\mathrm{supp}(a_i)| \geq |\mathrm{supp}(za_i)| = |\mathrm{supp}(b_i)| > i |\mathrm{supp}(a_i)|$$

and so $|\operatorname{supp}(z)| > i$ for every $i \geq 1$. This contradicts the finiteness of the support of the element $z$ and establishes the lemma.

**Lemma 6** Assume that $K_0[S]$ is right self-injective. Then $K_0[S]$ has no infinite set of independent right ideals generated by idempotents.

*Proof.* Suppose that there exists an infinite set of idempotents $\{f_1, f_2, \ldots\}$ of $K_0[S]$ such that $f_i K_0[S]$, $i \geq 1$, are independent right ideals. Let $\mathsf{N} = \bigcup_{j=1}^{\infty} N_j$ be a partition of the set $\mathsf{N}$ of natural numbers such that all $N_j$ are infinite sets and $N_i \cap N_j = \varnothing$ for $i \neq j$. Let $I_1 = \sum_{i \in N_1} f_i K_0[S]$, and let $I_1'$ be the maximal essential extension of $I_1$ in $K_0[S]$. If $B_1$ is the complement of $I_1'$ containing the set $\{f_i \mid i \notin N_1\}$, then $K_0[S] = I_1' \oplus B_1$ as right $K_0[S]$-modules; see [54], Chapter 2. Since $K_0[S]$ has a left identity by Lemma 1 (ii), then $I_1' = e_1 K_0[S]$ for some $e_1 = e_1^2 \in I_1'$. Consequently, $e_1 f_i = f_i$ for any $i \in N_1$ and $e_1 K_0[S] \cap \sum_{i \notin N_1} f_i K_0[S] = 0$. We claim that $K_0[S] e_1$ is infinite dimensional. If this is not the case, then there exists $n > 1$ such that $f_{i_n} e_1 \in \sum_{k < n} f_{i_k} K_0[S] e_1$ for some $i_1, \ldots, i_n \in N_1$. Thus, $f_{i_n} = f_{i_n}(e_1 f_{i_n}) \in \sum_{k < n} f_{i_k} K_0[S]$. This contradicts the fact that $f_i K_0[S]$, $i \geq 1$, are independent right ideals of $K_0[S]$ and establishes our claim.

Now, replacing the subset $\{f_i \mid i \in N_1\}$ of $\{f_i \mid i \in \mathsf{N}\}$ by the single idempotent $e_1$, we get again a set of independent right ideals $e_1 K_0[S]$, $f_i K_0[S]$ for $i \notin N_1$. Thus, iterating this procedure leads to a sequence of idempotents $e_1, e_2, \ldots$ such that the right ideals $e_i K_0[S]$ are independent and $\dim_K K_0[S] e_i = \infty$. This contradicts Lemma 5 and shows that $K_0[S]$ has no infinite set of independent right ideals generated by idempotents.

**Corollary 7** Assume that $K[S]$ is right self-injective. Then $K[S]$ is semilocal.

*Proof.* From the preceding lemma, we know that $K[S]$ has no infinite set of orthogonal idempotents. Since the idempotents of $K[S]/\mathcal{J}(K[S])$ may be lifted to idempotents in $K[S]$ by Lemma 1 (iii), then $K[S]/\mathcal{J}(K[S])$ has no infinite set of orthogonal idempotents. Now, again by Lemma 1 (iii), this is a regular algebra so that it must be artinian; see [55], 19.26. This means that $K[S]$ is semilocal.

From the above result and Theorem 13 in Chapter 14, it follows that $S$ has no infinite subgroups provided that $K[S]$ is right self-injective and $\operatorname{ch}(K) = 0$. The same is true in the case of a positive characteristic. However, since there is no obvious way to transfer the self-injectivity of $K[S]$ to $K[G]$, where $G$ is a subgroup of $S$, we cannot use Theorem 2 to deduce the finiteness of $G$. For this we exploit the notion of algebraically compact modules. Recall that a right module $M$ over a ring $R$ is algebraically compact if any set of equations

(a)   $\sum_{i \in I} x_i a_{ij} = m_j$, $j \in J$, where $a_{ij} \in R$, $m_j \in M$ and, for any $j \in J$,
     all but finite $a_{ij} = 0$ ($I,J$ are arbitrary nonempty sets )

such that any finite subsystem of equations has a solution in $M$, has itself a solution in $M$. It is well known that any injective $R$-module is algebraically compact [113]. We include the argument for completeness. Let $F$ be a free right $R$-module with basis $\{x_i \mid i \in I\}$. Define $P$ as the submodule of $F$ generated by all $\sum_{i \in I} x_i a_{ij}$, $j \in J$, appearing in the system of equations (a). The hypothesis on the solvability of finite subsystems of (a) in $M$ implies that the rules $f(\sum_i x_i a_{ij}) = m_j$ define a homomorphism of right $R$-modules $f : P \to M$. Since $M$ is injective, this extends to a homomorphism of right $R$-modules $f' : F \to M$. This shows that $(f(x_i))_{i \in I}$ is a solution of (a) in $M$, and so $M$ is algebraically compact.

The advantage of dealing with algebraic compactness is that this property is much better behaved than injectivity under some algebraic operations.

We will use the following analog of the result of Lawrence [135], asserting that countable rings (or countably dimensional algebras) that are right self-injective must be right artinian. Here, for any algebra $R$, $R^{(\mathbb{N})}$ and $R^{\mathbb{N}}$ stand for the direct sum and the direct product of copies of $R$ indexed by the set $\mathbb{N}$ of natural numbers. The result is due to Zimmermann [277].

**Lemma 8** Let $R$ be a unitary $K$-algebra such that $R$ is an algebraically compact right $R$-module. If $\dim_K R$ is countable, then $R$ is right perfect.

*Proof.* Suppose that there exists a strictly descending chain of left principal ideals $L_1 \supset L_2 \supset \cdots$ of $R$. Choose for any $i \geq 1$ an element $m_i \in L_i \setminus L_{i+1}$. Let $g : R^{(\mathbb{N})} \to R$ be the homomorphism of right $R$-modules defined by the rule $g((r_i)_{i \in \mathbb{N}}) = \sum_{i \in \mathbb{N}} r_i$. From the definition of algebraic compactness, it follows that $g$ extends to a homomorphism of right $R$-modules $g' : R^{\mathbb{N}} \to R$. Consider the $K$-linear mapping $\phi : K^{\mathbb{N}} \to R$ defined by the rule $\phi((\lambda_i)_{i \in \mathbb{N}}) = g'((\lambda_i m_i)_{i \in \mathbb{N}})$. It is enough to show that $\phi$ is an embedding since then $\dim_K R$ is not countable, contradicting our hypothesis. Let $g'((\lambda_i m_i)_{i \in \mathbb{N}}) = 0$ for some $\lambda_i \in K$, $i = 1, 2, \ldots$. If $n$ is the least integer with $\lambda_n \neq 0$, then put

$$t_i = 0 \quad \text{for } i \leq n, \qquad t_i = \lambda_i m_i \quad \text{for } i > n$$

We know that $L_{n+1} = Ra$ for some $a \in R$. This and the definition of the $t_i$, $i \in \mathbb{N}$, imply that $(t_i)_{i \in \mathbb{N}} = (x_i)_{i \in \mathbb{N}} a$ for some $x_i \in R$. Now,

$$0 = g'((\lambda_i m_i)_{i \in \mathbb{N}}) = \lambda_n m_n + g'((t_i)_{i \in \mathbb{N}})$$

and so

$$m_n = -\lambda_n^{-1} g'((t_i)_{i \in \mathbb{N}}) = -\lambda_n^{-1} g'((x_i)_{i \in \mathbb{N}}) a \in Ra = L_{n+1}$$

This contradicts the choice of $m_n$ and shows that $R$ satisfies d.c.c. on principal left ideals. It is well known that $R$ is right perfect in this case; see [55], 22.29.

**Lemma 9** Assume that $K[S]$ is right self-injective. Then $S$ has no infinite subgroups.

*Proof.* Let $G$ be a subgroup of $S$ with an identity $e$. Since $K[S]$ is an algebraically compact right $K[S]$-module, then any system of equations

(b)         $\sum_{i \in I} x_i a_{ij} = m_j, \; j \in J$       with $a_{ij} \in K[G]$, $m_j \in K[G]$

has a solution $(b_i)_{i \in I}$ in $K[S]$ whenever any finite subsystem of (b) has a solution in $K[G]$. Then, for any $j \in J$, $i \in I$, we have $em_j = m_j$, $ea_{ij} = a_{ij}$, and so $m_j = \sum_{i \in I} b_i a_{ij} = \sum_{i \in I} eb_i ea_{ij}$. Since $G$ is a group-like in $eSe$, then, from Lemma 14 in Chapter 4, it follows that $m_j = \sum_{i \in I} c_i a_{ij}, j \in J$, where $c_i = \pi_G(eb_i e)$ for the projection $\pi_G$ of $K[eSe]$ onto $K[G]$. Therefore, $(c_i)_{i \in I}$ is a solution of system (b) in $K[G]$. This shows that $K[G]$ is an algebraically compact right $K[G]$-module. From Lemma 8 and from Theorem 21 in Chapter 14, it then follows that any countable subgroup of $S$ is finite. Hence, any subgroup of $S$ must be finite, too.

**Lemma 10** Assume that $K_0[S]$ is right self-injective. Then $S$ satisfies d.c.c. on principal left ideals.

*Proof.* Since $K_0[S]$ has a left identity $e$ by Lemma 1 (ii), then, for any $s \in S$, we have $S^1 s = Ss$. Suppose that a chain of principal left ideals $I_0 \supset I_1 \supset I_2 \supset \cdots, I_i = Ss_i$ for some $s_i \in S$, is given. Since, for any $i \geq 0$, there exists $t_i \in S$ such that $s_{i+1} = t_i s_i$, then putting $z_{i+1} = t_i t_{i-1} \ldots t_0, i \geq 0$, we get a chain of left ideals $Sz_1 \supseteq Sz_2 \supseteq \cdots$. The rule $x \to (z_1 + \cdots + z_n)x$ defines a homomorphism of right $K_0[S]$-modules $\phi_n : r(K_0[Sz_{n+1}]) \to K_0[S]$. Since

$$(z_1 + \cdots + z_{n+1}) - (z_1 + \cdots + z_n) = z_{n+1} \in Sz_{n+1} \subseteq l(r(K_0[Sz_{n+1}]))$$

then these extend to a homomorphism $\phi : \sum_{n \geq 1} r(K_0[Sz_{n+1}]) \to K_0[S]$. The Baer condition implies that there exists $z \in \bar{K}_0[S]$ such that $\phi(x) = zx$ for any $x \in \sum_{n \geq 1} r(K_0[Sz_{n+1}])$. Therefore, for any $n \geq 1$, $z - (z_1 + \cdots + z_n) \in$

$l(r(K_0[Sz_{n+1}]))$. From Lemma 1 (iv) it follows that $[z - (z_1 + \ldots + z_n) - a_{n+1}]K_0[S] = 0$ for some $a_{n+1} \in K_0([Sz_{n+1}])$. In particular, we obtain $zs_0 = (z_1 + \cdots + z_n)s_0 + a_{n+1}s_0$. Since $z_is_0 = s_i$ for any $i \geq 1$, then $zs_0 = (s_1 + \cdots + s_n) + b_{n+1}$ for some $b_{n+1} \in K_0[Ss_{n+1}]$. The choice of the elements $s_i$ implies now that $|\operatorname{supp}(zs_0)| \geq |\operatorname{supp}(zs_0) \backslash Ss_{n+1}| = |\operatorname{supp}(s_1 + \cdots + s_n)| = n$. Because $n$ is arbitrary, this contradicts the finiteness of $\operatorname{supp}(zs_0)$ and completes the proof.

We are now in a position to summarize the preceding auxiliary facts in the following result due to Okniński [179].

**Theorem 11** Assume that $K[S]$ is right self-injective. Then $K[S]$ is right perfect.

*Proof.* Since $K[S]$ is semilocal by Corollary 7, $S$ has no infinite subgroups by Lemma 9, and $S$ has d.c.c. on principal left ideals by Lemma 10, then the assertion follows from Corollary 22 in Chapter 14.

Observe that the above theorem, together with Corollary 22 in Chapter 14, implies that $S$ must be locally finite whenever $K[S]$ is right self-injective.

Using Theorem 11 and a result of Osofsky see [55], Section 24, we are now able to solve the main problem for (right and left) self-injective semigroup algebras.

**Theorem 12** Assume that $K[S]$ is right and left self-injective. Then $S$ is finite.

*Proof.* From Theorem 11, Lemma 1 (ii), and their left-right symmetric analogs, we know that $K[S]$ is a right and left perfect algebra with an identity. Therefore, the right and left socles of $K[S]$ are essential as right, respectively left, ideals of $K[S]$, which shows that $K[S]$ is a right and left $PF$-algebra in the sense of [55], 24.32. This and the theorem of Osofsky (see [55], 24.33), imply that $K[S]$ is a quasi-Frobenius algebra, and so it, in particular, is artinian. Hence, $S$ is finite by Theorem 23 in Chapter 14.

Our next aim is to show that the assumptions of Theorem 12 may be weakened when semisimple semigroups only are considered. For this we need the following preparatory result.

**Lemma 13** Assume that $J = \mathfrak{M}^0(G, I, M, P)$ is a completely 0-simple ideal of $S$. If $K_0[S]$ is right self-injective, then $I$ is finite, $G$ is finite, and $K_0[J]$ has a left identity. Moreover, $K_0[S/J]$ is right self-injective provided that $J \neq S$.

*Proof.* Since $G$ is isomorphic to a subgroup of $S$, then it is finite by Lemma 9. Suppose that $|I| = \infty$. Choosing elements $s_1$, $s_2$, ... from distinct rows of $J$, we get a set of independent right ideals $K_0[s_i S] = K_0[s_i J]$ of $K_0[S]$, $i = 1, 2, \ldots$; see Chapter 1. Since $I$ is infinite, then any $Ss_i = Js_i$ is infinite. This contradicts Lemma 5 and shows that $I$ is finite. We know that $J = \bigcup_{i=1}^{n} e_i S$ for some $n \geq 1$, $e_i = e_i^2 \in J$, in this case. Thus, $K_0[J] = \sum_{i=1}^{n} K_0[e_i S]$ and Lemma 1 (vi) imply that $K_0[J] = e K_0[S]$ for some $e = e^2 \in K_0[J]$. Assume that $S \neq J$. Let $f$ be a left identity of $K_0[S]$. Then $K_0[S] = e K_0[S] \oplus (f - e)K_0[S]$ as right $K_0[S]$-modules. Consequently, $(f - e)K_0[S]$ is an injective right $K_0[S]$-module. Now $K_0[S/J] \cong K_0[S]/K_0[J] = K_0[S]/e K_0[S] \cong (f - e)K_0[S]$. Therefore, $K_0[S/J]$ is an injective right $K_0[S]$-module. Since the $K_0[S]$-module and $K_0[S/J]$-module structures on $K_0[S/J]$ coincide, then $K_0[S/J]$ is right self-injective.

Now, we can give a complete extension of Theorem 2 to the class of semigroup algebras of semisimple semigroups.

**Theorem 14** Assume that $S$ is a semisimple semigroup such that $K[S]$ is right self-injective. Then $S$ is finite.

*Proof.* From Theorem 11 and from Theorem 21 in Chapter 14, it follows that $S$ has a chain of ideals $S = S_n \supset S_{n-1} \supset \cdots \supset S_1$ such that $S_1$ and all $S_i/S_{i-1}$ are completely 0-simple. We proceed by induction on $n$. Let $\mathfrak{M}^0(G, I, M, P)$ be a Rees matrix presentation of $S_1$. Then $|G| < \infty$, $|I| < \infty$ by Lemma 13. Put $A = l_{K_0[S_1]}(K_0[S_1])$. From Lemma 3 in Chapter 5, it follows that $K_0[S_1]/A$ is a finite dimensional $K$-algebra. Thus, if $n = 1$, then Lemma 3 implies that $S = S_1$ is a finite semigroup. If $n > 1$, then $S/S_1$ admits a chain of ideals $S/S_1 \supset \cdots \supset S_2/S_1$ with completely 0-simple factors, so that Lemma 13 and the induction hypothesis imply that $S/S_1$ is finite. Hence, $K_0[S]/A$ is finite dimensional and, again by Lemma 3, $S$ is a finite semigroup.

It is not known whether the fact that $K[S]$ is right self-injective always implies that $S$ is a finite semigroup. When trying to establish this implication, we may assume, by Theorem 11 and by Theorem 21 in Chapter 14, that $S$ has a chain of ideals $S = S_n \supset S_{n-1} \supset \cdots \supset S_1$ such that $S_1$ and all factors $S_i/S_{i-1}$ are either completely 0-simple or right $T$-nilpotent semigroups. Further, Lemma 13 and an induction as in the proof of Theorem 14 allow us to assume that $S_1$ is a nil semigroup. We will show that $S_1$ must be nilpotent in this case.

**Lemma 15** Let $J$ be a nil ideal of $S$. If $K_0[S]$ is right self-injective, then $J$ is nilpotent.

*Proof.* Suppose that $J$ is not nilpotent. Then from Lemma 11 in Chapter 2, it follows that, for every $n \geq 1$, there exists $s_n \in J$ such that $J^{2n}s_n = \theta$ and $t_1^{(n)}s_n, t_2^{(n)}s_n, \ldots, t_n^{(n)}s_n$ are distinct nonzero elements of $S$ for some $t_i^{(n)} \in J^n$, $i = 1, \ldots, n$. Observe that $t_i^{(3^n)} \in J^{3^n} \subseteq l(r(J^{3^n})) \subseteq l(r(J^{3^{n-1}2}))$. This allows us to define a homomorphism of right $K_0[S]$-modules $\phi : \sum_{n=0}^{\infty} r(J^{3^n 2}) \to K_0[S]$ by the rule

$$\phi(x) = [t_1^{(1)} + (t_1^{(3)} + t_2^{(3)} + t_3^{(3)}) + \cdots + (t_1^{(3^n)} + \cdots + t_{3^n}^{(3^n)})]x$$

for any $x \in r(J^{3^n 2})$, $n \geq 0$. Then there exists an element $z \in K_0[S]$ such that $\phi(x) = zx$ for all $x \in \sum_{n=0}^{\infty} r(J^{3^n 2})$. Substituting $x = s_{3^n}$, in view of the fact that $s_{3^n} \in r(J^{3^n 2})$, we get $zs_{3^n} = \phi(s_{3^n}) = \sum_{i=0}^{n} \sum_{j=1}^{3^i} t_j^{(3^i)}s_{3^n}$. Then,

$$|\operatorname{supp}(zs_{3^n})| \geq \left| \operatorname{supp}\left( \sum_{j=1}^{3^n} t_j^{(3^n)}s_{3^n} \right) \right| - \left| \operatorname{supp}\left( \sum_{i=0}^{n-1} \sum_{j=1}^{3^i} t_j^{(3^i)}s_{3^n} \right) \right|$$

Since all $t_j^{(3^n)}s_{3^n}$ are distinct and nonzero, then it follows that $|\operatorname{supp}(zs_{3^n})| \geq 3^n - (1 + 3 + \cdots + 3^{n-1}) > 3^n/2$. While this holds for every $n \geq 1$, this contradicts the finiteness of the supports of elements in $K_0[S]$ and proves the assertion.

The above result shows that it is reasonable to expect that $K[S]$ is semiprimary whenever it is right self-injective; see Theorem 21 in Chapter 14 and its proof. Observe that no examples are known of right but not left self-injective semiprimary algebras; see [56]. This, in view of Theorem 12, supports the conjecture on the finiteness of $S$. We can provide additional evidence in this line. The following result is a direct consequence of the theorem of Lawrence [135] mentioned before Lemma 8 and our Theorem 23 in Chapter 14.

**Proposition 16** Let $S$ be a countable semigroup such that $K[S]$ is right self-injective. Then $S$ is finite.

Once we know that (right and left) self-injective semigroup algebras are finite dimensional, we may look for a description of these algebras in terms of the underlying semigroups. This may be given in the case of semisimple semigroups. The following result obtained in [179] shows that this class of algebras is, in fact, determined by the group algebras of finite groups.

**Theorem 17**  Assume that $S$ is a semisimple semigroup such that $K[S]$ has an identity. Then the following conditions are equivalent:

(i)  $K[S]$ is right (or left) self-injective.
(ii)  $K[S]$ is a quasi-Frobenius algebra.
(iii)  $S$ is a finite strongly $p$-semisimple semigroup, where $p = \text{ch}(K)$.
(iv)  $K[S] \cong M_{n_1}(K[G_1]) \oplus \cdots \oplus M_{n_r}(K[G_r])$ for some $r \geq 1$, $n_i \geq 1$, and some finite subgroups $G_i$ of $S$.

*Proof.*  If (i) holds, then $S$ is finite by Theorem 14, and so $K[S]$ is a quasi-Frobenius algebra; see [55], 24.4.

If (ii) holds, then, with the notation of the proof of Theorem 14, Lemma 13 and its left-right symmetric analog imply that $K_0[S_1]$ has an identity, and $K_0[S/S_1]$ is a quasi-Frobenius algebra if $S \neq S_1$. Then, an induction shows that every $K_0[S_i/S_{i-1}]$ has an identity. Since every principal factor of $S$ is isomorphic to one of the factors $S_i/S_{i-1}$ or $S_1$ by [26], § 2.6, then (iii) holds.

The implication (iii) $\Rightarrow$ (iv) is established in Corollary 27 of Chapter 5. Since group algebras of finite groups are quasi-Frobenius (see Theorem 2) and the class of quasi-Frobenius algebras is closed under the matrix algebra construction (see [55], 24.7), then (iv) implies (i).

The reason for assuming the existence of an identity element in $K[S]$ in the above theorem is to omit some pathologies when considering modules over nonunitary algebras. Apparently, Theorem 17 covers the following cases:

$K[S]$ is also left self-injective, by Lemma 1 (ii) and its left-right symmetric analog.
$S$ is an inverse semigroup—the involution $s \to s^{-1}$, $s \in S$, establishes the left-right symmetry of $K[S]$ (see [26], § 1.9), and the preceding case applies.
$K \neq \mathbb{Q}$—if $K[S]$ has no two-sided identity in this case, then it is easy to construct a homomorphism of right $K[S]$-modules that does not satisfy the Baer condition; see [179]. On the other hand, it is easy to check by direct computation that $\mathbb{Q}[T]$ is right self-injective, where $T = \{s,t\}$ is a semigroup of right zeros, i.e., $s^2 = s$, $t^2 = t$, $st = t$, $ts = s$.

If $S$ is not semisimple, then an intrinsic characterization of self-injective algebras $K[S]$ in terms of $S$ is not known. However, in certain cases, such a result can be obtained through the class of Frobenius algebras; see [32], Chapter IX. For this, we introduce the following notation.

Let $S = \{s_1, \ldots, s_n, \theta\}$ be a finite semigroup with zero. Let $\{\omega_1, \ldots, \omega_n\}$ be a set of parameters. Define the $n \times n$ matrix $\Gamma_S = (\gamma_{ij})$ by $\gamma_{ij} = \omega_k$ if $s_i s_j = s_k$ and by $\gamma_{ij} = 0$ if $s_i s_j = \theta, j = 1, \ldots, n$. $\Gamma_S$ is called the Nakayama matrix of parameters of $S$ (corresponding to the given ordering of $S$) [167]. By a specialization of $\Gamma_S$ in a field $L$, we mean the matrix obtained from $\Gamma_S$ by substituting some scalars $\alpha_k \in L$ in place of the corresponding parameters $\omega_k$ in $\Gamma_S$. Since we are interested only in the question of when a given specialization of $\Gamma_S$ is a nonsingular matrix, then the ordering of $S$ is of no importance (a permutation of $S$ results in the corresponding permutation of rows and columns of $\Gamma_S$). Observe also that the structural matrices $\mathbf{R}$, $\mathbf{L}$ defined for $K_0[S]$ in Chapter 14 are specializations of $\Gamma_S$.

The following result, coming essentially from the characterization of Frobenius algebras given in [32], 61.3, provides a criterion for determining Frobenius semigroup algebras.

**Proposition 18** Let $S = \{s_1, \ldots, s_n, \theta\}$ be a finite semigroup with zero. Then $K[S]$ is a Frobenius algebra if and only if the Nakayama matrix of parameters of $S$ becomes nonsingular under some specialization in $K$.

*Proof.* We know that, if $K_0[S]$ is a Frobenius algebra, then there exists a nondegenerate bilinear form $f: K_0[S] \times K_0[S] \to K$ that is associative, that is, $f(ab, c) = f(a, bc)$ for any $a$, $b$, $c \in K_0[S]$; see [32], 61.3. If $M = (m_{ij})$ is the matrix of $f$ in the basis $S = \{s_1, \ldots, s_n\}$ of $K_0[S]$, then $f(s_i, s_j) = m_{ij}$ for $i, j = 1, \ldots, n$. Moreover, if $s_i s_j = s_k s_l$, then,

$$m_{ij} = f(s_i, s_j) = f(s_i s_j, 1) = f(s_k s_l, 1) = f(s_k, s_l) = m_{kl}$$

where $1$ is the identity of $K_0[S]$. Hence, $M$ is a specialization of the Nakayama matrix of parameters of $S$ in $K$, and $M$ is nonsingular because $f$ is nondegenerate.

On the other hand, assume that $N = (n_{ij})$ is a nonsingular specialization of $\Gamma_S$ in $K$. It is straightforward that the bilinear form $g$ on $K_0[S]$, the matrix of which in the basis $\{s_1, \ldots, s_n\}$ is $N$, is associative. By [32], 61.3, it is then sufficient to show that $K_0[S]$ has an identity in this case. Since $N$ is nonsingular, then there exists a vector $\lambda = (\lambda_1, \ldots, \lambda_n) \in K^n$ such that, for any $k$ with $s_k \in S^2$, the $k$th entry of the vector $\lambda N$ equals $n_{ij}$, where $s_k = s_i s_j$. It then follows that $g(e, s_i s_j) = n_{ij} = g(s_i, s_j)$ for $e = \sum_{r=1}^{n} \lambda_r s_r$. Since $g$ is associative, then we get $0 = g(es_i, s_j) - g(s_i, s_j) = g(es_i - s_i, s_j)$. The fact that $i, j$ are arbitrary and $g$ is nondegenerate implies then that $es_i = s_i$ for all $i$, so that $e$ is a left identity of $K_0[S]$. A symmetric argument shows that $K_0[S]$ has a right identity, and the result follows.

The following observation shows that it is enough to examine the specializations of $\Gamma_S$ in the prime subfield of $K$ only in order to check whether $K[S]$ is a Frobenius algebra.

**Lemma 19** Let $L$ be a field extension of $K$. Then, for any $K$-algebra $R$, the following conditions are equivalent:

(i)   $R$ is a Frobenius algebra.
(ii)  $R \otimes_K L$ is a Frobenius algebra.

*Proof.* Since $R$ has an identity if and only if $R \otimes_K L$ has an identity, then we can assume that $R$ has an identity. Now, we use the fact that $\mathrm{Hom}_K(R,K) \otimes_K L \cong \mathrm{Hom}_L(R \otimes_K L, L)$ as $K$-spaces and even as left $R \otimes_K L$-modules under the natural module structures; see [32], § 60. Thus, $R \cong \mathrm{Hom}_K(R,K)$ as left $R$-modules if and only if $R \otimes_K L \cong \mathrm{Hom}_L(R \otimes_K L, L)$ as left $R \otimes_K L$-modules which, in view of [32], 61.1, establishes our assertion.

As the first application, we give a characterization of semisimple artinian semigroup algebras that is more intrinsic than that of Chapter 14. It is a direct consequence of Theorem 24 in Chapter 14, Theorem 17, Proposition 18, and Lemma 19 applied to $S^0$, because the semisimple artinian algebras are Frobenius.

**Corollary 20** The following conditions are equivalent:

(i)   $K[S]$ is semisimple artinian.
(ii)  $S$ is a finite semisimple semigroup such that the order of every subgroup of $S$ in invertible in $K$ and the Nakayama matrix of parameters of $S^0$ becomes nonsingular under a specialization in the prime subfield of $K$.

Before applying the above results to our original problem of describing semigroup algebras that are self-injective, let us note that there are known examples of quasi-Frobenius but not Frobenius semigroup algebras. For example, the linear semigroup $S \subseteq M_6(\mathbb{Q})$, consisting of matrices $e_{11} + e_{55}$, $e_{12} + e_{56}, e_{13}, e_{21} + e_{65}, e_{22} + e_{66}, e_{23}, e_{33} + e_{44}, e_{45}, e_{46}, 0$, is of this type [167]. We will apply the foregoing results to the case of commutative semigroups. It is a well-known and easy consequence of the appropriate definitions that the notions of "quasi-Frobenius" and "Frobenius" coincide in this case; see [32], 61.1. In fact, if $R$ is a commutative quasi-Frobenius finite dimensional algebra, then any indecomposable direct summand of the $R$-module $R$ has multiplicity 1, so that, by the definition, $R$ embeds into

$R^* = \text{Hom}_K(R, K)$ as left $R$-modules. Since $\dim_K R = \dim_K R^*$, then $R$, $R^*$ are isomorphic as left $R$-modules. This means that $R$ is a Frobenius algebra.

We will also exploit the fact that finite dimensional unitary right self-injective algebras are quasi-Frobenius; see [32], 58.6.

**Corollary 21**   Let $S$ be a commutative semigroup. Then,

(i)   $K[S]$ is right self-injective if and only if it is a Frobenius algebra. Moreover, in this case, $S$ is finite and $K_0[S] \cong K_0[S_1] \oplus \cdots \oplus K_0[S_n]$ for some $n \geq 1$ and monoids $S_1, \ldots, S_n$ with no idempotents distinct from $1$, $\theta$.

(ii)  If $S$ is a finite semigroup with zero, then $K[S]$ is a Frobenius algebra if and only if a specialization of the Nakayama matrix of parameters of $S$ in the prime subfield of $K$ is nonsingular.

*Proof.*   (i) The former follows from Theorem 12, Lemma 1 (ii), and the foregoing remarks. Assume first that $S$ has no nonzero and nonidentity idempotent. Since $K_0[S]$ has an identity, then $S$ is not nilpotent. But $S$ is finite; hence, $S$ has a nonzero idempotent. It then follows that $S$ is a monoid.

If $S$ is arbitrary, then let $e$ be a minimal nonzero idempotent of $S$. Then $eS$ is an ideal of $S$ that is a monoid with no nonzero and nonidentity idempotents and, by Corollary 8 in Chapter 4, $K_0[S] \cong K_0[eS] \oplus K_0[S/eS]$. Moreover, $K_0[eS]$, $K_0[S/eS]$ are Frobenius algebras as direct summands of $K_0[S]$. Thus, the proof may be easily completed by induction on the number of nonzero idempotents of $S$.

(ii) is a direct consequence of Proposition 18 and Lemma 19.

Several authors investigated some other types of conditions for $K[S]$ to be Frobenius in the apparently simplest case of finite commutative monoids; see [213], [214], [241], [272], and [273]. This was based on the reduction to the case where $S = U(S) \cup N$, $N$—a nilpotent ideal of $S$ or $N = \varnothing$, given in Corollary 21. Monoids of this type are called elementary (commutative) monoids. Many module theoretic necessary and sufficient conditions for such $K[S]$ to be Frobenius have been found. For example, a result of this type obtained in [213] establishes the equivalence of the fact that $K[S]$ is a Frobenius algebra and the fact that the $K[U(S)]$-modules $r_{K[S]}(\mathcal{J}(K[S]))$ and $K[G]$ are isomorphic, where $G$ denotes the complement of the Sylow $p$-subgroup of $U(S)$, $p = \text{ch}(K)$ (possibly 0).

In the special case where $S$ is an elementary monoid with trivial group of units, the criterion described in Corollary 21 simplifies considerably. In this case, $S = \{1\} \cup N$ and, to omit triviality, we may assume that $N \neq \varnothing$,

$\theta$. Observe first that, for any $s$, $t \in S$, $sS = tS$ if and only if $s = t$. In fact, if $\theta \neq s = tx$, $t = sy$ for some $x$, $y \in S$, then $s = syx = s(yx)^2 = \cdots$. The nilpotency of $N$ implies that $yx = 1$, so that $x = y = 1$ and $s = t$ as desired. Thus, $S$ has a minimal nonzero ideal $I$ of order 2. Write $t_1 = 1$, $N = \{t_2, \ldots, t_n, \theta\}$, where $\{t_n, \theta\} = I$. Define a matrix $A_S = (\alpha_{ij})$ by $\alpha_{ij} = 1$ if $t_i t_j = t_n$, $\alpha_{ij} = 0$ if $t_i t_j \neq t_n$ for $i, j \in \{1, 2, \ldots, n\}$. Clearly, $A_S$ is a specialization of the Nakayama matrix of parameters of $S$ in $K$. With this notation, we have the following result due to Wenger [272].

**Proposition 22** Let $S = \{t_1 = 1, t_2, \ldots, t_n, \theta\}$ be an elementary monoid with trivial group of units. Then $K[S]$ is a Frobenius algebra if and only if the matrix $A_S \in M_n(K)$ is nonsingular.

*Proof.* Suppose that $A_S$ is a singular matrix. Then there exist $\lambda_k \in K$, $k = 1$, $2$, $\ldots$, $n$, not all 0, such that $\sum_{k=1}^{n} \lambda_k R_k = 0$, where $R_k$ denotes the $k$th row of $A_S$. From the definition of the $\alpha_{ij}$, it follows that $\alpha_{j1} = 0$ for $j \neq n$ and $\alpha_{n1} = 1$ because $t_j t_1 = t_j$ for any $j = 1$, $2$, $\ldots$, $n$. Hence, $\lambda_n = 0$ and $\sum_{k=1}^{n-1} \lambda_k R_k = 0$. Put $a = \sum_{k=1}^{n-1} \lambda_k t_k \in K_0[S]$. Then, for any $j$, $a t_j = \sum_{k=1}^{n-1} \lambda_k t_k t_j = \sum_{k=1}^{n} \beta_k t_k$ for some $\beta_1$, $\ldots$, $\beta_n \in K$. Now, $\beta_n = \sum_{k \in Z_j} \lambda_k$, where $Z_j = \{k \mid t_k t_j = t_n\}$. Since $\sum_{k \in Z_j} \lambda_k$ coincides with the $j$th component of the zero vector $\sum_{k=1}^{n} \lambda_k R_k$, then $\beta_n = 0$. Thus, $a t_j \in \sum_{k=1}^{n-1} K t_k$ for any $j$, and so $a K_0[S] \cap K t_n = 0$. It follows that the socle of $K_0[S]$ is not an irreducible $K_0[S]$-module. But the socle of $K_0[S]$ is a module dual to $K_0[S]/\mathcal{J}(K_0[S]) \cong K$ by [32], 58.4. In view of [32], 58.6, this contradicts the fact that $K_0[S]$ is a Frobenius algebra.

The converse implication follows from Corollary 21.

Now, let $S$ be an arbitrary elementary monoid. Since $S \setminus U(S)$ is a nilpotent ideal of $S$, if nonempty, then it is easy to see (refer to the paragraph preceding Proposition 22) that, for any $s \in S$, the $\mathcal{J}$-class of $s$ in $S$ equals $sU(S)$, that is, $sS = tS$ if and only if $t \in sU(S)$. Moreover, the relation $\rho = \{(x, y) \in S \times S \mid x \in yU(S)\}$ is a congruence on $S$, and an element $u \in S/\rho$ is invertible if and only if it is the $\rho$-class of an invertible element of $S$. Therefore, $S/\rho$ is an elementary monoid with a trivial group of units. We will show in Theorem 25 that $K[S/\rho]$ is Frobenius whenever $K[S]$ is Frobenius. However, in general, the converse does not hold.

**Example 23** Let $S = \{1, a, b, \theta\}$ be a monoid with the multiplication rules $ab = ba = b$, $a^2 = 1$, $b^2 = \theta$. Clearly, $S$ is elementary with $U(S) = \{1, a\}$. Moreover, $T = S/\rho = \{1, b', \theta\}$ with $(b')^2 = 0$. It is easy to see that $K_0[T]$ is a Frobenius algebra for any field $K$ (e.g., use Proposition 22). On the other hand, $K_0[S]$ is Frobenius if and only if ch$(K) \neq 2$. In fact, the Nakayama matrix of parameters of $S$

$$\begin{pmatrix} \omega_e & \omega_a & \omega_b \\ \omega_a & \omega_e & \omega_b \\ \omega_b & \omega_b & 0 \end{pmatrix}$$

is singular for any $\omega_e, \omega_a, \omega_b \in K$ if ch$(K) = 2$ and is nonsingular for the specialization $\omega_e = 1$, $\omega_a = 0$, $\omega_b = 1$ if ch$(K) \neq 2$.

**Lemma 24** Let $R$ be a finite dimensional commutative unitary $K$-algebra, and let $I$ be an ideal of $R$. Then $R/I$ is a Frobenius algebra if $R$ is a Frobenius algebra and $r_R(I)$ a principal ideal.

*Proof.* We will use the fact that $R$ is Frobenius if and only if a linear function $f : R \to K$ exists such that ker$(f)$ does not contain nonzero ideals of $R$; see [32], 61.3. We know that $r(I) = cR$ for some $c \in R$. Consider the function $h : R \to K$ given by $h(x) = f(cx)$. Clearly, $h$ is linear and ker$(h) = \{x \in R \mid cx \in \text{ker}(f)\}$. If $J \neq 0$ is an ideal of $R$ contained in ker$(h)$, then $cJ \subseteq \text{ker}(f)$ and, by the choice of $f$, we get $cJ = 0$. Thus, $J \subseteq r(cR) = r(r(I))$, the latter equal to $I$ by [32], 61.3. Hence, $I$ is the largest ideal of $R$ contained in ker$(h)$. This shows that the function $h' : R/I \to K$ given by $h'(x + I) = h(x)$ is well defined and that ker$(h')$ does not contain nonzero ideals of $R/I$. Hence, again by [32], 61.3, $R/I$ is a Frobenius algebra.

Note that the converse of the above result, as well as a generalization to not necessarily commutative algebras, is given in [167].

We close this chapter with the following result of Wenger [272]. Here $\rho$ stands for the congruence introduced before Example 23.

**Theorem 25** Let $S$ be an elementary monoid. If $K[S]$ is a Frobenius algebra, then $K[S/\rho]$ is a Frobenius algebra.

*Proof.* If $S = U(S)$, then $K[S/\rho] \cong K$, and we are done. Thus, assume that $S$ has a zero element. We first claim that $T = r_{K_0[S]}(e) \cap S$ is zero (note that 0 and $\theta$ are identified in $K_0[S]$), where $e = \sum_{u \in U(S)} u$. Suppose the contrary. Put $S_1 = S \setminus U(S)$. Then the ideal series $S \supset S_1 \supseteq T$ may be refined to a principal series for $S$: $S = S_0 \supset S_1 \supset \cdots \supset S_p \supset S_{p+1} = T \supset \cdots \supset S_r \supset S_{r+1} = \theta$ for some $r \geq p \geq 0$. Choose $s_i \in S_i \setminus S_{i+1}$, $i = 0, 1, \ldots, r$. Then $S_i \setminus S_{i+1}$ is the $\mathcal{J}$-class of $s_i$, and so $S_i \setminus S_{i+1} = s_i U(S)$. Put $f = \sum_{s \in s_p U(S)} s$. We will show that $fS_1 = 0$ in $K_0[S]$. Since $U(S)$ acts transitively on $s_p U(S)$, then $es = |G|f$ for any $s \in s_p U(S)$, where $G = \{u \in U(S) \mid us_p = s_p\}$. Hence, $f = |G|^{-1}es$ (note that $|G| \neq 0$ in $K$ because $es \neq 0$ by the definition of

$p$). If $p > 0$, then $sS_1 \subseteq S_{p+1}$, and so $fS_1 = |G|^{-1}esS_1 \subseteq |G|^{-1}eS_{p+1} = 0$. If $p = 0$, then $e = f$, and $fS_1 = eS_1 = 0$ in this case, too.

It is now clear that

$$r_{K_0[S]}(f) = (r_{K_0[S]}(f) \cap K[U(S)]) + K_0[S_1]$$

Let $g = \sum_{t \in S_r U(S)} t$. Then, similarly,

$$r_{K_0[S]}(g) = (r_{K_0[S]}(g) \cap K[U(S)]) + K_0[S_1]$$

because $gS_1 = 0$ in $K_0[S]$. Let $a = \sum_{x \in S} \lambda_x x \in K[U(S)]$, $\lambda_x \in K$. Since $af = (\sum \lambda_x)f$, then $a \in r_{K_0[S]}(f)$ if and only if $\sum \lambda_x = 0$. Similarly, $a \in r_{K_0[S]}(g)$ if and only if $\sum \lambda_x = 0$, which implies that $r_{K_0[S]}(f) = r_{K_0[S]}(g)$. But $K_0[S]$ is Frobenius, so that we get $fK_0[S] = gK_0[S]$; see [32], 61.3. Thus, $p = r$ and $g = f$ since supp$(f) \subseteq S_p \setminus S_{p+1}$ and supp$(g) \subseteq S_r$. This means that $T = S_{r+1} = \theta$, that is, $r_{K_0[S]}(e) \cap S$ has no nonzero elements as claimed.

Let $b \in K_0[S]$ be such that $eb = 0$. Write $b = b_0 + \cdots + b_r$, where supp$(b_i) \subseteq S_i \setminus S_{i+1}$. Then $eb_i = 0$ for any $i$ because supp$(eb_i) \subseteq S_i \setminus S_{i+1}$. Fix some $i$, and let $b_i = \sum_{s \in S_i U(S)}(\beta_s s)$, $\beta_s \in K$. Since $S_i \setminus S_{i+1} = s_i U(S)$ and $eu = e$ for every $u \in U(S)$, then we get $0 = eb_i = e \sum_{s \in S_i U(S)}(\beta_s s) = (\sum_{s \in S_i U(S)} \beta_s)es_i$. Now $es_i \neq 0$ by the first part of the proof, so that $\sum \beta_s = 0$. Consequently, $b \in \omega(K[U(S)])K_0[S]$, and so $r_{K_0[S]}(e) \subseteq \omega(K[U(S)])K_0[S]$. Note that the converse inclusion is clear. Now, let $\phi : K_0[S] \to K_0[S/\rho]$ be the natural homomorphism. Then, by Corollary 2 in Chapter 4,

$$\begin{aligned}
\ker(\phi) = I(\rho) &= K\{s - t \mid (s,t) \in \rho, \; s,t \in S\} \\
&= K\{s - us \mid u \in U(S), \; s \in S\} \\
&= \omega(K[U(S)])K_0[S]
\end{aligned}$$

Therefore, we have shown that $\ker(\phi) = r_{K_0[S]}(e)$. Since $K_0[S]$ is a Frobenius algebra, then, again by [32], 61.3, $r(\ker(\phi)) = eK_0[S]$. Thus, the assertion follows from Lemma 24.

Finally, note that if, in the case described in Theorem 25, the order of the group $U(S)$ is invertible in $K$, then $a = |U(S)|^{-1} \sum_{u \in U(S)} u$ is an idempotent, and we have $K_0[S] \cong aK_0[S] \oplus (1-a)K_0[S] \cong aK_0[S] \oplus \ker(\phi)$, so that the assertion of Theorem 25 is straightforward.

## Comments on Chapter 16

A substantial part of the theory of Frobenius algebras used in this chapter was developed by Nakayama [167]. The study of the self-injectivity conditions for $K[S]$, beyond the group algebra case discussed in [203], § 3.2, was started by Ponizovskii [213] and Wenger [270] with the case of Frobenius algebras of finite commutative semigroups. The approach of Ponizovskii and Wenger continued in [214] and [272], respectively, has led to some module theoretic conditions on $K[S]$ not presented here. The case of finite inverse, and more generally regular, semigroups (covered by Theorem 17) was considered by Ponizovskii in [213a] and [216] and by Wenger in [270] and [271]. In [273], an error in the main result of [213] was pointed out, but some related results were obtained. Grassman [74] made an effort toward generalizing some of the above module theoretic conditions to the case of Frobenius semigroup algebras of not necessarily commutative semigroups over the field of complexes. A special class of Frobenius semigroup algebras was discussed by Sato in [241].

The first attempt to study self-injectivity in the case of not necessarily finite semigroups was given by Kozhukhov in [118], where the assertion of our Theorem 17 was obtained for the class of inverse semigroups. This was generalized to the case of regular self-injective semigroup algebras by Okniński in [178]. The special case of inverse semigroups was also settled by Shoji in [250]. The preparatory lemmas leading to Theorem 11 come from [179], while the use of Osofsky's theorem to derive Theorem 12 was suggested to the author by Menal. Zimmermann extended Theorem 2 to the class of algebraically compact group algebras in [277]. A systematic study of semigroup algebras with this property was begun by Kljushin and Kozhukhov in [115], from which Lemma 15 is extracted. Note that it was erroneously claimed in [135] that the semigroup algebra of a countable infinite semigroup of left zeros is right self-injective; see Lemma 1 (ii). However, this algebra is right algebraically compact.

Finally, observe that, from the results of Menal [153], it follows that, for any $K$-algebra $R$ and any nonalgebraic field extension $F \subseteq K$, the fact that $R \otimes_F K$ is right self-injective implies that $R$ is a right self-injective, semilocal, algebraic $K$-algebra. This is mentioned in connection with our Corollary 7 and the fact that $S$ is locally finite whenever $K[S]$ is right self-injective.

# 17

# Other Finiteness Conditions: A Survey

A number of finiteness conditions, other than those discussed in the previous chapters, have been investigated in the class of group and semigroup rings. The results obtained and the techniques used are very different from those presented here. The main reason is that they are expressed by some "qualitative" conditions on the set of ideals of a given algebra rather than the "quantitative" ones that we are concerned with. In particular, none of those conditions implies the periodicity of the underlying semigroup. We give a brief survey of the main results concerning the most important finiteness conditions of this type.

Recall that a unitary algebra $R$ is a right (left) fir if any right (left) ideal of $R$ is free and of unique rank, as a right (left) $R$-module. $R$ is a fir if it is both a right and left fir. If the same condition is imposed on the set of right finitely generated ideals of $R$ only, then $R$ is called a semifir. It is known that this condition is right-left symmetric [31]. If any right (left) ideal of $R$

is projective as an $R$-module, then $R$ is called right (left) hereditary, and $R$ is right (left) semihereditary if any finitely generated right (left) ideal of $R$ is projective. For the general theory of algebras of these types, we refer to [31]. If $R$ is a commutative domain, then, as is well known, $R$ is hereditary if and only if $R$ is a Dedekind domain, and $R$ is semihereditary if and only if $R$ is a Prüfer domain. It is clear that we have the following implications:

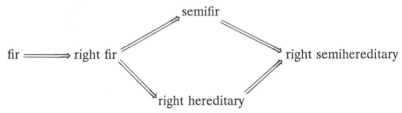

as well as their left-symmetric analogs. Observe that the class of regular algebras, studied in Chapter 15, is contained in the class of right (and left) semihereditary algebras.

Recall that, since in a group ring $K[G]$, a right-left symmetry is established by the involution given by $g \to g^{-1}$, $g \in G$, then there is no distinction between the right, and the appropriate left, properties of $K[G]$.

We start with the strongest of the above-considered conditions. The following result is due to Wong [274].

**Theorem 1**  Let $S$ be a monoid. Then $K[S]$ is a fir if and only if $S$ is a free product of a free group and a free monoid.

The sufficiency of the above condition is well known and was established by many authors; see [31]. Some examples of monoid algebras that are right firs but not left firs were given by Cohn in [31] and by Skornjakov in [253]. In particular, the monoid $S$, defined by generators $y, x_i, i = 1, 2, \ldots$, subject to the relations $yx_i = x_{i-1}$, leads to a semigroup algebra $K[S]$ that is of this type.

A semigroup $S$ is said to be rigid if, for every $s, t \in S$, the condition $sS \cap tS \neq \varnothing$ implies that either $sS \subseteq tS$ or $tS \subseteq sS$. It is known that this condition is right-left symmetric and that any cancellative rigid semigroup embeds into a group; see [31], Section 0.8. Cohn showed, in [31], Theorem 2.10.3, that $K[S]$ is a right fir whenever $S$ is a rigid monoid with a.c.c. on principal right ideals, satisfying the condition

(a)                                     $st = 1 \Longrightarrow s = t = 1$

Monoids satisfying the latter condition are called conical. This motivated the work done in [119], where one-sided firs were completely characterized.

That conditions (1–5) given below, together with the left analogs of (3) and (5), imply that $S$ is a free product of a free group and a free monoid was, in fact, used to derive the necessity in Theorem 1. The theorem below is due to Kozhukhov [119].

**Theorem 2** Let $S$ be a monoid. Then $K[S]$ is a right fir if and only if the following conditions hold:

(1)  $S$ is cancellative.
(2)  The group of units $U(S)$ is a free group.
(3)  $S$ has a.c.c. on principal right ideals.
(4)  $S$ is rigid.
(5)  For any $s \in S \setminus U(S)$, $t \in S$, and $u \in U(S)$, the equality $su = ts$ implies that $u = 1$.

It is worth noting that, if $S$ satisfies the above conditions, then there is an embedding $\phi : S \to G$ into a free group $G$ on a set $X$ such that $X \subseteq \phi(S)$.

The group algebras that are semifirs were completely described by Dicks and Menal in [43].

**Theorem 3** Let $G$ be a group. Then $K[G]$ is a semifir if and only if $G$ is locally free, that is, any finitely generated subgroup of $G$ is free.

It may be deduced from Theorem 2 that, if $S$ is a directed union of free products of free groups and free monoids, then $K[S]$ is a semifir. It was conjectured by Dicks that the converse also holds. This was confirmed in some special cases. First of all, the conditions of Theorem 2 imply that $S$ has the above property. This is also the case if $S$ is finitely generated over its group of units $U(S)$. (Then $S$ appears to be a free product of a locally free group $U(S)$ and a free monoid [152].) Some further evidence that the conjecture might be true was given by Dicks and Schofield in [44]. Namely, it was shown that, if $K[S]$ is a semifir, then the universal group of $S$ is locally free. On the other hand, in terms of the conditions listed in Theorem 2, Cohn asked whether $K[S]$ is a semifir whenever $S$ is conical, rigid, and irreflexive; that is, $usv = s$ for $u, v \in S$ and a nonunit $s \in S$ implies that $u = v = 1$. An example of a conical and rigid monoid $S$ such that $K[S]$ is not a semifir for every $K$ was constructed by Cedo in [18].

Hereditary group algebras were described by Dicks in [40]. For stating this result, we need some preparation. Let $X$ be an oriented graph with the nonempty set of vertices $V(X)$ and a set of edges $E(X)$. By $(i,t) : E(X) \to V(X) \times V(X)$, we denote the incidence map of $X$, so that, for any $e \in E(X)$, $i(e)$ and $t(e)$ are the initial and terminal vertices of $e$, respectively. We assume that $X$ is a connected graph. $X$ may be viewed as a small category with the object set $\mathrm{ob}(X) = E(X) \cup V(X)$ and nonidentity morphisms $i_e : e \rightarrow i(e)$, $t_e : e \rightarrow t(e)$ for $e \in E(X)$. Any functor $\mathfrak{g} : X \rightarrow$ *Groups* into the category of groups is called a connected graph of groups. The homomorphisms $\mathfrak{g}(i_e) : \mathfrak{g}(e) \rightarrow \mathfrak{g}(i(e))$ are denoted by $g \rightarrow g^{i_e}$, and a similar notation is used for $t_e$. Let $T$ be a spanning tree for $X$, that is, a subgraph with the same vertex set and with a minimal edge set so that the subgraph is still connected. The fundamental group of $\mathfrak{g}$ with respect to $T$ is defined as the group $\pi(\mathfrak{g}, T)$ universal with the following properties:

(1)  For each vertex $v$ of $X$, there is a group homomorphism $\mathfrak{g}(v) \rightarrow \pi(\mathfrak{g}, T)$.
(2)  For each edge $e$ of $X$, there is an element $q(e)$ of $\pi(\mathfrak{g}, T)$ such that $q(e)^{-1} g^{i_e} q(e) = g^{t_e}$ for all $g \in \mathfrak{g}(e)$, and if $e$ is an edge of $T$, then $q(e) = 1$.

Since the isomorphism class of $\pi(\mathfrak{g}, T)$ appears to be independent of the choice of $T$, one may speak of the fundamental group $\pi(\mathfrak{g})$ of $\mathfrak{g}$ with no reference to a spanning tree.

For more details, proofs, and applications, the reader is referred to [40] and [41].

**Theorem 4**  Let $G$ be a group. Then $K[G]$ is hereditary if and only if

(i)  $G$ is the fundamental group of a connected graph of finite groups with orders invertible in $K$.

Moreover, if $G$ is finitely generated, then the above is equivalent to any of the following conditions:

(ii)  $G$ has a free subgroup of finite index, and the orders of finite subgroups of $G$ are invertible in $K$.
(iii)  $G$ is a fundamental group of a finite connected graph of finite groups of orders invertible in $K$.

The more difficult "only if" part of the above theorem is, in fact, due to a result of Dunwoody, see [40], which establishes the assertion on $G$ if the augmentation ideal $\omega(K[G])$ of $K[G]$ is a projective $K[G]$-module. The converse is obtained by showing that whenever $G$ is of type (i), then $r \, \mathrm{gl} \, \dim_{K[G]} K \leq 1$, where $K$ is viewed as a $K[G]$-module with trivial $G$-action. (Observe that, for every group $G$, we have $r \, \mathrm{gl} \, \dim K[G] \leq r \, \mathrm{gl} \, \dim_{K[G]} K$; see [203], Theorem 10.3.6.) We note that it is known that a fundamental group $G$ of a connected graph of finite groups has no free subgroups of rank 2 if and only if either of the following holds:

$G$ is countable locally finite.

$G$ is finite-by-(cyclic infinite).

$G$ is finite-by-$Z \coprod Z$; see [41].

In the case where $G$ is a nilpotent group, the description simplifies considerably. This was first proved by Goursaud and Valette in [72], but it may also be easily derived from the above theorem and from remarks.

**Theorem 5**  Let $G$ be a nilpotent group. Then $K[G]$ is hereditary if and only if either of the following holds:

(i)   $G$ is finite-by-(cyclic infinite), and the order of the torsion subgroup of $G$ is invertible in $K$.

(ii)  $G$ is locally finite and countable, and the order of every element of $G$ is invertible in $K$.

Hereditary monoid rings were considered by Cheng and Wong in [21], with a restriction to a class of cancellative monoids. The following extension of Theorem 1 was obtained.

**Theorem 6**  The following conditions are equivalent for a monoid $S$:

(i)   $K[S]$ is a hereditary domain.

(ii)  $K[S]$ is hereditary, and $S$ is torsion-free and weakly cancellative.

(iii) $S$ is a free product of a free group and a free monoid.

(iv)  $K[S]$ is a fir.

Here, "torsion-free" means that, for any $a \in S$, we have $a = 1$ whenever $a^n = 1$ for some $n \geq 1$, and "weakly cancellative" means that either of the equalities $ab = a$, $ba = a$ implies that $b = 1$ and that $aub = ab$ for some $u \in U(S)$ implies that $u = 1$.

Semihereditary monoid rings have been extensively studied, but conclusive results are concerned with some special classes of semigroups [24], [44], [130]. The group algebra case was settled by Dicks and Schofield in [44].

**Theorem 7**  The group algebra $K[G]$ is semihereditary if and only if the following conditions hold:

(1)  The order of every finite subgroup of $G$ is invertible in $K$.

(2)  Every finitely generated subgroup of $G$ has a free subgroup of finite index.

The result in the special case of locally nilpotent groups was first established in [72].

The result in the special case of locally nilpotent groups was first established in [72].

**Theorem 8**  Let $G$ be a locally nilpotent group. Then $K[G]$ is semihereditary if and only if the order of every finite subgroup of $G$ is invertible in $K$, and $G/T$ embeds into the additive group of rational numbers, where $T$ denotes the torsion subgroup of $G$.

The next result is taken from the paper of Kuzmanovich and Teply [130].

**Theorem 9**  Let $S = \bigcup_{\alpha \in \Gamma} G_\alpha$ be a semilattice $\Gamma$ of torsion groups $G_\alpha$. Then $K[S]$ is right semihereditary if and only if each $G_\alpha$ is locally finite and has no elements of order noninvertible in $K$.

The last of the above-mentioned classes of semigroups for which the semihereditariness of $K[S]$ is entirely described in terms of $S$ consists of commutative semigroups. However, the description obtained by Chouinard et al. [24] is highly complex and not easy to state. Here, we formulate the result only in the fundamental case of cancellative semigroups (due to Hardy and Shores) and briefly sketch the procedure reducing the general case to this one.

**Theorem 10**  Let $S$ be a commutative cancellative monoid, and let $T$ be the torsion subgroup of the group of fractions of $S$. Then $K[S]$ is semihereditary if and only if the following conditions hold:

(1)  $\mathrm{ch}(K) = 0$ or $S$ is $p$-separative if $p = \mathrm{ch}(K) > 0$ (the latter equivalent to the fact that $T$ has no $p$-torsion).

(2)  $T \subseteq S$ and $S/{\sim}_T$ is isomorphic to a subgroup of the additive group of rational numbers or to the positive cone of such a group. (${\sim}_T$ is the congruence on $S$ determined by $T$.)

The general case involves a rather complicated procedure of deriving some cancellative semigroups from a given separative semigroup. Condition (4) in the next result refers to this class of semigroups. Recall that a commutative semigroup $S$ is a semilattice $\Gamma$ of its archimedean components $S_\alpha$, $\alpha \in \Gamma$, which are cancellative if $S$ is separative; see [26] or our Chapter 6. For any $\alpha \le \beta \in \Gamma$, we denote by $\phi_{\alpha,\beta}^{\mathrm{red}}$ the homomorphism $S_\alpha/{\sim}_{T_\alpha} \to S_\beta/{\sim}_{T_\beta}$, where $T_\alpha$, $T_\beta$ are the torsion subgroups of the groups of fractions $G_\alpha$, $G_\beta$ of $S_\alpha$, $S_\beta$, respectively, which is determined by the structural homomorphism $\phi_{\alpha,\beta} : G_\alpha \to G_\beta$. This notation is used below.

**Theorem 11**  Let $S$ be a commutative monoid. Then $K[S]$ is semihereditary if and only if the following conditions hold:

(1)   $\mathrm{ch}(K) = 0$ or $S$ is $p$-separative if $p = \mathrm{ch}(K) > 0$.
(2)   For any $\alpha \in \Gamma$, $\Gamma_{(\alpha)} = \{\beta \in \Gamma \mid \beta \leq \alpha$ and $\phi_{\alpha,\beta}^{\mathrm{red}}$ is trivial $\}$ is empty or
       is a finitely generated ideal of $\Gamma$.
(3)   For any $\alpha \in \Gamma$, the congruence $\tau_\alpha = \{(s,t) \in S \times S \mid se_\alpha = te_\alpha\}$, where
       $e_\alpha$ is the identity of $G_\alpha$, is finitely generated.
(4)   $K[T]$ is semihereditary for every $T$ belonging to a certain family of
       cancellative monoids derived from $S$.

Let us note the family of cancellative monoids arising in condition (4) contains all groups $G_\alpha$, $\alpha \in \Gamma$, so that every such group satisfies the conditions of Theorem 5. In particular, every $G_\alpha$ has torsion-free rank not exceeding 1.

We mention two particular cases, in which the description given in Theorem 11 simplifies considerably. The proofs were independently given in [130] as consequences of Theorem 9.

**Theorem 12** Let $S$ be a periodic commutative semigroup. Then $K[S]$ is semihereditary if and only if $S$ is a semilattice of groups and the order of every element of $S$ is invertible in $K$.

**Theorem 13** Let $S = \bigcup_{\alpha \in \Omega} G_\alpha$ be a commutative monoid that is a semilattice $\Omega$ of groups $G_\alpha$, $\alpha \in \Omega$, at least one of which is not torsion. Then $K[S]$ is semihereditary if and only if the following conditions hold:

(1)   The order of every element of $G_\alpha$, $\alpha \in \Omega$, is invertible in $K$.
(2)   $G_\alpha$ has torsion-free rank at most 1 for $\alpha \in \Omega$.
(3)   The set $\{\beta \in \Omega \mid \beta \leq \alpha, \ker(\phi_{\alpha,\beta})$ is not torsion$\}$ is either empty or
       else has a finite set of maximal elements such that every member of
       the set is exceeded by at least one maximal element; $\alpha \in \Gamma$.

For an extensive study of commutative monoid rings that are Bezout, arithmetical, Prüfer, principal ideal rings, or that belong to some related classes of rings, we refer to Gilmer's monograph [64].

Various conditions for semigroup rings of torsion-free commutative and cancellative semigroups have been studied in a series of papers by Matsuda [144]. A generalization of the class of Krull rings was investigated for semigroup rings by Jespers and Wauters in [104].

Finally, we note that most of the results of this chapter were proved in the more general setting of monoid rings with an arbitrary coefficient ring.

# Part IV

# Semigroup Algebras Satisfying Polynomial Identities

Associative algebras satisfying polynomial identities have been very extensively studied and, in fact, form one of the main fields of research in ring theory. They have proved to be a class of well-behaved algebras (particularly with respect to structure theorems, the theory of dimensions, and properties of the radical) and useful for the general theory of algebras and their representations. Our starting point in this part is the beautiful description of group algebras of this type. On the other hand, we tend to generalize some of the results on the commutative semigroup algebras presented in Gilmer's monograph [64].

After recalling the basic general results used in the sequel, we start in Chapter 19 with a natural semigroup theoretic consequence of the $PI$-property for $K[S]$—the so-called permutational property for semigroups. Then we turn to the problem of determining necessary and sufficient conditions for $K[S]$ to satisfy a polynomial identity. As was the case for several conditions studied in Part III, the $PI$-property for $K[S]$ can sometimes be studied, and characterized, "locally", via the algebras of the principal factors of $S$. While the general problem is entirely solved for semigroups of some important types only, a complete description of the special class of Azumaya semigroup algebras may be expected. This topic is presented in Chapter 25. The purpose of the remaining chapters is to show that much more information may be given on some specific problems concerning $PI$-algebras, when restricted to semigroup algebras. This is especially the case with the theory of dimensions and the Jacobson radical studied in Chapters 21 and 23. The class of so-called monomial algebras, arising independently from various problems in ring theory, is discussed in Chapter 24.

# 18

# Preliminaries on PI-Algebras

For any integer $m \geq 1$, let $K\langle x_1,\ldots,x_m \rangle$ denote the free $K$-algebra in $m$ free generators $x_1$, ..., $x_m$. A $K$-algebra $R$ is said to satisfy a polynomial identity (shortly, $R$ is a $PI$-algebra) if there exists an integer $n$ and a nonzero element $f = f(x_1,\ldots,x_n) \in K\langle x_1,\ldots,x_n \rangle$ such that $f(a_1,\ldots,a_n) = 0$ for every $a_1$, ..., $a_n \in R$. Commutative algebras and nilpotent algebras are the basic trivial examples of $PI$-algebras. It is also well known, and not hard to prove, that any finite dimensional $K$-algebra satisfies a polynomial identity; see [203], Lemma 5.1.6.

In Part IV, we discuss the results known for $PI$-semigroup algebras. Our main aim is to extend some facts on $PI$-group algebras and, on the other hand, on semigroup algebras of arbitrary commutative semigroups. Thus, the monographs of Passman [203] and Gilmer [64] are the basic references throughout Part IV.

We will repeatedly use the following fundamental results on the semi-simple and radical parts of a $PI$-algebra.

**Theorem 1**   Let $R$ be a $K$-algebra satisfying a polynomial identity of degree $n$. Then,

   (i)   For every prime ideal $P$ of $R$, the localization of $R/P$ with respect to its center is isomorphic to the matrix algebra $M_r(D)$ over a division $K$-algebra $D$ such that $\dim_{Z(D)} M_r(D) \leq (n/2)^2$. Moreover, $R/P$ embeds into $M_N(L)$ for a field $L \supseteq K$, and an integer $N \leq n/2$.
  (ii)   If $P$ is a right primitive ideal of $R$, then $R/P \cong M_r(D)$ with $D$, $r$ as above.
 (iii)   The algebra $M_r(L)$ satisfies an identity of degree $2r$.
 (iv)   For any nil ideal $I$ of $R$, we have $\mathcal{J}(I) = \mathcal{B}(I)$ and, if $R$ is a finitely generated $K$-algebra, then $\mathcal{J}(R)$ is a nilpotent ideal of $R$.

The first assertion in (i) is the strong version of the theorem of Posner; see [203], Theorem 5.4.10. Part (ii), together with the latter assertion in (i), is due to Kaplansky; see [203], Theorem 5.3.4. (Here $L$ is the splitting field of $D$.) (iii) is the Amitsur–Levitzki theorem; see [203], Theorem 5.1.9. Finally, (iv) is [203], Lemma 5.4.7, and the celebrated theorem of Braun [17].

The fact that every irreducible representation of $S$ over a division algebra may be replaced by an irreducible representation of $S$ over a field will be often used. This is a refinement of statement (i) above, which may be formulated as follows.

**Proposition 2**   Assume that $K[S]$ is a $PI$-algebra and $M$ a primitive ideal in $K[S]$. Then there exists an irreducible representation $\phi : S \to M_n(L)$ such that $(K[S]/M) \otimes_F L \cong M_n(L)$, where $F$ is the center of $K[S]/M$, $L$ is a splitting field for $K[S]/M$, and $\phi$ factors through the natural homomorphism $K[S] \to K[S]/M$.

The problem of describing group algebras satisfying polynomial identities has been completely settled. The following result due to Isaacs and Passman if $\mathrm{ch}(K) = 0$, and to Passman if $\mathrm{ch}(K) > 0$, is the starting point for our considerations.

**Theorem 3**   Let $G$ be a group. Then,

   (i)   $K[G]$ is a $PI$-algebra if and only if:

(1)  $G$ is abelian-by-finite, that is, $G$ has an abelian normal subgroup $A$ of finite index, if $\mathrm{ch}(K) = 0$.

(2)  $G$ is $p$-abelian-by-finite, that is, $G$ has a normal subgroup $A$ of finite index such that the commutator subgroup $A'$ is a $p$-group, if $\mathrm{ch}(K) = p > 0$.

In this case, the $FC$-subgroup $\Delta(G)$ of $G$ has finite index in $G$, and the commutator subgroup $\Delta(G)'$ is finite. Moreover, there is a bound (dependent on the degree of the identity satisfied in $K[G]$ only) on the product $[G : A]|A'|$.

(ii)  If $K[G]$ is a $PI$-algebra, then $\mathcal{J}(K[G]) = \mathcal{B}(K[G])$ and $\mathcal{J}(K[G]) = 0$ if $\mathrm{ch}(K) = 0$ or $\mathrm{ch}(K) = p > 0$ and $G$ has no normal subgroups of order divisible by $p$. If $\mathrm{ch}(K) = p > 0$ and the maximal normal $p$-subgroup $G_p$ of $G$ is finite, then $\mathcal{J}(K[G])$ is a nilpotent ideal.

(iii)  For all elements $g$, $h \in G$, we have $(g,h) \in \rho_{\mathcal{J}(K[G])}$ if and only if $g$, $h$ are in the same coset of the maximal normal $p$-subgroup $G_p$ of $G$, where $p = \mathrm{ch}(K) > 0$.

For the reader's convenience, we briefly mention the results of [203] from which Theorem 3 may be extracted. The assertions in (i) are established in Theorems 5.3.7, 5.3.9, and 5.2.14 in [203].

Assume that $K[G]$ is a $PI$-algebra. If $\mathrm{ch}(K) = 0$, then $\mathcal{J}(K[G]) = \mathcal{J}(K[A])K[G]$ by Theorem 7.2.7, and since $\mathcal{J}(K[A]) = 0$, by Corollary 7.3.2, then $\mathcal{J}(K[G]) = 0$. (The references in this and the following paragraph are to [203].)

Assume that $\mathrm{ch}(K) = p > 0$. If $(g,h) \in \rho_{\mathcal{J}(K[G])}$, $g$, $h \in G$, then $gP = hP$, where $P = \{x \in G \mid 1 - x \in \mathcal{J}(K[G])\}$ is a normal $p$-subgroup of $G$ by Lemma 10.1.13. On the other hand, since $[G : \Delta(G)] < \infty$, and $G_p \cap \Delta(G)$ is a locally finite group by Lemma 4.1.5, then $G_p$ is locally finite. Thus, $G_p$ coincides with the least normal locally finite $p$-group $H \subseteq G$ such that $\Delta(G/H)$ has no nonidentity finite normal $p$-subgroups. From Theorem 8.4.16 it then follows that $\omega(K[G_p]) \subseteq \mathcal{B}(K[G])$, so that the elements of $G_p$ lie in the same class of the congruence $\rho_{\mathcal{B}(K[G])} \subseteq \rho_{\mathcal{J}(K[G])}$, and (iii) holds. Now, by (i), $G/G_p$ has an abelian subgroup of finite index $A/(A \cap G_p)$ that has no elements of order $p$. Thus, again, Theorem 7.2.7 and Corollary 7.3.2 imply that $\mathcal{J}(K[G/G_p])$ is a nilpotent ideal. This and the inclusion $\omega(K[G_p]) \subseteq \mathcal{B}(K[G])$ imply that $\mathcal{B}(K[G]) = \mathcal{J}(K[G])$. If $G_p$ is finite, then the nilpotency of $\mathcal{J}(K[G])$ follows because $\omega(K[G_p])K[G]$ is nilpotent by Lemma 3.1.6 (see Corollary 8.1.14). Finally, if $G$ has no normal subgroups of order divisible by $p$, then the above, together with Connell's theorem, implies that $\mathcal{J}(K[G]) = 0$.

We note that the equality $\mathcal{J}(K[G]) = \mathcal{B}(K[G])$ may also be derived from Theorem 1 (iv) because, for every $a \in \mathcal{J}(K[G])$, we have $a \in \mathcal{J}(K[G]) \cap K[\langle\!\langle \mathrm{supp}(a) \rangle\!\rangle] \subseteq \mathcal{J}(K[\langle\!\langle \mathrm{supp}(a) \rangle\!\rangle])$ (see Corollary 16 in our Chapter 4) or directly from Corollary 5 in Chapter 11.

It is well known that, if an algebra $R$ satisfies an identity of degree $n$, then it satisfies a multilinear identity of degree $n$; see [203], Lemma 5.1.1. This, and the standard argument as in [203], Lemma 5.1.3, allows us to derive the following consequence.

**Proposition 4** Let $F$ be the prime subfield of the field $K$, and let $S$ be a semigroup. Then the algebras $K[S]$ and $F[S]$ satisfy the same multilinear polynomial identities.

The above shows that, as in the group algebra case, one may expect a characterization of the $PI$-property for $K[S]$ in terms of $S$ and the characteristic of $K$ only.

We will often use the following observation.

**Proposition 5** Assume that $S$ has a zero element. Then $K[S]$ is a $PI$-algebra if and only if $K_0[S]$ is a $PI$-algebra. Moreover, if $S$ is not a nil semigroup, then the algebras $K[S]$, $K_0[S]$ satisfy the same multilinear identities.

*Proof.* The former assertion follows from the fact that $K[S] \cong K_0[S] \oplus K$; see Corollary 9 in Chapter 4. Let $f(x_1, \ldots, x_n)$ be a nonzero multilinear polynomial satisfied in $K_0[S]$. Then, for every $s_1, \ldots, s_n \in S$, $f(s_1, \ldots, s_n) = \alpha\theta$ in $K[S]$ for some $\alpha \in K$. Obviously, $\alpha$ is the sum of all coefficients in $f(x_1, \ldots, x_n)$. Let $s \in S$ be a non-nilpotent element. Then $f(s, \ldots, s) = \alpha s^n$, and so $\alpha = 0$. This shows that $f(s_1, \ldots, s_n) = 0$ in $K[S]$ and, since $S$ is a $K$-basis of $K[S]$ and $f$ is multilinear, then $f$ is an identity in $K[S]$.

Let us also note the following standard connection between the $PI$-properties in distinct characteristics.

**Proposition 6** Assume that $\mathbb{Q}[S]$ satisfies a polynomial identity of degree $n$. Then, for every prime $p$, $\mathbb{F}_p[S]$ satisfies an identity of degree $n$.

*Proof.* Obviously $\mathbb{Q}[S]$ satisfies an identity $f$ of degree $n$ with integral coefficients. Since we may assume that these coefficients are relatively prime, then $f \pmod{p}$ is a desired identity in $\mathbb{F}_p[S]$.

## Comments on Chapter 18

We refer to the monographs of Jacobson [94], Procesi [222], and Rowen [239] for the general theory of $PI$-algebras. However, we send the reader to [203] whenever possible because this is the main reference to the results on group algebras as well. A presentation of the deep theorem of Braun on the radical of a $PI$-algebra, together with the intermediate steps leading to it (due to other authors), can only be found in the original paper [17]. The study of the $PI$-property for semigroup algebras of semigroups other than groups was begun by Domanov in [46], where Proposition 5 was first used.

# 19

# Semigroups Satisfying Permutational Property

Let $n \geq 2$ be an integer. We say that a semigroup $S$ has the property $\mathfrak{P}_n$ if, for any elements $s_1, \ldots, s_n \in S$, there exists a nontrivial permutation $\sigma$ in the symmetric group $\mathcal{S}_n$ such that $s_1 \ldots s_n = s_{\sigma(1)} \ldots s_{\sigma(n)}$. Obviously, $\mathfrak{P}_2$ is the commutativity and $\mathfrak{P}_2, \mathfrak{P}_3, \ldots$ a chain of successively weaker properties. It is easy to construct examples of semigroups satisfying $\mathfrak{P}_{n+1}$ but not $\mathfrak{P}_n$. For example, $S = X/I$, where $X$ is a free semigroup on free generators $x_1, x_2, \ldots, x_n$ and $I$ the ideal consisting of all words of length $n + 1$, is of this type. We say that $S$ has the permutational property (shortly, $S$ has $\mathfrak{P}$) if some $\mathfrak{P}_n$, $n \geq 2$, is satisfied in $S$.

Our interest in the class of semigroups satisfying the permutational property comes from the following simple observation.

**Proposition 1** Assume that $K[S]$ satisfies a polynomial identity of degree $n$. Then $S$ has the property $\mathfrak{P}_n$.

*Proof.* Since we know that $K[S]$ satisfies a multilinear polynomial identity of degree $n$, then there exist scalars $\alpha_\sigma \in K$ such that $x_1 \ldots x_n = \sum_{1 \neq \sigma \in S_n} \alpha_\sigma x_{\sigma(1)} \ldots x_{\sigma(n)}$ for all $x_1, \ldots, x_n \in K[S]$. Choosing the $x$ from $S$ only, we see that $\alpha_\sigma \neq 0$ for some $\sigma$. Thus, the $K$-independence of the elements of $S$ implies that $S$ has the property $\mathfrak{P}_n$.

Since, in view of Proposition 4 in Chapter 18, if one of the algebras $K[S]$, $K_0[S]$ satisfies a polynomial identity, so does the other, the permutational property of $S$ also follows if we assume that $K_0[S]$ is a $PI$-algebra.

Observe that, if the multiplicative semigroup of an algebra $R$ has the property $\mathfrak{P}_n$, then $R$ is a $PI$-algebra since it satisfies the identity $\prod_{1 \neq \sigma \in S_n} (x_1 \ldots x_n - x_{\sigma(1)} \ldots x_{\sigma(n)})$ (the product taken over any fixed ordering of $S_n \setminus \{1\}$), with all coefficients equal to $\pm 1$, hence nontrivial. This suggests the question: To what extent is the property $\mathfrak{P}$ (imposed on the multiplicatively closed basis $S$ of the algebra $K[S]$ only) responsible for the $PI$-property of $K[S]$?

An analogy between the $PI$-property for algebras and the property $\mathfrak{P}$ for semigroups results from the celebrated Shirshov theorem on factorization of words in a finitely generated semigroup; see [17], Appendix, and [252], 5.§.2. This may be stated as follows.

**Theorem 2** Let $T = \langle s_1, \ldots, s_r \rangle$ be a finitely generated semigroup satisfying the property $\mathfrak{P}_n$, $n \geq 2$. Then there exist $p \geq 1$ and elements $v_1, \ldots, v_p \in T$, which are words of length $\leq n$ in $s_1, \ldots, s_r$, and there exist integers $k$, $N \geq 1$ such that any element $s \in S$ of length $\geq N$ (as a word in $s_1, \ldots, s_r$) may be written in the form

(a) $\qquad w(s) = v_{i_1}^{j_1} \ldots v_{i_k}^{j_k} \qquad$ where $j_m \geq 0$, $m = 1, \ldots, k$

Moreover, all $p$, $N$, and $k$ are bounded by a function of $n$, $r$, and the decomposition of each element $w(s)$ of the form (a) with respect to $s_1, \ldots, s_r$ is a permutation of $s$ with respect to $s_1, \ldots, s_r$.

As a consequence, two important analogs of results on general $PI$-algebras may be derived. We first state the Burnside-type result obtained by Restivo and Reutenauer in [233], which corresponds to the Shirshov theorem on $PI$-algebras with algebraic bases; see [239], Theorem 4.2.8.

**Theorem 3** Let $S$ be a periodic semigroup satisfying the permutational property. Then $S$ is locally finite.

*Proof.* If $T = \langle s_1, \ldots, s_r \rangle$ is a finitely generated subsemigroup of $S$, then it has some $\mathfrak{P}_n$, $n \geq 2$. Since $T$ is periodic, then, with the notation of Theorem 2, there exist finitely many elements of the form $v_{i_1}^{j_1} \ldots v_{i_k}^{j_k}$ in $T$. Consequently, $T$ is finite.

Observe also that the following analog of the result on the growth of finitely generated $PI$-algebras (see [121], Corollary 10.7) is a direct consequence of Theorem 2 and the appropriate definitions; see Chapter 8 or [121].

**Theorem 4** Let $S$ be a finitely generated semigroup satisfying the property $\mathfrak{P}$. Then $S$ has polynomial growth, and so the algebra $K[S]$ has finite Gelfand–Kirillov dimension for every field $K$.

*Proof.* We use the notation of Theorem 2. By the type of a word $v_{i_1}^{j_1} \ldots v_{i_k}^{j_k}$ of the form (a) with $i_l \neq i_{l+1}$ for $l = 1, \ldots, k-1$, we mean the $k$-tuple $(i_1, \ldots, i_k)$. Clearly, there are finitely many possible types. Let $m > N$. As in the case of the polynomial algebra (see [121], Example 1.6), one shows that the number of words in $v_1, \ldots, v_p$ of length $\leq m$ of a given type is bounded by a polynomial in $m$ of degree $k$. Since any element of $S$ that is a word of length $m$ in $s_1, \ldots, s_r$ is a word of length $\leq m$ in $v_1, \ldots, v_p$, then the number of elements of $S$ that are words of length $\leq m$ in $s_1, \ldots, s_r$ is bounded by a polynomial in $m$ of degree $k$. This means that $S$ has polynomial growth of exponent not exceeding $k$, establishing the assertion.

Before studying the property $\mathfrak{P}$ for the most important special classes of semigroups, let us state two auxiliary general lemmas. The former uses another nice combinatorial result on so-called repetitive mappings; see [139], Chapter 4.

**Lemma 5** Let $S$ be a semigroup that is a union of its subsemigroups $S_1, \ldots, S_n$. If every $S_i$, $i = 1, \ldots, n$, has the property $\mathfrak{P}$, then $S$ has the property $\mathfrak{P}$.

*Proof.* We know that there exists $m \geq 2$ such that any $S_i$ has $\mathfrak{P}_m$. Let $f : T \to S$ be a homomorphism of a free semigroup $T$ on a set of free generators $X$ onto $S$. Define a function $\phi : T \to \mathbb{N}$ by the rule $\phi(x_1 x_2 \ldots x_k) = $ the least integer $r$ such that $f(x_1) \ldots f(x_k) \in S_r$, $x_i \in X$. Since $\phi(T)$ is a finite set, then, from [139], Theorem 4.1.1, we know that $\phi$ is a repetitive function, that is, for every $j \geq 1$ there exists $N(j) \geq 1$ such that any word $x = x_1 x_2 \ldots x_{N(j)}$, $x_i \in X$, has a factorization $x = u y_1 y_2 \ldots y_j v$, $u, v, y_i \in T$, such that $\phi(y_1) = \phi(y_2) = \cdots = \phi(y_j)$.

Therefore, with $j = m$, we get

$$f(x_1)f(x_2)\ldots f(x_{N(m)}) = f(x) = f(u)f(y_1)\ldots f(y_m)f(v)$$
$$= f(u)f(y_{\sigma(1)})f(y_{\sigma(2)})\ldots f(y_{\sigma(m)})f(v)$$

for some $1 \neq \sigma \in \mathbf{S}_m$. It then follows that $S$ has the property $\mathfrak{P}_{N(m)}$.

**Lemma 6** Let $I$ be an ideal in a semigroup $S$. Then $S$ has the permutational property if and only if the semigroups $I$, $S/I$ have the permutational property.

*Proof.* The necessity is clear. Thus, assume that $I$ has $\mathfrak{P}_r$ and $S/I$ has $\mathfrak{P}_m$ for some $r, m \geq 2$. Put $n = rm$, and let $x_1, \ldots, x_n$ be any elements of $S$. If there exists $i$, $0 \leq i \leq n - m$, such that $x_{i+1}\ldots x_{i+m} \notin I$, then, by the assumption on $S/I$, $x_{i+1}\ldots x_{i+m} = x_{i+\sigma(1)}\ldots x_{i+\sigma(m)}$ for some $1 \notin \sigma \in \mathbf{S}_m$. Consequently,

$$x_1\ldots x_n = x_1\ldots x_i x_{i+\sigma(1)}\ldots x_{i+\sigma(m)} x_{i+m+1}\ldots x_n$$

and we are done in this case. On the other hand, if every product of the subsequent $m$ elements $x_j$ lies in $I$, then putting $y_i = x_{(i-1)m+1}\ldots x_{im}$, $i = 1, \ldots, r$, we may use the property $\mathfrak{P}_r$ in $I$. Thus, $x_1\ldots x_n = y_1\ldots y_r = y_{\tau(1)}\ldots y_{\tau(r)}$ for some $1 \neq \tau \in \mathbf{S}_r$, yielding the assertion in this case.

Our first main result in this chapter will give a characterization of cancellative semigroups with the property $\mathfrak{P}$. We start with the following observation.

**Lemma 7** Let $S$ be a cancellative semigroup satisfying the permutational property. Then $S$ has a two-sided group of fractions.

*Proof.* It is clear that $S$ cannot have noncommutative free subsemigroups. Thus, the assertion follows from Lemma 1 in Chapter 7.

The theorem below was formerly proved for groups by Curzio et al. [34]. Then, using their techniques and Proposition 9 in Chapter 7, Okniński extended it to the present form in [186].

**Theorem 8** The following conditions are equivalent for any cancellative semigroup $S$:

(i) $S$ has the permutational property.
(ii) $S$ has a two-sided group of fractions that is finite-by-abelian-by-finite.

The proof will be split into several lemmas. We start with the simpler implication (ii) $\Rightarrow$ (i).

**Lemma 9** Let $H$ be a subgroup of a group $G$. Assume that $H$ has the property $\mathfrak{P}_r$ and that $H$ has a finite index $m$ in $G$. Then $G$ has the property $\mathfrak{P}_{rm}$.

*Proof.* Put $n = rm$, and consider any elements $x_1, \ldots, x_n \in G$. Also, put $x_0 = 1$. Since $[G : H] = m$, then, among the cosets $H = x_0 H, x_0 x_1 H, \ldots$, $x_0 x_1 \ldots x_n H$, there are $r + 1$ equal cosets, say,

$$x_0 x_1 \ldots x_{i_1} H = x_0 x_1 \ldots x_{i_2} H = \cdots = x_0 x_1 \ldots x_{i_{r+1}} H$$

for some $0 \leq i_1 < i_2 < \cdots < i_{r+1} \leq n$. Then the elements $y_j = x_{i_j+1} \ldots x_{i_{j+1}}, j = 1, \ldots, r$, belong to $H$. The property $\mathfrak{P}_r$ for $H$ allows us to choose a nontrivial permutation $\sigma \in S_r$ such that $y_1 \ldots y_r = y_{\sigma(1)} \ldots y_{\sigma(r)}$. Consequently,

$$
\begin{aligned}
x_1 x_2 \ldots x_n &= x_1 x_2 \ldots x_{i_1} y_1 y_2 \ldots y_r x_{i_{r+1}+1} x_{i_{r+1}+2} \ldots x_n \\
&= x_1 x_2 \ldots x_{i_1} y_{\sigma(1)} y_{\sigma(2)} \ldots y_{\sigma(r)} x_{i_{r+1}+1} x_{i_{r+1}+2} \ldots x_n
\end{aligned}
$$

which shows that $G$ has $\mathfrak{P}_n$.

**Lemma 10** Assume that the commutator subgroup $G'$ of a group $G$ has finite order $m$. Then $G$ has the property $\mathfrak{P}_{m+1}$.

*Proof.* Put $n = m + 1$, and let $x_1, \ldots, x_n$ by any elements of $G$. Since, for any $2 \leq j \leq n$, the elements $x_1 \ldots x_n$, $(x_n x_{n-1} \ldots x_j)(x_1 x_2 \ldots x_{j-1})$ have the same image under the natural homomorphism $G \to G/G'$, then there exist elements $z_j \in G'$ with

$$x_1 x_2 \ldots x_n = (x_n x_{n-1} \ldots x_j)(x_1 x_2 \ldots x_{j-1}) z_j$$

If some $z_j = 1$, $j = 2, 3, \ldots, n$, then we are finished. Thus, assume the contrary. Since $|G' \setminus \{1\}| = m - 1 = n - 2$, then $z_k = z_l$ for some integers $k, l$, satisfying $2 \leq k < l \leq n$. Then,

$$
\begin{aligned}
(x_n x_{n-1} \ldots x_l)(x_1 x_2 \ldots x_{l-1}) z_l &= x_1 x_2 \ldots x_n \\
&= (x_n x_{n-1} \ldots x_{l+1} x_l \ldots x_{k+1} x_k)(x_1 x_2 \ldots x_{k-1}) z_k
\end{aligned}
$$

which implies that $x_1 x_2 \ldots x_{l-1} = (x_{l-1} \ldots x_{k+1} x_k)(x_1 x_2 \ldots x_{k-1})$. Multiplying by $x_l x_{l+1} \ldots x_n$ on the right, we derive the desired assertion in this case, too.

The two foregoing lemmas may be summarized in the following result, which establishes the implication (ii) $\Rightarrow$ (i) in Theorem 8.

**Proposition 11** Let $G$ be a group with a normal subgroup $N$ such that the groups $G/N$ and $N'$ are finite with orders $l, m$, respectively. Then $G$ has the property $\mathfrak{P}_n$, where $n = (m + 1)l$.

The following lemma is the key to proving the converse in Theorem 8.

**Lemma 12**  Let $S$ be a semigroup with the property $\mathfrak{P}_n$, and let $G$ be a group of fractions of $S$. Then the index of the $FC$-center $\Delta(G)$ of $G$ is finite, and it is bounded by a function of $n$.

*Proof.*  Let $F = \Delta(G)$. Obviously, we may assume that $[G : F] > 1$. Consider a sequence of elements $x_1, x_2, \ldots, x_m$, $m \geq 1$, of $S$ such that

(1) $x_{i_1} x_{i_2} \ldots x_{i_k} \notin F$ whenever $i_1, i_2, \ldots, i_k$ are distinct integers in the set $\{1, 2, \ldots, m\}$.

(2) $x_1 x_2 \ldots x_m \neq x_{\sigma(1)} x_{\sigma(2)} \ldots x_{\sigma(m)}$ for all nontrivial $\sigma \in \boldsymbol{S}_m$.

Since $G \neq F$, then $S$ is not contained in $F$. Thus, such a sequence exists for $m = 1$ because one can simply choose $x_1 \in S \setminus F$. Further, while $S$ has $\mathfrak{P}_n$, then sequences of this type do not exist for $m \geq n$. We let $m$ denote the maximal integer subject to the existence of a sequence $x_1, \ldots, x_m \in S$ satisfying (1) and (2). Let $Z$ be the set consisting of $1_G$ and all the elements of the form $x_{i_1}^{-1} x_{i_2}^{-1} \ldots x_{i_k}^{-1}$, where $i_1, \ldots, i_k$ are distinct integers in the set $\{1, \ldots, m\}$. If $x_{m+1}$ is any element of $S \setminus ZF$, then it is easy to see that the sequence $x_1, x_2, \ldots, x_{m+1}$ satisfies (1). Consequently, it does not satisfy (2), and there exists $\sigma \neq 1$ in $\boldsymbol{S}_{m+1}$ such that $x_1 x_2 \ldots x_{m+1} = x_{\sigma(1)} x_{\sigma(2)} \ldots x_{\sigma(m+1)}$. From the choice of $x_1, \ldots, x_m$, it follows that $\sigma(m + 1) \neq m + 1$.

For each $\sigma$ in $\boldsymbol{S}_{m+1}$ such that $\sigma(m + 1) \neq m + 1$, define $A_\sigma$ to be the set (possibly empty) of all $x_{m+1} \in S \setminus ZF$ such that $x_1 x_2 \ldots x_{m+1} = x_{\sigma(1)} x_{\sigma(2)} \ldots x_{\sigma(m+1)}$. Then,

$$S \subseteq \bigcup_{z \in Z} zF \cup \bigcup_\sigma A_\sigma$$

If $A_\sigma \neq \varnothing$, then fix an element $s_\sigma \in A_\sigma$. If $\sigma(i) = m + 1$, then, for any $t \in A_\sigma$, we have

$$x_1 x_2 \ldots x_m s_\sigma = x_{\sigma(1)} x_{\sigma(2)} \ldots x_{\sigma(i-1)} s_\sigma x_{\sigma(i+1)} \ldots x_{\sigma(m+1)}$$
$$x_1 x_2 \ldots x_m t = x_{\sigma(1)} x_{\sigma(2)} \ldots x_{\sigma(i-1)} t x_{\sigma(i+1)} \ldots x_{\sigma(m+1)}$$

Multiplying by $s_\sigma^{-1}$ and $t^{-1}$ on the right, respectively, we deduce that $s_\sigma u_\sigma s_\sigma^{-1} = t u_\sigma t^{-1}$, where $u_\sigma = x_{\sigma(i+1)} x_{\sigma(i+2)} \ldots x_{\sigma(m+1)}$. This shows that $t \in s_\sigma C_G(u_\sigma)$. Consequently,

$$S \subseteq \bigcup_{z \in Z} zF \cup \bigcup_\sigma s_\sigma C_G(u_\sigma)$$

Since $u_\sigma \notin F$ by (1), then, from the right-left symmetric analog of Proposition 9 in Chapter 7, it follows that

$$G = SS^{-1} = \bigcup_{z \in Z} zF$$

Therefore, $[G : F] < |Z|$. For any $r \geq 1$, define $f(r) = \sum_{j=0}^{r} \binom{r}{j} j!$. Then $|Z| \leq f(m) < f(n)$ since $m < n$, and the result follows.

We are now able to complete the proof of Theorem 8. Since $S$ has the property $\mathfrak{P}$, then, by Lemma 7, $S$ has a group of fractions $G$. Lemma 12 implies that the $FC$-center $F$ of $G$ has finite index in $G$. It is then enough to show that $F$ itself is finite-by-abelian-by-finite. Since, in view of Lemma 5 in Chapter 7, $F$ is the group of fractions of $S \cap F$, then we may assume that $G = F$. If $S$ is commutative, then so is $G$, and the result follows. Thus, we may suppose that $S$ has $\mathfrak{P}_n$ but not $\mathfrak{P}_{n-1}$ for some $n > 2$. Then there exist elements $x_1, \ldots, x_{n-1} \in S$ such that $\sigma = 1$ whenever $x_1 x_2 \ldots x_{n-1} = x_{\sigma(1)} x_{\sigma(2)} \ldots x_{\sigma(n-1)}$, $\sigma \in \mathcal{S}_{n-1}$. Since $G$ is an $FC$-group, then the subgroup $N = \bigcap_{i=1}^{n-1} C_G(x_i)$ has finite index in $G$; see [203], Lemma 4.1.5. Take any elements $s, t \in S \cap N$, and define $y_i = x_i$ for $i = 1$, $2, \ldots, n - 2$, and $y_{n-1} = x_{n-1}s$, $y_n = t$. Since $S$ has $\mathfrak{P}_n$, then there exists $1 \neq \sigma \in \mathcal{S}_n$ such that

(b) $\qquad\qquad y_1 y_2 \ldots y_n = y_{\sigma(1)} y_{\sigma(2)} \ldots y_{\sigma(n)}$

Set $\sigma(i) = n$, $\sigma(j) = n - 1$. Suppose that $j < i$. Since $s$, $t$ commute with all the elements $x_1, x_2, \ldots, x_{n-1}$, then the above equality yields $x_1 x_2 \ldots x_{n-1} = x_{\tau(1)} x_{\tau(2)} \ldots x_{\tau(n-1)}$, where $\tau \in \mathcal{S}_{n-1}$ is the permutation defined by $\tau(k) = \sigma(k)$ for $k = 1, 2, \ldots, i - 1$, and $\tau(k) = \sigma(k + 1)$ for the remaining $k$ in $\{1, \ldots, n - 1\}$, if any. Then the choice of the $x_k$ implies that $\tau = 1$. Consequently, $i = n$ and $\sigma = 1$, a contradiction. Thus, we must have $j > i$, and the equality (b) shows that $sts^{-1}t^{-1} \in \langle\langle x_1, \ldots, x_{n-1} \rangle\rangle$. From Lemma 5 and Corollary 8, both in Chapter 7, it then follows that

$$N' = ((S \cap N)(S \cap N)^{-1})' \subseteq \langle\langle x_1, x_2, \ldots, x_{n-1} \rangle\rangle$$

Now, $N'$ is periodic, being the commutator subgroup of an $FC$-group; see [203], Lemma 4.1.5. Moreover, the elements of finite order in a finitely generated $FC$-group form a finite subgroup; see [203], Lemma 4.1.5. Therefore, $N'$ is finite and, since $[G : N]$ is finite, then $G$ is finite-by-abelian-by-finite. This completes the proof of Theorem 8.

In the case where $S$ is finitely generated, we derive the following consequence.

**Corollary 13**   Assume that $S$ is a finitely generated cancellative semigroup. Then the following conditions are equivalent:

(i)   $S$ has the permutational property.
(ii)  $S$ has an abelian-by-finite group of fractions.

(iii)   $K[S]$ is a $PI$-algebra.

*Proof.*   In view of Theorem 3 in Chapter 18 and Proposition 1, it is enough to show that (i) implies (ii). Obviously, the group of fractions $G$ of $S$ is finitely generated. Since Lemma 12 implies that the $FC$-subgroup $F$ of $G$ has finite index in $G$, then $F$ is finitely generated, too. From Lemma 7 in Chapter 7, we know that the center of $F$ has a finite index in $F$, and the result follows.

From the proof of the implication (i) $\Rightarrow$ (ii) in Theorem 8, it follows that any $FC$-group with the property $\mathfrak{P}$ has a finite commutator subgroup. Thus, Lemma 10 implies that the class of groups with finite commutator subgroups coincides with the class of $FC$-groups satisfying $\mathfrak{P}$.

Corollary 13 may be used to derive a characterization of noetherian $PI$-algebras $K[S]$ of cancellative semigroups. We first state an application of Theorem 6 in Chapter 12 to the case of $PI$-algebras.

**Theorem 14**   Assume that $S$ is a semigroup such that $K[S]$ is a right noetherian $PI$-algebra. Then $S$ is finitely generated.

*Proof.*   Every cancellative subsemigroup of a homomorphic image of $S$ has a finite-by-abelian-by-finite group of fractions by Proposition 1 and Theorem 8. Thus, the assertion follows from Theorem 6 in Chapter 12.

**Theorem 15**   Let $S$ be a cancellative monoid. Then the following conditions are equivalent:

(i)    $K[S]$ is a right noetherian $PI$-algebra.
(ii)   $K[S]$ is a right and left noetherian $PI$-algebra.
(iii)  $S$ is finitely generated, has a.c.c. on right ideals, and satisfies the permutational property.
(iv)   $S$ is a finitely generated subsemigroup of an abelian-by-finite group, and $S$ has a.c.c. on right ideals.

*Proof.*   (i) implies, in view of Theorem 14, that $S$ is finitely generated. Thus, the implications (i) $\Rightarrow$ (iii) $\Rightarrow$ (iv) follow; see Corollary 13.

Assume that (iv) holds. Then $S$ is a subsemigroup of a finitely generated group $G = \langle\langle S \rangle\rangle$ with an abelian normal subgroup $H$ of finite index. Since $G$ is polycyclic-by-finite, then $K[S]$ is right noetherian by Theorem 9 in Chapter 11. From Lemma 1 in Chapter 11, it follows that $K[S \cap H]$ is noetherian. Thus, $K[S]$ is also left noetherian by the left-right symmetric analog of Lemma 1 in Chapter 11. Therefore, (iv) implies (ii). The remaining implication (ii) $\Rightarrow$ (i) is trivial.

Observe that none of the conditions in (iii) above may be omitted. For the property $\mathfrak{P}$ and finite generacy, appropriate examples can be easily found within the class of groups. Let $G$ be the group defined by generators $x$, $y$, $z$, subject to the relations $zxz^{-1} = y$, $zyz^{-1} = x$, $xy = yx$. Then $H = \langle\langle x,y,z^2 \rangle\rangle$ is an abelian subgroup of index 2 in $G$. Let $S = \langle x,z \rangle^1$. Since $G$ is a semidirect product of $\langle\langle x,y \rangle\rangle$ and $\langle\langle z \rangle\rangle$, it is easy to see that the algebra $K[S]$ is not noetherian. For example, $x^n z \notin x^m zS$ for all $n \neq m$, so that $S$ has an infinite ascending chain of right ideals $\bigcup_{i=1}^{n} x^i zS$. In particular, the assertion of Theorem 10 in Chapter 11 cannot be extended to finite extensions of $FC$-groups.

We note that there are groups with the property $\mathfrak{P}$ that are not abelian-by-finite, so that, in general, the permutation property does not imply the $PI$-property for the corresponding group algebra; see Theorem 3 in Chapter 18.

**Example 16**    Let $G$ be the group with generators $z, x_1, x_2, \ldots$, subject to the relations $[x_i,z] = 1$, $x_i^p = z^p = 1$, $[x_i,x_{i+1}] = z$, $[x_i,x_j] = 1$ for $i \geq 1, j > i + 1$, where $p$ is a fixed prime. Clearly, $z \in Z(G)$, and $G/\langle\langle z \rangle\rangle$ is abelian. It then follows that $G' = \langle\langle z \rangle\rangle$. In particular, $G$ is an $FC$-group with the property $\mathfrak{P}$ (in fact, $S$ has $\mathfrak{P}_{p+1}$ by Lemma 10). From the relations, it is clear that $x_i Z(G) \neq x_j Z(G)$ for $i \neq j$. Hence, $[G : Z(G)] = \infty$. Consequently, $G$ is not abelian-by-finite since abelian-by-finite $FC$-groups are center-by-finite; see Lemma 7 of Chapter 7.

Now, we turn to another fundamental class of semigroups for which the permutational property relates a given semigroup to some groups derived from it. This is the class of semisimple semigroups; see Chapter 1. The following result of Domanov [48] plays a crucial role.

**Theorem 17**    Assume that $S$ is a 0-simple semigroup satisfying the permutational property. Then $S$ is completely 0-simple.

*Proof.*    Suppose that $S$ is not completely 0-simple. We consider three cases.

Case 1: $S$ has a nonzero idempotent $e$.

It is well known that $S$ contains a bicyclic subsemigroup in this case, that is, the monoid $\langle p,q \rangle$ with the identity $e$ given by two generators $p$, $q$, subject to the relation $pq = e$; see Lemma 1 in Chapter 1. Put $x_i = q^{i-1}p^i$, $i = 1, 2, \ldots$. Then, for any $n \geq 2$, we have

$$x_1 x_2 \ldots x_n = p(qp^2)(q^2 p^3) \ldots (q^{n-1}p^n) = p^n$$

On the other hand, if $1 \neq \sigma \in S_n$, then take the least integer $j$ such that $\sigma(j) \neq j$. Then,

$$x_{\sigma(1)}x_{\sigma(2)}\ldots x_{\sigma(n)} = p^{j-1}(q^{\sigma(j)-1}p^{\sigma(j)})x_{\sigma(j+1)}\ldots x_{\sigma(n)} = q^{\sigma(j)-j}s$$

for some $s \in \langle p,q \rangle$. Since $\sigma(j) - j > 0$ and any element of $\langle p,q \rangle$ has a unique presentation in the form $q^r p^t$, for some non-negative integers $r$, $t$, then $x_{\sigma(1)}x_{\sigma(2)}\ldots x_{\sigma(n)} \neq p^n$. This shows that, for any $n \geq 2$, $S$ does not satisfy $\mathfrak{P}_n$, which contradicts our assumption.

Case 2: $S$ has no nonzero idempotents, and there exist elements $s$, $f \in S$ such that $sf = s$, $s \neq \theta$.

Since $S$ is 0-simple, then $f = xsq$ for some $x, q \in S^1$. Moreover, $q \neq 1$ since otherwise $f = xs = xsf = f^2$, contradicting the fact that $S$ has no nonzero idempotents. Put $p = xs$. Then,

(c)     $$pf = p, \ pq = f, \ pf^k = p, \ p^k q^k = f \qquad \text{for any } k \geq 1$$

The above equalities imply that the elements $p, f$ are not nilpotent because $p^k = \theta$ or $f^k = \theta$ yields $p = pf^k = \theta$ or $f = p^k q^k = \theta$, contradicting the fact that $s \neq \theta$. Put $x_i = q^{i-1}p^i$, $i = 1, 2, \ldots$. For any $n \geq 1$, we have $x_1 x_2 \ldots x_n = p(qp^2)(q^2 p^3)\ldots(q^{n-1}p^n) = f^{n-1}p^n$.

Suppose that the latter equals $x_{\sigma(1)}x_{\sigma(2)}\ldots x_{\sigma(n)}$ for some $1 \neq \sigma \in S_n$. If $j$ is the least integer with $\sigma(j) \neq j$, then,

$$x_{\sigma(1)}x_{\sigma(2)}\ldots x_{\sigma(n)} = f^j q^{\sigma(j)-j}p^{\sigma(j)}x_{\sigma(j+1)}\ldots x_{\sigma(n)}$$

In view of (c), the element $p^{\sigma(j)} x_{\sigma(j+1)} \ldots x_{\sigma(n)}$ may be written as $f^{j_1}q^{i_1} f^{j_2}q^{i_2} \ldots f^{j_m}q^{i_m}p^l$ for some non-negative integers $m, l, j_k, i_k$. Thus, we get

(d)     $$f^{n-1}p^n = f^{j_0}q^{i_0}f^{j_1}q^{i_1}\ldots f^{j_m}q^{i_m}p^l \qquad \text{where } j_0 = j, i_0 = \sigma(j) - j$$

Put $r = \sum_{k=1}^m i_k$, and multiply the above equality by the element $p^{r+1}$ on the left. Then, through (c), we come to $p^{n+r+1} = p^{l+1}$. Since $p$ is not nilpotent and $S$ has no nonzero idempotents, then $p$ is not a periodic element. This implies that $n + r = l$. Thus, multiplying (d) by $p^r$ on the left, we get

$$p^l = p^{n+r} = f^t p^l$$

where $t = 1$ if $i_m \neq 0$ and $t = j_m + 1$ if $i_m = 0$. Then $f = p^l q^l = f^t p^l q^l = f^{t+1}$ and, since $t \neq 0$, $f$ is periodic. This, in view of the assumption on $S$, contradicts the fact that $f$ is not nilpotent and proves the assertion in this case.

Case 3: For any $x, y \in S$, we have $y = \theta$ whenever $yx = y$.

Define a relation $\leq$ on $S$ by $x \leq y$ if $x \in yS^1$. If $x \leq y$ and $y \leq x$, then $x = yu, y = xv$ for some $u, v \in S^1$, which implies that $xvu = x$. Consequently, either $x = \theta$ and $y = \theta$, or $v = 1 = u$, and also $x = y$, which shows that the relation $\leq$ is antisymmetric. Thus, it is clear that $\leq$ is a partial order on $S$, and we write $x < y$ if $x \leq y$ and $x \neq y$. For any $n > 1$, we will construct a sequence $x_n, x_{n-1}, \ldots, x_1$ of nonzero elements of $S$ such that $x_{i+1} \leq x_i x_{i+1} \ldots x_n$ for $i = 1, 2, \ldots, n - 1$. To this end, let us establish two properties of the order $\leq$.

(i)  If $y < z$, and $xy \neq \theta$, then $xy \neq xz$. In fact, we know that $y = zu$ for some $u \in S$. Then $xy = xzu$, and the assumption in case 3 implies that $xy \neq xz$.

(ii)  If $x < y$, $x \neq \theta$, then there exists $z \in S$ with $y \leq zx$. For this, observe that the 0–simplicity of $S$ implies that $y = zxv$ for some $z, v \in S^1$. Then $y \leq zx$, and $z \neq 1$ since otherwise $y \leq x$, which is impossible.

Take any $x, y \in S$ with $xy \neq \theta$. From (ii) it follows that there exists $z \in S$ such that $x \leq zxy$. Then $x \leq zx$, and we put $x_n = x$, $x_{n-1} = z$. Assume that the elements $x_n, x_{n-1}, \ldots, x_{i+1}$ have been chosen. Then $x_{i+2} \leq x_{i+1} x_{i+2} \ldots x_n \neq \theta$, and $x_{i+1} x_{i+2} \ldots x_n < x_{i+1}$ by our assumption in this case. Thus, by (ii), there exists $x_i \in S$ with $x_{i+1} \leq x_i x_{i+1} \ldots x_n$, which establishes our claim.

Observe that $x_{i+1} \leq x_i \ldots x_n < x_i$ for any $i = 1, 2, \ldots, n - 1$, and so $x_l < x_k$ for $k < l$. Now, for any $1 \neq \sigma \in S_n$ and for the minimal integer $j$ with $\sigma(j) \neq j$, we have

$$x_{\sigma(j)} x_{\sigma(j+1)} \ldots x_{\sigma(n)} < x_{\sigma(j)} \leq x_{j+1} \leq x_j x_{j+1} \ldots x_n$$

This and (i) imply that

$$\begin{aligned} x_{\sigma(1)} x_{\sigma(2)} \ldots x_{\sigma(n)} &= x_1 x_2 \ldots x_{j-1} \bigl( x_{\sigma(j)} \ldots x_{\sigma(n)} \bigr) \\ &< x_1 x_2 \ldots x_{j-1} \bigl( x_j \ldots x_n \bigr) \end{aligned}$$

again contradicting the fact that $S$ has the permutational property and completing the proof of the theorem.

In view of the above result, we will now turn to the question of when a completely 0-simple semigroup $S = \mathfrak{M}^0(G, I, M, P)$ has the permutational property. Clearly, the group $G$, being isomorphic to a subgroup of $S$, satisfies $\mathfrak{P}$ in this case, too. It seems that some finiteness conditions on the sandwich matrix $P$ might be found that are necessary and sufficient (under the assumption that $G$ has $\mathfrak{P}$) for $S$ to have the property. Conditions of this type may be given with respect to the $PI$-property of the corresponding algebras $K[S]$, $K[G]$. They will be discussed in Chapter 20. Here, we start with a surprising example showing that the rank of $P$ is not responsible for the property $\mathfrak{P}$ of $S$ (as might be suggested by Proposition 6 in Chapter 20).

**Example 18**   Let $S = \mathfrak{M}^0(\{1\}, \mathbb{Z}, \mathbb{Z}, P)$ be the completely 0-simple semigroup, with the sandwich matrix $P$ defined by $p_{ii} = \theta$, $p_{ij} = 1$ for $i \neq j$, $i, j \in \mathbb{Z}$. Then $S$ may be identified with the semigroup $\{s_{ij} \mid i, j \in \mathbb{Z}\} \cup \{\theta\}$, with the multiplication rules

$$s_{ij}s_{kl} = \begin{cases} s_{il} & \text{if } j \neq k \\ \theta & \text{if } j = k \end{cases}$$

We claim that $S$ has the property $\mathfrak{P}_{10}$. Thus, let $x_1, \ldots, x_{10} \in S$ be given. The following obvious properties of $S$ will be used:

(i)   If $y_1 y_2 \ldots y_k = \theta$ for some $y_1, y_2, \ldots, y_k \in S$, then there exists $i \in \{1, 2, \ldots, k-1\}$, with $y_i y_{i+1} = \theta$.

(ii)   For any $x, y \in S$, $xy = \theta$ if and only if there exists $n \in \mathbb{Z}$ such that $x$ lies in the $n$th column and $y$ lies in the $n$th row of $S$.

(iii)   If $y_1 y_2 \ldots y_k \neq \theta$, $z_1 z_2 \ldots z_r \neq \theta$ for some $k, r \geq 1$, $y_i, z_j \in S$, such that $y_1, z_1$ lie in the same row of $S$ and $y_k, z_r$ lie in the same column of $S$, then $y_1 y_2 \ldots y_k = z_1 z_2 \ldots z_r$.

If $x_1 x_2 \ldots x_{10} = \theta$, then (i) allows us to find a nontrivial permutation of the $x$ with the zero product. Thus, suppose that $x_1 x_2 \ldots x_{10} \neq \theta$. Assume first that, for some $n \in \mathbb{Z}$, there exist three integers $r, s, t \in \{1, 2, \ldots, 10\}$, $r < s < t$, such that the elements $x_r, x_s, x_t$ lie in the same row of $S$. Then $x_1 x_2 \ldots x_{10} = x_1 x_2 \ldots x_{r-1} y_1 y_2 y_3$, where $y_1 = x_r x_{r+1} \ldots x_{s-1}$, $y_2 = x_s x_{s+1} \ldots x_{t-1}$, $y_3 = x_t x_{t+1} \ldots x_{10}$. Hence, $y_1, y_2, y_3$ lie in the same row of $S$ and, consequently, $(x_1 x_2 \ldots x_{r-1}) y_2 \neq \theta$, $y_2 y_1 \neq \theta$, $y_1 y_3 \neq \theta$ by (ii). Thus, again by (ii) and (iii), $x_1 x_2 \ldots x_{10} = x_1 x_2 \ldots x_{r-1} y_2 y_1 y_3$, and we are done in this case. Hence, we may assume that, for any $n \in \mathbb{Z}$, there are at most two $x$ lying in the $n$th row of $S$. A similar argument allows us to assume that there are at most two $x$ lying in the $n$th column of $S$.

Let $j < 10$ be maximal with respect to the property $x_j x_2 \neq \theta$. From the foregoing, it follows that $j \geq 7$, since at most two $x$ may annihilate $x_2$ on the left. Let us consider all elements of the form

(e)                          $x_1 x_i x_{i+1} \ldots x_j x_2 x_3 \ldots x_{i-1} x_{j+1} x_{j+2} \ldots x_{10}$

where $i \in \{3, \ldots, j\}$. By (i) we always have $x_i x_{i+1} \ldots x_j x_2 x_3 \ldots x_{i-1} \neq \theta$. Further, $x_1 x_i = \theta$ for at most two $i$, and $x_{i-1} x_{j+1} = \theta$ for at most two $i - 1$, $i \in \{3, \ldots, j\}$. Since $j \geq 7$, then at least one element of the form (e) is nonzero. Now $x_1 x_2 \ldots x_{10}$ also is nonzero, so (iii) implies that the product $x_1 x_2 \ldots x_{10}$ is "permutable." This shows that $S$ satisfies $\mathfrak{P}_{10}$.

It may be shown that the semigroup $S$ constructed above satisfies $\mathfrak{P}_6$ but not $\mathfrak{P}_5$. This requires more computation.

On the other hand, the algebra $K_0[S]$ is the union of its subalgebras $K_0[S_i]$, where $S_i = \{s_{jk} \mid -i \leq j \leq i,\ -i \leq k \leq i\}, i \geq 1$. Since the submatrix $P_i$ of $P$ corresponding to $S_i$ (see Chapter 1), is nonsingular over $K$ whenever $2i \neq 0$ in $K$, then Corollary 26 in Chapter 5 implies that $K_0[S_i] \cong M_{2i+1}(K)$ in this case. Consequently, $K_0[S]$ contains matrix algebras of arbitrarily big dimensions. (Note also that $K[S] = \bigcup_{i \in \mathbb{Z}} K[S_i]$ is a regular algebra.) Therefore, by the Amitsur–Levitzki theorem, $K[S]$ is not a $PI$-algebra.

Let $S = \mathfrak{M}^0(G, I, M, P)$ be a completely 0-simple semigroup. We say that $S$ is triangularizable if, for every finite subsets $M' \subseteq M$, $I' \subseteq I$, there exist finite sets $N$, $J$ such that $M' \subseteq N \subseteq M$, $I' \subseteq J \subseteq I$, $|N| = |J|$, and there exists a permutation of the rows and a permutation of the columns of the $|N| \times |N|$ submatrix $P_{NJ}$ of $P$ (see Chapter 1) such that the resulting $|N| \times |N|$ matrix is triangular and all its diagonal entries are nonzero. For example, any completely 0-simple inverse semigroup is isomorphic to a triangularizable semigroup by [26], Theorem 3.9 ($|M| = |I|$ and $P$ may be chosen to be an identity $I \times I$ matrix). From the results on the isomorphisms of completely 0-simple semigroups [26], §3.4, it follows that by permuting rows and columns of $P$ we get a semigroup isomorphic to $S$.

Using an argument similar to that of the proof of the "easy" implication in the Amitsur–Levitzki theorem, we show that, for semigroups of this type, in contrast to the preceding example, the cardinality of the sets $I$, $M$ controls the property $\mathfrak{P}$.

**Proposition 19**  Let $S = \mathfrak{M}^0(G, n, n, P)$, $n \geq 1$, be a completely 0-simple semigroup that is triangularizable. Then $S$ does not satisfy $\mathfrak{P}_{2n-1}$.

*Proof.*  By the preceding remarks, we may assume that $P$ is a triangular matrix, say lower triangular, with nonzero diagonal. Let $s_{i,j} = (1, i, j) \in S$, for $i, j \in \{1, 2, \ldots, n\}$. It is clear that

$$s_{n,n}s_{n,n-1}s_{n-1,n-1}s_{n-1,n-2}s_{n-2,n-2}\cdots s_{2,2}s_{2,1}s_{1,1}$$
$$= (p_{n,n}(p_{n-1,n-1})^2(p_{n-2,n-2})^2 \cdots (p_{2,2})^2 p_{1,1}, n, 1)$$

Since all $p_{i,i}$ are nonzero, the above product is a nonzero element of $S$. On the other hand, it is easy to see that any reordering of the factors leads to a zero product. The assertion follows.

As mentioned above, a completely 0-simple inverse semigroup $S = \mathfrak{M}^0(G, I, I, P)$ is triangularizable. Thus, Proposition 19 implies that $|I| \leq m/2$ (that is, $|E(S) \setminus \theta| \leq m/2$) whenever $S$ has the property $\mathfrak{P}_m$. This will be used in the proof of the next result.

**Proposition 20**  A completely 0-simple inverse semigroup $S$ has the property $\mathfrak{P}$ if and only if $S$ has finitely many idempotents and the maximal subgroups of $S$ have the property $\mathfrak{P}$.

*Proof.*  It remains to prove sufficiency of these conditions. If $S = \mathfrak{M}^0(G,I,I,P)$ and $I$ is finite, then $S = \bigcup_{i,j \in I} S_{(j)}^{(i)}$ is a finite union of semigroups defined in Chapter 1. Any $S_{(j)}^{(i)}$ is either a group with zero or a semigroup with zero multiplication. Since the latter always has the property $\mathfrak{P}_2$, then the assertion follows from Lemma 5.

Examples 16 and 18 show that, in general, the property $\mathfrak{P}$ is not equivalent to the $PI$-property of the corresponding semigroup algebra. However, if $S$ is finitely generated, then the property $\mathfrak{P}$ of $S$ is equivalent to the $PI$-property of $K[S]$ in either of the following cases:

(1)  $S$ is cancellative, by Corollary 13.
(2)  $S$ is periodic, by Theorem 3.

The latter motivated the question asked in [233] whether it is always the case for finitely generated semigroups. The class of completely 0-simple semigroups provides more examples of semigroups of this type.

**Proposition 21**  Let $S = \mathfrak{M}^0(G,I,M,P)$ be a finitely generated completely 0-simple semigroup with the property $\mathfrak{P}$. Then $K[S]$ is a $PI$-algebra.

*Proof.*  Obviously, a finite set of generators of $S$ lies in a finite number of rows and columns of $S$, so that the sets $I$, $M$ must be finite. Moreover, if $S = \langle x_1, x_2, \ldots, x_n \rangle$, $x_j = (g_j, i_j, m_j)$ for some $g_j \in G^0$, $i_j \in I$, $m_j \in M$, then $G$ must be generated by the finite set $\{g_j, p_{mi} \mid j = 1, 2, \ldots, n, i \in I, m \in M\} \setminus \{\theta\}$. Thus, by Corollary 13, $K[G]$ is a $PI$-algebra. In view of Lemma 3 in Chapter 5, $K_0[S]/l(K_0[S])$ embeds into a matrix algebra over $K[G]$. It follows that it is a $PI$-algebra; see [222], Theorem 3.3.2. Then $K[S]$ is a $PI$-algebra, too.

Finitely generated inverse semigroups with the property $\mathfrak{P}$ admit a nice description, which shows that the implication "$S$—with $\mathfrak{P} \Rightarrow K[S]$—$PI$" also holds for this class of semigroups [188]. To state this result, we first extend one of the assertions of Proposition 20.

**Lemma 22**  Let $S$ be a finitely generated inverse semigroup. If $S$ has the permutational property, then $S$ has finitely many idempotents.

*Proof.* Assume that $\mathfrak{P}_k$ holds in $S$, and let $n$ be the integral part of $k/2$. From Proposition 19, we know that, for any $x \in S$, the cardinality $n_x$ of the set $E_x$ of nonzero idempotents of the principal factor $S_x$ does not exceed $n$. The algebra $K_0[S_x]$ may be identified with the matrix algebra $M_{n_x}(K[G_x])$, where $G_x$ is a maximal subgroup of $S_x$; see Corollary 26 in Chapter 5. Here, a nonzero element $t \in S_x$ is identified with a matrix in $M_{n_x}(G_x^0)$ with exactly one nonzero entry. Proposition 28 in Chapter 5 implies that $K_0[S]$ is a subdirect product of all contracted principal factor algebras $M_{n_x}(K[G_x])$, $x \in S$, where any $s \in S$ is treated as $(s_x)_{x \in S} \in \prod_{x \in S} M_{n_x}(K[G_x])$ with $s_x = \phi_x(s \sum_{e \in E_x} e)$, and $\phi_x$ is the natural homomorphism $K[SxS] \to K_0[S_x]$. Observe that if $\phi_x(se)$, $\phi_x(sf)$ are nonzero for some $e, f \in E_x$, $e \neq f$, then the nonzero entries of the matrices $\phi_x(se)$, $\phi_x(sf)$ do not lie in the same column of the matrix algebra $M_{n_x}(K[G_x])$. Therefore, $s_x$ is a matrix with at most $n_x$ nonzero entries, any two of which lie in distinct columns. Further, $s_x = \phi_x(\sum_{e \in E_x} es)$ because $\phi_x(\sum_{e \in E_x} e)$ is the identity of $M_{n_x}(K[G_x])$. Hence, similarly, every two entries of $s_x$ lie in distinct rows of $M_{n_x}(K[G_x])$.

It is clear that $\rho = \{(s,t) \in S \times S \mid s_x - t_x \in M_{n_x}(\omega(K[G_x]))$ for all $x \in S\}$, $\omega(K[G_x])$ denoting the augmentation ideal of $K[G_x]$, is a congruence on $S$. Moreover, $S/\rho$ embeds into the algebra

$$\prod_{x \in S} (M_{n_x}(K[G_x])/M_{n_x}(\omega(K[G_x]))) \cong \prod_{x \in S} M_{n_x}(K)$$

From the above observations, it follows that the image $s'$ of $s \in S$ under the natural homomorphism $S \to S/\rho$ is of the form $s' = (s'_x)_{x \in S}$, where $s'_x \in M_{n_x}(\{0,1\})$ has at most $n_x$ nonzero entries, any two of which lie in distinct rows and distinct columns of $M_{n_x}(K)$. From Lemma 14 in Chapter 3 and the fact that $n_x \leq n$, it then follows that $(s'_x)^{n!}$ is an idempotent matrix. Hence $(s')^{n!}$ is an idempotent, too. Therefore, $S/\rho$ is a finitely generated periodic semigroup with the permutational property, and Theorem 3 implies that it is finite.

Assume that two idempotents $e, f \in S$ with $(e,f) \in \rho$ are given. Since $S$ is inverse, then, for any $y \in E_e$, $fy$ is an idempotent. Hence, $\phi_e(fy) \in M_{n_e}(\{0,1\})$ is an idempotent matrix with at most one nonzero entry. Therefore, $\phi_e(f \sum_{y \in E_e} y)$ is a diagonal idempotent in $M_{n_e}(K)$. Since the same applies to the idempotent matrix $\phi_e(e)$ of rank 1, then the fact that $(e,f) \in \rho$ implies that $\phi_e(f \sum_{y \in E_e} y) = \phi_e(fx)$ for some $x \in E_e$, and $\phi_e(e) = \phi_e(fx)$. In view of the definition of $\phi_e$, this shows that $e = fx$.

Consequently, $e = fx = ffx = fe$. A similar argument shows that $f = ef$ and, since the idempotents of $S$ commute, then $e = f$. It then follows that $(e,f) \notin \rho$ for any distinct idempotents $e, f \in S$. Thus, the finiteness of $S/\rho$ implies that $S$ has finitely many idempotents.

**Theorem 23** Let $S$ be an inverse semigroup. Then the following conditions are equivalent:

(i)   $S$ is finitely generated and satisfies the permutational property.
(ii)  $S$ has finitely many idempotents, and all subgroups of $S$ are finitely generated and abelian-by-finite.
(iii) $K[S]$ is a right and left noetherian $PI$-algebra.

*Proof.* If (i) holds, then, as in Lemma 22, any principal factor algebra $K[S_x] \cong M_{n_x}(K[G_x])$ of $K[S]$ is a homomorphic image of $K[S]$. Therefore, it is finitely generated, and so any $G_x$ is a finitely generated group. Now, (ii) follows from Lemma 22 and Corollary 13.

If (ii) holds, then $K[S] \cong M_{n_1}(K[G_1]) \oplus \cdots \oplus M_{n_k}(K[G_k])$ for some $k$, $n_j \geq 1$, and some subgroups $G_j$ of $S$; see Corollary 27 in Chapter 5. Thus, $K[G_j]$ is a noetherian $PI$-algebra by Theorem 15. Then so is the algebra $K[S]$.

The implication (iii) $\Rightarrow$ (i) was established in Proposition 1 and Theorem 14. (Observe that Theorem 14 is not really needed here— the noetherian property implies that $K[S]$ cannot have infinite sets of commuting idempotents, and so $E(S)$ must be finite, and one may proceed on the basis of Corollary 27 in Chapter 5 and the fact that noetherian group algebras are finitely generated.)

We now show that a counterexample to the question: $S$—finitely generated with $\mathfrak{P} \Rightarrow K[S]$—a $PI$-algebra? may be found in a class very close to completely 0-simple semigroups. This will be a completely semisimple semigroup with two principal factors constructed in [189].

**Example 24** Let $S = \{ s_{ij} \mid i,j \in \mathbb{Z} \} \cup \{ \theta \}$ be the completely 0-simple semigroup described in Example 18. Define an extension $T$ of $S$ as the monoid generated by $S$ and two elements $y, z$ subject to the rules

$$zy = yz = 1, \; zs_{ij} = s_{i+1,j}, \; s_{ij}z = s_{i,j-1}, \; ys_{ij} = s_{i-1,j}, \; s_{ij}y = s_{i,j+1}$$

for any $i, j \in \mathbb{Z}$. [When $S$ is viewed as the semigroup of $\mathbb{Z} \times \mathbb{Z}$ matrices over $\{1\}^0$ with at most one nonzero row, in which a row of the sandwich matrix $P$ lies (see Lemma 3 in Chapter 5), $y$ and $z$ may be identified with the matrices $Y = (y_{ij})$, $y_{i,i+1} = 1$, $y_{ij} = 0$ for $j \neq i + 1$, $Z = (z_{ij})$, $z_{i,i-1} = 1$, $z_{ij} = 0$ for $j \neq i - 1$, $i, j \in Z$, respectively.] It is clear that $T$ is finitely generated, for example, $T = \langle y, z, s_{00} \rangle$. Moreover, $S$ is an ideal of $T$, with the factor semigroup $T/S$ isomorphic to an infinite cyclic group with zero adjoined. Thus, from Example 18 and Lemma 6, it follows that $T$ has the permutational property and that $K[T]$ is not a $PI$-algebra.

Observe also that a countable infinite direct sum of copies of the field $K$ is, in a natural way, a right (and left) $K_0[T]$-module. It is easy to see that this module is faithful (use the above embedding of $K_0[T]$ into $Z \times Z$ matrices) and irreducible; see Lemma 16 in Chapter 5. Thus, $K_0[T]$ is a right, and similarly left, primitive algebra.

Let us note that it will be shown in Chapter 24 that examples of this type cannot be constructed in some further class (besides those mentioned above) of semigroup algebras.

Example 24 shows that there exist primitive non-$PI$ homomorphic images of finitely generated semigroup algebras of semigroups satisfying the permutational property. We close this chapter with an observation on irreducible skew linear representations of finitely generated semigroups satisfying $\mathfrak{P}$. Any such representation determines, and is determined by, a homomorphism of the corresponding semigroup algebra onto a simple artinian algebra. Thus, the following result asserts, in fact, that any artinian homomorphic image of $K[S]$ is a $PI$-algebra.

**Proposition 25** Let $S$ be an irreducible subsemigroup of the matrix algebra $M_n(D)$, $n \geq 1$, $D$—a division algebra. Assume that $S$ is finitely generated and satisfies $\mathfrak{P}$. Then $M_n(D)$ is a $PI$-algebra, and $D$ has finite dimension over its center.

*Proof.* From Proposition 8 in Chapter 1, we know that there exists $r \leq n$ such that $M_r(D)$ is isomorphic to (and will be identified with) the $Z(D) = K$-linear space $K\{T\}$ generated by a cancellative subsemigroup $T$ of $S$. Since $S$ is finitely generated, then $M_n(D)$ is a finitely generated $K$-algebra. Thus, $D$ and $M_r(D)$ also are finitely generated $K$-algebras. If $M_r(D) = K[x_1, x_2, \ldots, x_k]$, then any $x_i$ may be written as $\sum_j \alpha_{ij} h_{ij}$ for some $\alpha_{ij} \in K$, $h_{ij} \in T$. Let $H$ be the subsemigroup of $T$ generated by the $h_{ij}$. Then $K\{H\} = M_r(D)$, and it is a homomorphic image of the semigroup algebra $K[H]$. Since $K[H]$ is a $PI$-algebra by Corollary 13, then $M_r(D)$ is also. By the Kaplansky theorem (see Theorem 1 in Chapter 18), $D$ is finite over its center in this case, and $M_n(D)$ also is a $PI$-algebra.

## Comments on Chapter 19

The permutational property, in fact, first appeared in connection with Shirshov's results on $PI$-algebras being applications of his height theorem; see [252],5.§2; [17], Appendix; and [239], Theorem 4.2.7. Then, it appeared explicitly only in the 1980s: in [33] in the context of groups and in [233] with respect to the Burnside problem for semigroups. Since then, it has been an object of extensive study. We list the papers not referred to in the text. Groups with $\mathfrak{P}$ were first studied by Curzio et al. in [33]. Among other things, they showed that $\mathfrak{P}_3$ is equivalent to the fact that the commutator subgroup has order $\leq 2$. Some further results on $\mathfrak{P}_3$ for groups and, more generally, for regular semigroups, were given by Garzon and Zalcstein in [60]. Groups with the property $\mathfrak{P}_4$ were studied by Bianchi et al. [10], as well as Longobardi and Maj [137]. In particular, they were shown to be metabelian. In [11] and [12], Blyth considered a formally weaker property: "there exists $n \geq 2$ such that for any $s_1, s_2, \ldots, s_n \in S$ we have $s_{\sigma(1)} \cdots s_{\sigma(n)} = s_{\tau(1)} \cdots s_{\tau(n)}$ for some $\sigma, \tau \in \mathcal{S}_n, \sigma \neq \tau$" in the class of groups, showing that it is equivalent to $\mathfrak{P}$. He proved also that, for every $n \geq 3$, there exists a bound, depending on $n$ only, on the order of semisimple groups satisfying $\mathfrak{P}_n$. Some nice general results on $\mathfrak{P}$ and its above generalization (not equivalent to $\mathfrak{P}$ in the class of semigroups) were given by Justin and Pirillo in [106], (see also [211]), from which Lemma 5 is taken. We note that the notion of repetitive mappings used in this lemma is a special case of the celebrated Ramsey theorem; see [139], Chapter 4. Some results on a related permutation condition considered with respect to the theory of languages are surveyed in [234]. It was claimed in [61] that the implication "$S$—finitely generated with $\mathfrak{P} \Rightarrow K[S]$—$PI$" also holds for regular linear semigroups over $K$ if $\mathrm{ch}(K) = 0$. While there is a gap in the proof of this fact, a recent result of Okniński and Putcha shows that all linear semigroups with $\mathfrak{P}$ lead to $PI$-semigroup algebras [193a]. Irving [91] constructed a finitely generated semigroup algebra $K[S]$, all of which primitive homomorphic images are finite dimensional but of arbitrarily big dimensions. In particular, $K[S]$ is not a $PI$-algebra. Finally, we note that a simple cancellative linear semigroup that is not a group is given in Example 15 in Chapter 3. Thus, the assertion of Theorem 17 cannot be extended to 0-simple linear semigroups.

# 20

# PI-Semigroup Algebras

In this chapter, we discuss the basic classes of semigroups for which necessary and sufficient conditions for the $PI$-property of the corresponding semigroup algebras may be given. Some results of this type were proved in Chapter 19 by means of the permutational property. We start with the case of cancellative semigroups, which is completely settled. The following result will be crucial for the considerations in the subsequent chapters.

**Theorem 1** Let $S$ be a cancellative semigroup such that $K[S]$ satisfies a polynomial identity. Then $S$ has a group of (two-sided) fractions $G$, and $K[G]$ is a $PI$-algebra satisfying the same multilinear identities as $K[S]$.

*Proof.* Since $S$ has the permutational property by Proposition 1 in Chapter 19, then Lemma 7, also in Chapter 19, implies that $S$ has a group of fractions $G$. Let $f(x_1,\ldots,x_n)$ be a multilinear polynomial in noncommuting indeterminates $x_1$, ..., $x_n$, which vanishes on $K[S]$. Then, for any $s_2$, ...,

$s_n \in S, f(x_1, s_2, \ldots, s_n) = 0$ is a linear generalized identity on $S$. Thus, from Lemma 17 in Chapter 7, it follows that $f(x_1, s_2, \ldots, s_n) = 0$ holds for all $x_1 \in G$. Now, for every $g_1 \in G$, $f(g_1, x_2, s_3, \ldots, s_n) = 0$ is an identity on $S$, and so, again by Lemma 17 in Chapter 7, it is an identity on $G$. Proceeding this way, we come, after $n$ steps, to $f(g_1, \ldots, g_n) = 0$ for all $g_1, \ldots, g_n \in G$. Since $f$ is multilinear, then it is an identity on $K[G]$. The result follows.

The above assertion was first established, using a different proof, by Zelmanov in [276]. It exploited the fact that, for every $PI$-algebra $R$, every central localization $R'$ of $R$ also is a $PI$-algebra, and it satisfies the same multilinear identities as $R$; see [203], Lemma 5.1.3. A reduction to the case where $G$ is an $FC$-group was used in [276]. From Corollary 8 in Chapter 7, we know that $G = SZ(S)^{-1}$ in this case. Therefore, $K[G]$ is a central localization of $K[S]$, and the assertion of Theorem 1 follows from the above-mentioned general result on central localizations.

As observed in Example 24 in Chapter 19, the permutational property alone does not seem to be the right way to attack the problem of characterizing the $PI$-property of semigroup algebras of finitely generated semisimple semigroups that are not groups. Since, in view of Theorem 17 in Chapter 19, we know that semisimple semigroups with $PI$-semigroup algebras must be completely semisimple, we propose in this chapter another way of relating the $PI$-property of $K[S]$ to the $PI$-property of some group algebras $K[H]$. Here $H$ will run through the set of all maximal subgroups of $S$. We start with the basic case of a completely 0-simple semigroup $S = \mathfrak{M}^0(G, I, M, P)$ such that $K[G]$ is a $PI$-algebra. By Lemma 7 and Corollary 18 both in Chapter 5, $\mathcal{J}(K[S])^3 = 0$ for every field $K$ with $\mathrm{ch}(K) = 0$ because $\mathcal{J}(K[G]) = 0$; see Theorem 3 in Chapter 18. Similarly, if $\mathrm{ch}(K) = p > 0$ and $G$ has no normal subgroups of order divisible by $p$, then also $\mathcal{J}(K[S])^3 = 0$. Thus, $K[S]$ is a $PI$-algebra if and only if $K[S]/\mathcal{J}(K[S])$ is a $PI$-algebra. This extends the result known for group rings; see Theorem 3 in Chapter 18. This and the Kaplansky theorem allows us to deal with the $PI$-property for $K[S]$ through the linear representations of $K[G]$, and $K[S]$.

Recall that if $\hat{R} = \mathfrak{M}(R, I, M, P)$ is a Munn algebra over an algebra $R$ and if $\varphi : R \to R'$ is a homomorphism of algebras, then, by $\varphi(P)$ we denote the matrix $(\varphi(p_{mi}))_{m \in M, i \in I}$. If $R = M_n(D)$ for an algebra $D$ and some $n \geq 1$, then $\overline{P}_D$ stands for the $(M \cdot n) \times (I \cdot n)$ matrix over $D$ obtained from $P$ by erasing the matrix brackets in all entries $p_{mi} \in M_n(D)$ of $P$. In particular, if $p_{mi}$ are endomorphisms of a linear space $V$ of dimension $n$ over a division algebra $D$, then, replacing $p_{mi}$ by their matrices in a fixed basis of $V$, we come to an $(M.n) \times (I.n)$ matrix over $D$, which will be further denoted by $\overline{P}$.

Our first result characterizes the $PI$-property of $K[S]$ in terms of the ranks of the irreducible representations of the maximal subgroup of $S$. It was obtained by Domanov in [46].

**Theorem 2** Let $S = \mathfrak{M}^0(G, I, M, P)$ be a completely 0-simple semigroup such that $G$ has no normal subgroups of order divisible by $p$ if $\mathrm{ch}(K) = p > 0$. Then $K_0[S]$ is a $PI$-algebra if and only if the following conditions are satisfied:

(1)  $K[G]$ is a $PI$-algebra.
(2)  There exists $k \geq 1$ such that, for every irreducible representation $\varphi$ of $G$ over a field $L$ with $\mathrm{ch}(L) = \mathrm{ch}(K)$, we have $\rho(\overline{\varphi(P)}) \leq k$.

In this case, if $k$ is the minimal integer for which (2) holds, and $n$ is the minimal degree of an identity satisfied in $K_0[S]$, then $2k \leq n \leq 2k + 2$.

*Proof.*  Let $\varphi : G \to M_r(L)$ be an irreducible representation of $G$ over a field $L$ with $\mathrm{ch}(L) = \mathrm{ch}(K)$. Then we have the natural homomorphisms of algebras

(a)        $L_0[S] \cong \mathfrak{M}(L[G], I, M, P) \longrightarrow \mathfrak{M}(M_r(L), I, M, \varphi(P))$

From Proposition 23 in Chapter 5 (with $\hat{R} = \mathfrak{M}(M_r(L), I, M, \varphi(P))$), we know that if $\rho(\overline{\varphi(P)}) < \infty$, then,

(b)        $\mathfrak{M}(M_r(L), I, M, \varphi(P))/\mathfrak{B}(0) \cong M_{\rho(\overline{\varphi(P)})}(L)$

If $K[S]$ is a $PI$-algebra, then so is $K[G]$ because $G$ embeds into $S$. Moreover, from (a), (b), and the Amitsur-Levitzki theorem, it follows that $L[S]$ does not satisfy identities of degree less than $2 \cdot \rho(\overline{\varphi(P)})$ if $\rho(\overline{\varphi(P)}) < \infty$. On the other hand, if $\rho(\overline{\varphi(P)}) = \infty$, then, for every $k \geq 1$, we may choose a submatrix $P_k$ of $P$ consisting of finitely many rows of $P$ such that $k \leq \rho(\overline{\varphi(P_k)})$. Then also $\rho(\overline{\varphi(P_k)}) < \infty$. The subalgebra $L_0[S_k]$ of $L_0[S]$ generated by the $\mathcal{L}$-classes of $S$ corresponding to the rows of $P$ lying in $P_k$ does not satisfy identities of degree $\leq 2k$. Consequently, $K_0[S]$ is not a $PI$-algebra by Proposition 4 in Chapter 18, which contradicts the hypothesis and completes the proof of the necessity.

Assume now that (1) and (2) hold. From Theorem 3 in Chapter 18, it follows, in view of the hypothesis on $G$, that $\mathcal{J}(K[G]) = 0$. Therefore, $K_0[S] \cong \mathfrak{M}(K[G], I, M, P)$ embeds into $\prod_J \mathfrak{M}(K[G]/J, I, M, P)$, where $J$ runs over the set of primitive ideals of $K[G]$. Since $K[G]$ is a $PI$-algebra, then, by Proposition 2 in Chapter 18, there exist $n \geq 1$, a field

$L$, and an irreducible representation $\psi : G \to M_n(L)$ such that $K[G]/J$ embeds into $M_n(L)$ and $\psi$ factors through the natural homomorphism $\varphi : K[G] \to K[G]/J$. From (2) it follows that $\rho(\overline{\psi(P)}) \leq k$ so that the Amitsur–Levitzki theorem implies that $\mathfrak{M}(M_n(L),I,M,\psi(P))/\mathfrak{B}(0) \cong M_{\rho(\overline{\psi(P)})}(L)$ satisfies the standard identity $[x_1,\ldots,x_{2k}] = 0$. From Lemma 7 in Chapter 5, it then follows that the identity $x_{2k+1}[x_1,\ldots,x_{2k}]x_{2k+2} = 0$ is satisfied in $\mathfrak{M}(M_n(L),I,M,\psi(P))$. Thus, $\mathfrak{M}(K[G]/J,I,M,\overline{\varphi(P)}) \subseteq \mathfrak{M}(M_n(L),I,M,\psi(P))$, and, since $J$ is arbitrary, $K_0[S]$ also satisfies this identity. The result follows.

While the above result completely settles the characteristic zero case, it is not known whether it may be extended to the case of an arbitrary maximal subgroup $G$ of $S$, if $\text{ch}(K) = p > 0$. A partial result in this direction is given below.

**Corollary 3** Assume that $\text{ch}(K) = p > 0$. Let $S = \mathfrak{M}^0(G,I,M,P)$ be a completely 0-simple semigroup such that $G$ has no infinite $p$-subgroups. Then $K[S]$ is a $PI$-algebra if and only if conditions (1) and (2) of Theorem 2 are satisfied.

*Proof.* We can assume that $K[G]$ is a $PI$-algebra. From Theorem 3 in Chapter 18, it follows that $G$ has a normal subgroup $H$ of finite index and a finite normal $p$-subgroup $T \subseteq H$ such that $H/T$ is abelian and has no nontrivial $p$-subgroups. Moreover, we know that $\mathcal{J}(K[G])$ is a nilpotent ideal (Theorem 3 in Chapter 18) and, by Corollary 18 and Lemma 7 both in Chapter 5, $\mathcal{J}(K[S])$ also is nilpotent. Since, as we saw in the proof of Theorem 2, condition (2) is, in fact, responsible for the $PI$-property of $K[S]/\mathcal{J}(K[S])$, the result follows easily.

Our next aim is to find an intrinsic description of the $PI$-property for $K_0[S]$ not referring to a complete set of irreducible representations of $G$. As suggested in the preceding chapter, a reasonable finiteness condition on the matrix $P$ should be involved. By Theorem 3 in Chapter 18, we know that $G$ has an abelian subgroup $H$ of finite index whenever $K[G]$ is a $PI$-algebra and $G$ has no nontrivial normal $p$-subgroups if $\text{ch}(K) = p > 0$. In this case, $K[G]$ is a free left $K[H]$-module with a basis consisting of the coset representatives of $H$ in $G$. Under a fixed such basis $X = \{x_1,\ldots,x_n\}$, $n = [G : H]$, we have the natural embedding $\chi : K[G] \to \text{End}_{K[H]} K[G] \cong M_n(K[H])$. Let $P(H,X) = (q_{ij})$ be the $M \times I$ matrix such that $q_{ij}$ is the matrix of the endomorphism $\chi(p_{ij})$ in the basis $X$. Then $\overline{P(H,X)}$ is an $(M.n) \times (I.n)$ matrix over the commutative algebra $K[H]$. For every $M \times I$

matrix $Q$ over a commutative algebra $A$, the rank of $Q$ is defined as the maximal integer $k$ such that $Q$ has a $k \times k$ submatrix $Q'$ with $\det(Q') \neq 0$, or infinity if such an integer does not exist. Clearly, this notion extends that of the ordinary rank of matrices over a field. Therefore, the rank of $Q$ will be denoted by $\rho(Q)$, the notation used in Chapter 1, and in Chapter 5 for matrices over division algebras. The rank $\rho(\overline{P(H,X)})$ of $\overline{P(H,X)}$ may be defined in this way. Using this notation, we prove another result obtained by Domanov in [46]. The proof employs the following special case of a general Maschke-type result.

**Lemma 4** Let $H$ be an abelian normal subgroup of finite index in a group $G$. Assume that $\mathrm{ch}(K) = 0$ and $M$ is the kernel of an onto homomorphism $K[H] \to K$. Then $V = K[G]/MK[G]$ is a completely reducible right $K[G]$-module. Moreover, the image $\bar{X}$ in $V$ of any set $X$ of coset representatives of $H$ in $G$ is a basis of $V$ over $K$.

*Proof.* If $\bar{M} = \bigcap_{x \in X} xMx^{-1}$, then $\bar{M}K[G]$ is an ideal in $K[G]$. It is enough to show that $R = K[G]/\bar{M}K[G]$ is a completely reducible $R$-module because then the modules $V_R$, $V_{K[G]}$ also must be completely reducible. Let $W$ be a right ideal of $R$. Since $\bar{M}K[G] \cap K[H] = \bar{M}$ (see Corollary 17 in Chapter 4), then $W$ is a submodule of the $A$-module $V$, where $A = K[H]/\bar{M}$. But $A$, being a direct product of some $K[H]/(xMx^{-1})$, is semisimple artinian, so that $W_A$ is a direct summand of $V_A$. Then $W_{K[H]}$ is a direct summand of $V_{K[H]}$. Since $K[G]$ is a crossed product $K[H] * (G/H)$, Maschke's theorem implies that $W$ is a direct summand of $V$ as a $K[G]$-module. This proves that $R_R$ is completely reducible. If $\sum \lambda_i x_i \in MK[G]$ for some $\lambda_i \in K$, $x_i \in X$, then every $\lambda_i x_i$ is in $MK[G]$, so that $\lambda_i = 0$. It is clear that $\sum_{x \in X} Kx + MK[G] = K[G]$. The result follows.

**Theorem 5** Let $K$ be a field of characteristic zero, and $S = \mathfrak{M}^0(G,I,M,P)$ be a completely 0-simple semigroup.

(i)   If $K_0[S]$ satisfies a polynomial identity of degree $m$, then $G$ has an abelian normal subgroup $H$ of finite index $n$ bounded by a function of $m$. Moreover, $\rho(\overline{P(H,X)}) \leq mn/2$ for any basis $X \subseteq G$ of the left $K[H]$-module $K[G]$.

(ii)  Assume that $H$ is an abelian normal subgroup of $G$ with $[G : H] = n < \infty$. If, for some $X$, $\rho(\overline{P(H,X)}) = k < \infty$, then the algebra $K_0[S]$ satisfies a polynomial identity of degree $2k + 2$.

*Proof.* (i) Since $K[G]$ is a subalgebra of $K[S]$, then the former assertion follows from Theorem 3 in Chapter 18. Assume that $[G : H] = n$ and that

$\rho(\overline{P(H,X)}) \geq k$ for some $k < \infty$. We shall show that there exists a field $L \supseteq K$ and an irreducible representation $\lambda$ of $G$ over $L$ with $\rho(\overline{\lambda(P)}) \geq k/n$. Then, in view of Theorem 2, $K[S]$ cannot satisfy polynomial identities of degree less than $2k/n$. It will follow that $k \leq mn/2$, which yields the assertion. Let $Q$ be a $k \times k$ submatrix of $P$ such that $\det(\overline{Q(H,X)}) \neq 0$. Since $\mathrm{ch}(K) = 0$ and $H$ is abelian, then $\mathcal{J}(K[H]) = 0$. Thus, there exists a homomorphism $\varphi$ of $K[H]$ onto a field $L$ such that $\det \varphi(\overline{Q(H,X)}) = \det \overline{\varphi(Q(H,X))} \neq 0$. This extends to a homomorphism $L[H] \to L$. We can assume that $L$ is algebraically closed passing to the algebraic closure of $L$ if necessary. Now, we get an extension of the above homomorphism to a homomorphism of right $L[G]$-modules $L[G] \to V = L[G]/\ker(\varphi)L[G]$, defined by $\sum b_i x_i \to \sum \varphi(b_i) x_i$ for $b_i \in L[H]$, and the coset representatives $x_i \in X$ of $H$ in $G$. By Lemma 4, $V$ is the direct sum $V_1 \oplus \cdots \oplus V_s$ of some irreducible right $L[G]$-modules $V_i$. Let $\psi : L[G] \to \mathrm{End}_L R$ be the homomorphism defined by $\psi(r)(v) = vr$ for $v \in V$, $r \in L[G]$, and let $\psi_i(r) = \psi(r)|_{V_i}$ for any $r \in L[G]$, $i = 1, \ldots, s$. Clearly, $\psi_i : L[G] \to \mathrm{End}_L(V_i)$ is a finite dimensional representation of $L[G]$. Since $V_i$ is an irreducible $L[G]$-module, then its centralizer $D$ is a division algebra, finite dimensional over $L$. Now, $L$ is algebraically closed, so that $D = L$ and, from the density theorem, it follows that $\psi_i(L[G]) = \mathrm{End}_L(V_i)$. Thus, $\psi_i$ is an irreducible representation of $G$. It is then enough to show that $\rho(\overline{\psi_i(P)}) \geq k/n$ for some $i$. Since $\psi$ represents right multiplication, then it is easy to see that $\overline{\psi(P)} = \overline{\varphi(P,X)}$, where $\overline{\psi(P)}$ is obtained from $\psi(P)$ by replacing every entry by its matrix in the basis $\varphi(x_1), \ldots, \varphi(x_n)$ of $V$ over $L$. Let $T$ be the $L$-subspace of $V^I$ generated by all images of $V$ under the action of rows of the matrix $\psi(P)$, treated as linear transformations $V \to V^I$. Since every row of $\psi(P)$ maps $V_i$ onto a subspace $T_i$ of $T$ contained in $V_i^I$, $i = 1, \ldots, s$, then $T$ is the direct sum of $T_1, \ldots, T_s$. Moreover, $\rho(\overline{\psi_i(P)}) = \dim_L T_i$, so that we come to

$$\sum_{i=1}^{s} \rho(\overline{\psi_i(P)}) = \sum_{i=1}^{s} \dim_L T_i = \dim_L T = \rho(\overline{\psi(P)}) = \rho(\overline{\varphi(P(H,X))}) \geq k$$

Since $s \leq n$, this shows that, for some $i$, we have $\rho(\overline{\psi_i(P)}) \geq k/n$. This completes the proof of (i).

(ii) Assume that $\rho(\overline{P(H,X)}) = k$. We have a natural embedding $\mathfrak{M}(\mathbb{Q}[G], I, M, P) \subseteq \mathfrak{M}(\mathbb{Q}[H], I \cdot n, M \cdot n, \overline{P(H,X)})$ coming from the embedding $\mathbb{Q}[G] \to M_n(\mathbb{Q}[H])$, and it is enough to show that the latter satisfies the identity $x_{2k+1}[x_1, \ldots, x_{2k}]x_{2k+2} = 0$. Since $H$ is abelian, then

$\cap_\varphi \ker(\varphi) = 0$, where $\varphi$ runs over the set of irreducible representations of $Q[H]$. Thus, in view of the fact that $\overline{\varphi(P(H,X))} = \varphi(\overline{P(H,X)})$ for any $\varphi$, we get $\max_\varphi \{\rho(\varphi(\overline{P(H,X)}))\} = \rho(\overline{P(H,X)}) = k$. Now, as in the proof of Theorem 2, $K_0[S]$ is a subdirect product of algebras satisfying the identity $x_{2k+1}[x_1,\ldots,x_{2k}]x_{2k+2} = 0$, and so it satisfies the same identity.

We note that the second assertion of the above theorem also holds in the case where $\mathrm{ch}(K) = p > 0$ and $G$ has no nontrivial $p$-subgroups. However, when (i) was proved, Lemma 4 was used, the analog of which may not be true in this case.

In the case where $\mathrm{ch}(K) = p > 0$ and there is no restriction on the maximal subgroup $G$ of $S$, the problem comes from the fact that the rank controls $K_0[S]/\mathcal{J}(K_0[S])$ only. One might expect that the role of $P(H,X)$ should be played by the matrix $P(H/H',X)$, where $H$ has finite index in $G$ and $H'$ is a finite $p$-group. The following question arises: How does $\mathcal{J}(K_0[S])$ affect the rank of $\overline{P(H/H',X)}$?

We state some partial results, not dependent on the characteristic of $K$, which are based on another notion of rank of the sandwich matrix $P$. This notion of rank does not refer to the normal abelian subgroups of $G$ and to linear representations of $G$. Let $p$ be a prime or zero. Let $P$ be a $M \times I$ matrix over a group with zero $G^0$, and let $\mathrm{row}_p(P)$ denote the left $F_p[G]$-submodule of $F_p[G]^I$ generated by the rows of $P$, where $F_p$ is the prime field with $\mathrm{ch}(F_p) = p$. We say that $P$ has finite row $p$-rank equal to $n$ if there exists a set of $n$ rows of $P$ that is left $F_p[G]$-independent and if $n$ is maximal with respect to this property. Clearly, $P$ has finite row $p$-rank if the Goldie dimension of $\mathrm{row}_p(P)$ is finite. Moreover, it is easy to see that if $F_p[G]$ is a domain, then the row $p$-rank of $P$ coincides with the Goldie dimension of $\mathrm{row}_p(P)$. If $\mathrm{row}_p(P)$ is a finitely generated $F_p[G]$-module, then we say that $P$ has finite strong row $p$-rank, and we define it as the least number of rows of $P$ generating this module. Similarly, using right $F_p[G]$-modules $\mathrm{col}_p(P)$, we can define column analogs of the above notions. We note that $p$ is not necessarily a prime here.

**Proposition 6** Let $S = \mathfrak{M}^0(G,I,M,P)$ be a semigroup of matrix type. Assume that $\mathrm{ch}(K) = p$, not necessarily nonzero. Then,

(i)   If the algebra $K_0[S]$ satisfies a polynomial identity of degree $n$, then the row $p$-rank of $P$ is less than $n$.

(ii)  If the strong row $p$-rank of $P$ is finite and $K[G]$ satisfies a polynomial identity, then $K_0[S]$ satisfies a polynomial identity.

*Proof.* Since the algebras $K[S]$, $F_p[S]$ and also $K[G]$, $F_p[G]$ satisfy the same multilinear identities, then we may assume that $K = F_p$.

(i) We proceed by induction on $n$. Clearly $n > 1$. Let $n = 2$, and let $f(x,y)$ be a multilinear polynomial satisfied in $K_0[S]$. If $f(x,y) = xy$, then $P$ is a zero matrix, and so its row $p$-rank is zero. Thus, assume that $f(x,y) = xy + \lambda yx$ for some $0 \neq \lambda \in K$. Then, for any $m$, $n \in M$, $j \in I$, we have $(p_{mj}, j, n) = (1,j,m)(1,j,n) = -\lambda(1,j,n)(1,j,m) = -\lambda(p_{nj}, j, m)$. If $m \neq n$, then $p_{mj} = p_{nj} = \theta$. Thus, if $|M| > 1$, then the $j$th column of $P$ is zero. Since $j \in I$ is arbitrary, then $P$ is a zero matrix. Similarly, if $|I| > 1$, then $P$ is a zero matrix. Hence, if $P$ is nonzero, then $|M| = |I| = 1$ and the row $p$-rank of $P$ does not exceed 1.

Suppose, now, that $n > 2$ and the row $p$-rank of $P$ is not less than $n$. Let $\bar{S} = \mathfrak{M}^0(G, I, \bar{M}, \bar{P})$ be a subsemigroup of $S$ consisting of $n$ columns of $S$ corresponding rows of $P$ are left $K[G]$-independent. Further, let $\bar{\bar{S}} = \mathfrak{M}^0(G, I, \bar{\bar{M}}, \bar{\bar{P}})$ be a subsemigroup of $\bar{S}$ consisting of some $n-1$ columns of $\bar{S}$. Then the rows of $\bar{\bar{P}}$ are left $K[G]$-independent and, by the induction hypothesis, $K_0[\bar{\bar{S}}]$ does not satisfy any multilinear identity of degree less than $n$. On the other hand, $K_0[\bar{S}]$ satisfies a multilinear polynomial $f = \sum_{i=1}^{k} f_i x_i$, where $k \leq n$, and $f_i \in K[x_1, \ldots, x_{i-1}, x_{i+1}, \ldots, x_n]$ is a multilinear polynomial of degree $n - 1$. Let $a_1, \ldots, a_{n-1} \in K_0[\bar{\bar{S}}]$, $a_n \in K_0[J]$, where $J$ is the subsemigroup of $\bar{S}$ such that $\bar{\bar{S}} \cap J = \{\theta\}$, $\bar{S} = \bar{\bar{S}} \cup J$ (i.e., $J$ is the remaining column of $\bar{S}$). Then $f_1(a_1, \ldots, a_{n-1})a_n = 0$ since $f(a_1, \ldots, a_n) = 0$ and $K_0[\bar{\bar{S}}] \cap K_0[J] = 0$. This shows that $f_1(a_1, \ldots, a_{n-1})K_0[J] = 0$ and, by Lemma 2 in Chapter 5, it follows that $f_1(a_1, \ldots, a_{n-1}) = 0$. Thus, $f_1$ is an identity in $K_0[\bar{\bar{S}}]$, which contradicts the fact that $K_0[\bar{\bar{S}}]$ does not satisfy any multilinear polynomial of degree less than $n$.

(ii) Let $\bar{S}$ be the subsemigroup of $S$ consisting of a set $\bar{M} \subseteq M$ of $n$ columns of $S$ such that the corresponding $n$ rows of $P$ form a generating set of the left $K[G]$-module $\text{row}_p(P)$. From the right-left symmetric version of Lemma 5 in Chapter 5, it follows that $K_0[S]$ and $K_0[\bar{S}]$ coincide modulo the left annihilator of $K_0[S]$. Now, if $\varphi : K_0[\bar{S}] \to M_n(K[G])$ is the homomorphism given by $\varphi(a) = \bar{P} \circ a$ ($K_0[\bar{S}]$ being identified with the Munn algebra $\mathfrak{M}(K[G], I, \bar{M}, \bar{P})$), then, by the hypothesis and [149], Theorem 13.4.8, $M_n(K[G])$ is a PI-algebra. Since $\ker(\varphi)$ is the right annihilator of $K_0[\bar{S}]$ (see Lemma 2 in Chapter 5), then (ii) follows.

Notice that defining the column $p$-rank, and the strong column $p$-rank of $P$ through the right $F_p[G]$-submodule of $F_p[G]^M$ generated by the columns of $P$, we can get a column analog of Proposition 6.

**Corollary 7**   Assume that $S = \mathfrak{M}^0(G,I,M,P)$ is a semigroup of matrix type such that $K[G]$ is a $PI$-algebra. Let $p = \mathrm{ch}(K)$, not necessarily nonzero. If the strong row $p$-rank of $P$ is finite, then row $p$-rank and column $p$-rank of $P$ are finite.

*Proof.*   This is immediate through Proposition 6 and its column analog.

It is easy to construct examples showing that the converse fails.

**Example 8**   Let $G$ be a group that is not finitely generated. It is well known that the augmentation ideal $\omega(F_p[G])$ is not finitely generated as a left ideal of $F_p[G]$ because $F_p[G]\omega(F_p[H]) \neq \omega(F_p[G])$ for every proper subgroup $H$ of $G$; see [203], the proof of Lemma 10.2.2. Thus, there exists a sequence of elements $g_1, g_2, \ldots \in G$ such that $g_{n+1} - 1 \notin \sum_{i=1}^{n} F_p[G](g_i - 1)$ for any $n \geq 1$. Let $P = (p_{ij})$ be the $\mathbb{N} \times 2$ matrix over $G$ given by $p_{n1} = 1, p_{n2} = g_n$ for $n \geq 1$. It is straightforward that the $n$th row of $P$ is not a left $F_p[G]$-combination of the first $n - 1$ rows for any $n > 1$. Hence, the strong row $p$-rank of $P$ is infinite, while the strong column $p$-rank of $P$ is obviously finite. In particular, if $F_p[G]$ is a $PI$-algebra, then the row $p$-rank of $P$ is finite by the column analog of Corollary 7.

Our next aim is to study the special case where the maximal subgroup $G$ of $S$ is finite and to connect the $PI$-property of $K_0[S]$ to the results of Chapter 14.

**Lemma 9**   Let $P$ be a $M \times I$ matrix over a field $K$ such that $\{p_{ij} \mid i \in M, j \in I\}$ is a finite set. If $\rho(P) < \infty$, then $P$ has finitely many distinct rows.

*Proof.*   We may assume that $M$, $I$ are infinite and that even $M, I = \mathbb{N}$. Suppose that $\rho(P) = n < \infty$, and choose an $n \times n$ submatrix $Q$ of $P$ with $\det(Q) \neq 0$. We may assume that $Q = (p_{ij})_{i,j=1,\ldots,n}$. Let $m > n$, and $P_{(m)}$ be the $m$th row of $P$. Then there exist $\lambda_1^{(m)}, \ldots, \lambda_n^{(m)} \in K$ such that $(\lambda_1^{(m)}, \ldots, \lambda_n^{(m)})Q = (p_{m1}, \ldots, p_{mn})$. Since, for distinct $m \in M$, there are finitely many possible vectors $(p_{m1}, \ldots, p_{mn}) \in K^n$ by the assumption on $P$, then there are finitely many possible vectors $(\lambda_1^{(m)}, \ldots, \lambda_n^{(m)}) = (p_{m1}, \ldots, p_{mn})Q^{-1} \in K^n$. Thus, there are finitely many linear combinations of the rows $P_{(1)}, \ldots, P_{(n)}$ that coincide with a row of $P$. Now, since $\rho(P) = n$, then any row of $P$ is a linear combination of $P_{(1)}, \ldots, P_{(n)}$, and so $P$ has finitely many distinct rows.

**Theorem 10**   Let $S = \mathfrak{M}^0(G,I,M,P)$ be a completely 0-simple semigroup over a finite group $G$. Then the following conditions are equivalent:

(1)   $K[S]$ is a $PI$-algebra.

(2)  $K[S]$ is semiprimary.

(3)  $K[S]$ is semilocal.

Moreover, if $G$ has no nontrivial normal $p$-subgroups, then these are equivalent to:

(4)  $P$ has finitely many distinct rows.

(5)  $P$ has finitely many distinct columns.

*Proof.* Let $N$ be the largest normal $p$-subgroup of $G$, where $p = \mathrm{ch}(K)$ ($N$ is trivial if $p = 0$). From Chapter 5, we know that the kernel of the natural homomorphism $K_0[S] \to K_0[\bar{S}]$, $\bar{S} = \mathfrak{M}^0(G/N, I, M, \bar{P})$, is a nilpotent ideal. Therefore, it is enough to establish the equivalence of (1–5) in the case where $G$ has no nontrivial normal $p$-subgroups.

Assume that $K[S]$ is a $PI$-algebra. Then $L[S]$ is a $PI$-algebra for an algebraically closed field $L \supseteq K$. Since $G$ has no normal $p$-subgroups, then, by Theorem 3 in Chapter 18 and Lemma 5 in Chapter 4, $G \cong G/\rho_{\mathcal{J}(L[G])} \subset L[G]/\mathcal{J}(L[G]) \cong M_{n_1}(L) \oplus \cdots \oplus M_{n_r}(L)$ for some $r, n_i \geq 1$. Suppose that $P$ has infinitely many distinct rows. Then one of the matrices $\varphi_k(P) = (\varphi_k(p_{ij}))_{i \in M, j \in I}$ has infinitely many distinct rows, where $\varphi_k$ is the projection $L[G] \to M_{n_k}(L)$, $k = 1, \ldots, r$. Therefore, the matrix $\varphi_k(P) \in M_{(n_k \cdot M) \times (n_k \cdot I)}(L)$ has infinitely many distinct rows. From Lemma 9 if follows that $\rho(\varphi_k(P)) = \infty$ which, in view of Theorem 2, contradicts our assumption. Hence, (1) implies (4).

(4) $\Rightarrow$ (2) follows from Theorem 21 and Corollary 22, both in Chapter 14.

(2) $\Rightarrow$ (3) is obvious.

(3) $\Rightarrow$ (1) Since $S$ is locally finite by Corollary 22 in Chapter 14, then $K[S]/\mathcal{J}(K[S])$ is finite dimensional by Corollary 11 in Chapter 14. Hence, $K[S]$ is a $PI$-algebra.

The equivalence of (1–3) and (5) follows by a symmetric argument or through a direct observation that (4) $\Leftrightarrow$ (5) since $G$ is finite.

**Remark 11**  (1) If $\mathrm{ch}(K) = 0$, then, in view of Theorem 13 in Chapter 14, the implications (3) $\Leftrightarrow$ (4) $\Leftrightarrow$ (5) $\Rightarrow$ (1) hold with no assumption that $G$ is finite.

(2) If $|G| < \infty$, then the strong row 0-rank of $P$ is finite if and only if $P$ has finitely many distinct rows. (For example, use Proposition 6 and Theorem 10.)

(3) Assume that $\mathrm{ch}(K) = p > 0$, and let $G$ be a finite $p$-group. Let $P = (p_{ij})$, $p_{ij} = 1$ for $i \neq j$, $p_{ii} = g$, for $i, j = 1, 2, \ldots$, where $1 \neq g \in G$.

Put $S = \mathfrak{M}^0(G, \mathbb{N}, \mathbb{N}, P)$. Then $K[S]$ is a semiprimary $PI$-algebra by the above theorem and Theorem 13 in Chapter 14. Moreover, it is easy to see that the row $q$-rank of $P$ equals 1 because $1 - g$ is a zero divisor in $L[G]$ for any $q$ and any field $L$ with $\mathrm{ch}(L) = q$. Obviously, $P$ has infinitely many distinct rows; hence, $\mathbb{Q}[S]$ is not a $PI$-algebra. Observe that $S$ has the permutational property since $K[S]$ is a $PI$-algebra.

Our last aim is to show that, in certain cases, the $PI$-property of a semigroup algebra $K[S]$ may be verified "locally," i.e., through an investigation of the principal factor algebras of $S$. Since, for any free semigroup $X$, the principal factor algebras of $X$ satisfy the polynomial $xy - yx$, then it is clear that we will be mainly concerned with the class of semigroups with no null principal factors, i.e., with semisimple semigroups or semigroups close to semisimple.

Recall that, for $t \in S$, we denote by $J_t$ the principal ideal $S^1 t S^1$, while $I_t$ stands for the ideal of nongenerators of $J_t$. Moreover, $S_t$ denotes the principal factor $J_t/I_t$. We prove an auxiliary result of independent interest.

**Lemma 12**   Assume that $s - t \in \mathcal{J}(K[S])$ for some $s, t \in S$. Then there exists $n \geq 1$ such that $s^n \in J_t$. Moreover, if $S_s$ is a 0-simple semigroup, then $J_s \subseteq J_t$ and, if $t = s^2$, then $s$ is a periodic element.

*Proof.*   Let $\varphi : K[S] \to K_0[S/J_t]$ be the natural homomorphism. As in Lemma 1 in Chapter 13, it follows that $\varphi(s) = \varphi(s - t) \in \mathcal{J}(K_0[S/J_t])$ is a periodic element of $S/J_t$. Since it lies in the radical, it must be nilpotent because it is not transcendental over $K$. The same argument applied to any $xsy - xty \in \mathcal{J}(K[S])$, $x, y \in S$, implies that $(xsy)^k \in J_t$ for some $k \geq 1$. If $J_s \nsubseteq J_t$, then $S_s$ is a homomorphic image of $J_s/(J_s \cap J_t)$. Thus, $S_s$ is a nontrivial nil semigroup, which implies that it is not 0-simple. Finally, if $t = s^2$, then Lemma 1 in Chapter 13 shows that $s$ is periodic.

**Theorem 13**   Assume that $K[S]$ is a $PI$-algebra. Then,

(i)  If $S$ is a semisimple semigroup, then it is completely semisimple and strongly $\pi$-regular of bounded index.

(ii)  If $S$ is a weakly periodic semigroup, then it is strongly $\pi$-regular.

*Proof.*   We know that $\bar{S} = S/\rho_{\mathcal{J}(K[S])}$ is a subdirect product of a family of linear semigroups $S^{(i)} \subseteq M_N(L_i)$, $i \in I$, for some fixed $N \geq 1$ and some fields $L_i \supseteq K$; see Theorem 1 in Chapter 18 and Lemma 5 in Chapter 4.

(i)  The first assertion follows from Proposition 1 and Theorem 17 both in Chapter 19. Since $S$ is regular, then so is any $S^{(i)}$ and, from Corollary 11 in Chapter 3, it follows that $S^{(i)}$ is strongly $\pi$-regular of bounded index $n_i$

and there exists $k \geq 1$ such that $n_i \leq k$ for all $i \in I$. Let $t = (t_i)_{i \in I} \in \bar{S}$. Then, for some $n \geq 1$, every $t_i^n$ lies in a subgroup $G_i$ of $S^{(i)}$, and let $e_i$ denote the identity of $G_i$. The regularity of $\bar{S}$ implies that $t^{3n}yt^{3n} = t^{3n}$ for some $y = (y_i)_{i \in I} \in \bar{S}$. Thus, $t_i^n y_i t_i^{2n} \in e_i S^{(i)} e_i$, and the above equality implies that $t_i^n y_i^n t_i^{2n} = e_i \in G_i$. Let $e = (e_i)_{i \in I}$. Then $t^n y t^{2n} = e$, and so $e \in \bar{S}$. Moreover, the element $t^n$ lies in the group of units of the submonoid $e\bar{S}e$ of $\bar{S}$. This shows that $\bar{S}$ is strongly $\pi$-regular of bounded index. Let $\varphi : S \to \bar{S}$ be the natural homomorphism. Assume that $\varphi(s) \in G$ for some $s \in S$ and a subgroup $G$ of $\bar{S}$. Suppose that $s^2 \in I_s$. If $\varphi(I_s) = J_{\varphi(s)}$, then $\varphi(u) = \varphi(s)$ for some $u \in I_s$, so that $s - u \in \mathcal{J}(K[S])$. This contradicts Lemma 12. Therefore, $\varphi(s^2) \in \varphi(I_s) \subseteq I_{\varphi(s)} \subsetneqq J_{\varphi(s)}$. This contradicts the fact that $\varphi(s^2) = \varphi(s)^2 \in G \subseteq J_{\varphi(s)} \setminus I_{\varphi(s)}$. Therefore, $s^2 \in J_s \setminus I_s$. Since $S$ is completely semisimple, this shows that $s$ lies in a subgroup of $J_s/I_s$, hence in a subgroup of $S$. Thus, the inverse image of a subgroup of $\bar{S}$ is a subgroup of $S$, and so $S$ is strongly $\pi$-regular of bounded index (the same as the index of $\bar{S}$). This completes the proof of (i).

(ii) Since all 0-simple principal factors of $S^{(i)}$, $i \in I$, are completely 0-simple by (i), then $S^{(i)}$ is $\pi$-regular. An argument similar to that given above shows that $\bar{S}$ is a strongly $\pi$-regular semigroup. If $s \in S$ is such that $\varphi(s)$ lies in a subgroup $G$ of $\bar{S}$, then, let $e \in S$ be such that $\varphi(e)$ is the identity of $G$. Lemma 12 allows us to assume that $e = e^2$. Since $se - s \in \mathcal{J}(K[S])$, Lemma 12 implies also that $J_{s^k} \subseteq J_{se} \subseteq J_e$ for some $k \geq 1$. Now, if $t \in S$ is such that $\varphi(t)$ is the inverse of $\varphi(s^k)$ in $G$, then $s^k t - e \in \mathcal{J}(K[S])$, and so $J_e \subseteq J_{s^k t} \subseteq J_{s^k}$. It follows that $J_e = J_{s^k}$. Similarly, one shows that $J_{s^{2k}} = J_e$. Since $S_e$ is a completely 0-simple semigroup by (i), it then follows that $s^k$ lies in a subgroup of $S$. This establishes the strong $\pi$-regularity of $S$.

For semigroups $S^{(i)}$ arising from $S$ as in Theorem 13, we obtain a "local" criterion for the $PI$-property.

**Proposition 14** Let $S \subseteq M_n(L)$ be a linear semigroup, where $L$ is a field. If $S$ is weakly periodic, then $K[S]$ is a $PI$-algebra if and only if there is a polynomial identity satisfied in all principal factor algebras $K_0[S_t]$, $t \in S$.

*Proof.* From Theorem 10 in Chapter 3 and Theorem 13, it follows that there exists a chain of ideals $T_1 \subseteq T_2 \subseteq \cdots \subseteq T_r = S$ of $S$ such that $T_1$ and all $T_i/T_{i-1}$, $i \geq 2$, are completely 0-simple or nilpotent. If one of these semigroups is completely 0-simple, then it coincides with some principal factor $S_t$, $t \in S$. Therefore, the algebra $K[S]$ has a chain of ideals $0 \subseteq K[T_1] \subseteq K[T_2] \subseteq \cdots \subseteq K[T_r]$, with $K_0[T_1]$ and all factor algebras nilpotent or coinciding with some principal factor algebras of $K[S]$; see

Lemma 7 in Chapter 4. Thus, the result follows easily through the fact that the class of $PI$-algebras is closed under ideal extensions.

Unfortunately, the 0-simple principal factors of the semigroups $S^{(i)}$, $i \in I$, arising in the proof of Theorem 13, need not be images of principal factors of $K[S]$ (which is the case for the principal factors of $S/\rho_{\mathcal{J}(K[S])}$ by Lemma 12). Thus, Proposition 14 cannot be applied to derive the $PI$-property of $K[S]/\mathcal{J}(K[S])$ once we know that $K[S]$ is a "locally $PI$-algebra"; that is, all principal factor algebras of $S$ are $PI$. However, this may be achieved in the case of finitely generated semigroups.

**Theorem 15** Let $S$ be a finitely generated semigroup that is weakly periodic. Then $K[S]$ is a $PI$-algebra if and only if $S$ has a chain of ideals $S = T_r \supseteq T_{r-1} \supseteq \cdots \supseteq T_1$ with all $T_1$, $T_i/T_{i-1}$ nilpotent or completely 0-simple and with the corresponding semigroup algebras being $PI$-algebras. Moreover, in this case, $K[S]/\mathcal{J}(K[S])$ satisfies any identity that holds in all principal factor algebras of $S$.

*Proof.* The sufficiency of the listed conditions follows as in Proposition 14. Thus, assume that $K[S]$ is a $PI$-algebra. Since $\mathcal{J}(K[S]) = \mathcal{B}(K[S])$ by Theorem 12 in Chapter 21, then it is an intersection of finitely many prime ideals of $K[S]$; see [222], Section 5.2. We know that $\bar{S} = S/\rho_{\mathcal{J}(K[S])}$ embeds into a product $\prod_{i=1}^{n} M_{m_i}(L_i)$ for some field extensions $L_i$ of $K$, $m_i \geq 1$, in this case. Hence, $\bar{S} \subseteq M_N(L)$ for a field $L \supseteq K$ and $N \geq 1$. Since $\bar{S}$ is $\pi$-regular by Theorem 17 in Chapter 19, then Proposition 8 in Chapter 3 implies that $\bar{S}$ has finitely many $\mathcal{J}$-classes with idempotents. From Lemma 12 it follows that, for any $e = e^2, f = f^2 \in S$, we have $J_e = J_f$ whenever $J_{\varphi(e)} = J_{\varphi(f)}$, where $\varphi : S \to \bar{S}$ is the natural homomorphism. Therefore, $S$ also has finitely many $\mathcal{J}$-classes with idempotents and so, by Theorem 3 in Chapter 3, there exists a chain of ideals of $S$, $S = T_r \supset T_{r-1} \supset \cdots \supset T_1$, with $T_1$ and all $T_i/T_{i-1}$, $i \geq 2$, nil or 0-simple. Assume that $T_1$ or some $T_i/T_{i-1}$ is a nil semigroup. Then $K_0[T_1]$, or $K_0[T_i/T_{i-1}]$, is contained in the radical of the algebra $K_0[S]$, or $K_0[S/T_{i-1}]$, respectively, because it is locally nilpotent by Theorem 3 in Chapter 19. From Theorem 1 in Chapter 18, it then follows that $T_1$, or $T_i/T_{i-1}$, respectively, is a nilpotent semigroup. If one of $T_i/T_{i-1}$ or $T_1$ is a 0-simple semigroup, then it is completely 0-simple by Theorem 13.

Let $\psi : K[S] \to R$ be a homomorphism onto a right primitive algebra $R$, and let $j$ be minimal with the property $\psi(K[T_j]) \neq 0$. Since $\psi(K[T_j])$ is an ideal of $R$, then $R = \psi(K[T_j])$ by Theorem 1 in Chapter 18. Moreover,

$\psi(K[T_{j-1}]) = 0$ if $j \geq 2$, and so $\psi(K[T_j])$ is a homomorphic image of $K[T_j]/K[T_{j-1}] \cong K_0[T_j/T_{j-1}]$ or $K[T_1]$ if $j = 1$. The result follows.

Observe that, as in the above proof, it is enough to assume that $K[S]/\mathcal{J}(K[S])$ is a Goldie $PI$-algebra in order to show that $S$ has finitely many $\mathcal{J}$-classes containing idempotents, whenever $S$ is any weakly periodic semigroup.

An important case where the above result may be applied is given below.

**Corollary 16**  Assume that $S$ is a weakly periodic semigroup such that $K[S]$ is right noetherian. Then $K[S]$ is a $PI$-algebra if and only if every principal factor algebra $K_0[T]$ of $S$ satisfies a polynomial identity.

*Proof.*  This is a direct consequence of Theorem 6 in Chapter 12 and Theorem 15.

We close this chapter with another particular case, in which the $PI$-property may be checked locally—through the principal factor algebras.

**Theorem 17**  Let $K$ be a field of characteristic $p$ (not necessarily $\neq 0$), and let $S$ be a strongly $p$-semisimple semigroup. Then $K[S]$ is a $PI$-algebra if and only if the following conditions are satisfied:

(1)  $S$ is completely semisimple with a bound on the number of rows and columns of all principal factors.

(2)  There is a polynomial identity satisfied in all group algebras $K[G]$, where $G$ is a subgroup of $S$.

*Proof.*  Assume that $K[S]$ is a $PI$-algebra. Since $S$ is semisimple, then it must be completely semisimple by Theorem 13. Hence, Corollary 26 in Chapter 5 and the Amitsur–Levitzki theorem imply that (1) holds. Assertion (2) is obvious.

The converse is a direct consequence of Proposition 28 and Corollary 26 both in Chapter 5, and of the fact that if an algebra $R$ satisfies an identity, then the matrix algebra $M_n(R)$ also satisfies an identity; see [149], Theorem 13.4.8. Specifically, $K[S]$ is a subdirect product of matrix algebras of bounded size over group algebras of maximal subgroups of $S$.

**Remark 18**  In view of Theorem 3 in Chapter 18, condition (2) in the above theorem may be restated as follows: there exists $N \geq 1$ such that every subgroup $G$ of $S$ has a normal subgroup $H$ such that $H$ is abelian if $\mathrm{ch}(K) = 0$ ($H'$ is a finite $p$-group if $\mathrm{ch}(K) = p > 0$) and $H' \cdot [G:H] \leq N$.

Note also that the above conditions mean that there exists an identity that is satisfied in all principal factor algebras $K_0[S_t]$, $t \in S$.

The most important application of Theorem 17 is concerned with the case of inverse semigroups.

**Theorem 19**  Let $S$ be an inverse semigroup. Then $K[S]$ is a $PI$-algebra if and only if the following conditions are satisfied:

(1)  There is a bound on the number of idempotents of the principal factors of $S$.
(2)  There is a polynomial identity satisfied in all group algebras $K[G]$, where $G$ is a subgroup of $S$.

*Proof.*  Since $S$ is inverse, then condition (1) is equivalent to the respective condition in Theorem 17. Since the strong $p$-semisimplicity of $S$ is a consequence of (1) (see Corollary 26 in Chapter 5), as well as of the $PI$-property of $K[S]$ (see Proposition 20 in Chapter 19 and Theorem 13), then the result follows from Theorem 17.

Finally, we list one more situation where strongly $p$-semisimple semigroups are involved.

**Proposition 20**  Let $K[S]$ be a regular $PI$-algebra. Then $S$ is a strongly $p$-semisimple semigroup, $p = \text{ch}(K)$ not necessarily $\neq 0$, which is locally finite.

*Proof.*  From Chapter 15 we know that any principal factor algebra $K_0[S_t]$, $t \in S$, is a regular $PI$-algebra and $S_t \cong \mathfrak{M}^0(G,I,M,P)$ is completely 0-simple. From Proposition 6 it follows that there exists $n \geq 1$ and a set of $n$ rows of $P$ such that any larger set of rows is left $K[G]$-dependent. Hence, the number of rows does not exceed $n$ since otherwise $K_0[S_t]$ has a nonzero left annihilator. Now, from Theorem 6 and Lemma 4 both in Chapter 15, it follows that the number of columns of $S_t$ is also finite and that $K_0[S_t]$ has an identity. Since $S$ is periodic by Proposition 2 in Chapter 15, then it is locally finite by Theorem 3 in Chapter 19.

## Comments on Chapter 20

The proofs of Theorems 2 and 5 follow the paper of Domanov [46], where the study of the $PI$-property of semigroup algebras of completely 0-simple semigroups was started. The auxiliary Lemma 4 is well known for more general algebras—we refer to [168] for another proof and a discussion of this topic. Row and strong row ranks of sandwich matrices were introduced by Okniński in [184], where Proposition 6 was proved. Also, Theorem 13 is taken from [184]. Domanov [49] showed that the algebra of a completely 0-simple semigroup over a finite group is left perfect if and only if it satisfies a polynomial identity, while the connection of conditions (2) and (4) in Theorem 10 was noted by Finkelstein and Kozhukhov in [58]; see Chapter 14. A description of $PI$-semigroup-algebras of inverse semigroups was first given by Domanov in [47]. Condition (1) of our Theorem 19 is there replaced by an alternative finiteness condition on the set of idempotents of $S$. We note, in view of Proposition 20, that Chekanu showed in [19] that every locally finite regular $PI$-algebra is locally semisimple, that is, every finite subset lies in a finite dimensional semisimple subalgebra.

# 21

# The Radical

In this chapter, we study the Jacobson radical of semigroup algebras satisfying polynomial identities. One of the main general results establishes the equality $\mathcal{J}(K[S]) = \mathcal{B}(K[S])$ for algebras of this type. This may be compared to the Nullstellensatz for $PI$-algebras proved by Braun in [17]: the Jacobson radical of a finitely generated $PI$-algebra is a nilpotent ideal. Thus, the hypothesis on the existence of a multiplicative basis in an algebra $R$ leads to a consequence related to that obtained in the finitely generated case. However, we are not able to prove the above equality using Braun's theorem through a reduction to finitely generated subalgebras. Our second main aim in this chapter is to find a description of elements lying in $\mathcal{J}(K[S])$. This is particularly readable in the cases of cancellative semigroups and commutative semigroups and, more generally, in the so-called permutative semigroups.

We start with the basic class of commutative semigroup algebras. While the second assertion of the following proposition directly follows from Corollary 5 in Chapter 11, we offer a simpler proof here. It refers to the semisimplicity of certain commutative group algebras only.

**Proposition 1** Let $S$ be a commutative cancellative semigroup, and let $K$ be a field of characteristic $q$. If $q > 0$, then assume also that $S$ is $q$-separative. Then $K[S]$ has no nonzero Jacobson radical subalgebras. In particular, $\mathcal{J}(K[S]) = 0$.

*Proof.* Let $G$ be the group of fractions of $S$. If $q > 0$, then we know that $G$ has no nontrivial $q$-subgroups. It is enough to show that $K[G]$ has no nonzero radical subalgebras.

Suppose that $R$ is a subalgebra of $K[G]$ and $0 \neq a \in \mathcal{J}(R) = R$. From Corollary 16 in Chapter 4, if follows that $a \in \mathcal{J}(R \cap K[H])$, where $H$ is the subgroup of $G$ generated by supp($a$). Therefore, replacing $G$ by $H$, we can assume that $G$ is finitely generated. It is well known that $\mathcal{J}(K[G]) = 0$; see [107], Section 3.4. Therefore, $K[G]$ is a subdirect product of finite field extensions $K_\alpha$, $\alpha \in \mathcal{A}$, of $K$. Then $R$ is a subdirect product of some $K$-subalgebras $D_\alpha \subseteq K_\alpha$, $\alpha \in \mathcal{A}$. Since $\dim_K D_\alpha < \infty$, then $\mathcal{J}(D_\alpha) = 0$ (in fact, $D_\alpha$ is a field) and, hence, $\mathcal{J}(R) = 0$. This contradicts the supposition and proves the assertion.

The following result was first established by Munn in [159].

**Theorem 2** Let $S$ be a commutative semigroup. Then $\mathcal{J}(K[S]) = \mathcal{B}(K[S]) = I(\rho)$, where

$$\rho = \begin{cases} \text{the least separative congruence on } S & \text{if } \mathrm{ch}(K) = 0 \\ \text{the least } p\text{-separative congruence on } S & \text{if } \mathrm{ch}(K) = p > 0 \end{cases}$$

*Proof.* We will first show that $I(\rho)$ is a nil ideal. Choose $x, y \in S$ with $(x,y) \in \rho$. If $\mathrm{ch}(K) = 0$, then, for some $n \geq 1$, we have

$$xy^n = y^{n+1}, \; yx^n = x^{n+1}$$

and so

$$(x - y)x^n = 0 = (x - y)y^n$$

Thus,

$$(x - y)^{2n+1} = (x - y) \sum_{i=0}^{2n} \binom{2n}{i} x^{2n-i} y^i = 0$$

If $\mathrm{ch}(K) = p > 0$, then $x^{p^k} = y^{p^k}$ for some $k \geq 1$, and so $(x-y)^{p^k} = 0$. Hence, in both cases, $I(\rho)$ is spanned by nilpotents (see Lemma 1 in Chapter 4), and, consequently, $I(\rho)$ is a nil ideal of $K[S]$. Since $K[S]/I(\rho) \cong K[S/\rho]$, then it remains to show that $\mathcal{J}(K[S/\rho]) = 0$. Now $S/\rho$ is separative and, by Theorem 5 in Chapter 6, $S/\rho$ is a semilattice $\mathcal{A}$ of cancellative semigroups $S_\alpha$, $\alpha \in \mathcal{A}$, so that $S/\rho$ embeds into a semigroup $G$ that is a semilattice $\mathcal{A}$ of the groups of fractions $G_\alpha$ of $S_\alpha$, $\alpha \in \mathcal{A}$. Moreover, by the hypothesis, every $S_\alpha$ is $p$-separative if $\mathrm{ch}(K) = p > 0$. Proposition 8 in Chapter 6 implies that $K[G]$ is a subdirect product of all group algebras $K[G_\alpha]$, $\alpha \in \mathcal{A}$. From Proposition 1, it follows that $K[S/\rho] \subseteq K[G]$ is a subdirect product of semisimple algebras. Therefore, $\mathcal{J}(K[S/\rho]) = 0$. This completes the proof.

**Remark 3**  Let $S$ be a commutative semigroup that is separative, and $p$-separative if $\mathrm{ch}(K) = p > 0$. From Theorem 2 it follows that $S$ is a subdirect product of semigroups $T_\beta$, $\beta \in \mathcal{B}$, determined by the homomorphisms $\varphi_\beta : K[S] \rightarrow K_\beta$ onto some fields $K_\beta \supseteq K$ such that $K[S]$ is a subdirect product of $K_\beta$, $\beta \in \mathcal{B}$. Moreover, $\bigcap_{\beta \in \mathcal{B}} I(\rho_\beta) = 0$, where $\rho_\beta$ are congruences on $S$ determined by $\varphi_\beta$ and, clearly, every $T_\beta \setminus \{0\} \subseteq K_\beta$ is a cancellative semigroup that is $p$-separative if $\mathrm{ch}(K) = p > 0$.

Once the proof of Theorem 2 has been reduced to the case of the separative ($p$-separative if $\mathrm{ch}(K) = p > 0$) semigroup $S/\rho$, we might also use an alternative technique coming from so-called $K$-complete semigroups (see [203], [251]) instead of using Proposition 1. We note that the original proof of Theorem 2 involves some more semigroup theoretic machinery [159]. Observe also that Theorem 2 extends the well-known description of the radical of commutative group algebras [107], Section 3.4. Here the least $p$-separative congruence plays the role of the largest $p$-subgroup.

The assertion of Theorem 2 may be extended to a wider class of semigroups. We say that $S$ is a permutative semigroup if there exists $n \geq 2$ and a nontrivial permutation $\sigma$ in the symmetric group $\mathcal{S}_n$ such that

(a)                    $x_1 x_2 \ldots x_n = x_{\sigma(1)} x_{\sigma(2)} \ldots x_{\sigma(n)}$

for every $x_1, \ldots, x_n \in S$. Clearly, this property is stronger than that considered in Chapter 19. In fact, since (a) is a multilinear identity in $S$, then the following is straightforward.

**Lemma 4**  Assume that $S$ is a permutative semigroup. Then $K[S]$ is a $PI$-algebra.

The basic auxiliary result on this class of semigroups was obtained by Putcha and Yaqub in [228].

**Lemma 5** Assume that $S$ is a permutative semigroup. Then there exists $m \geq 1$ such that, for any $x, y \in S$, $s, t \in S^m$, we have $sxyt = syxt$.

*Proof.* We know that

(b)  $x_1 \ldots x_n = x_{\sigma(1)} \ldots x_{\sigma(n)}$ for any $x_1, \ldots, x_n \in S$, and a fixed nontrivial permutation $\sigma \in \mathcal{S}_n$, $n \geq 2$.

We can assume that $n > 2$.

Suppose first that $\sigma(1) \neq 1$. Let $i$ and $j$ be such that $\sigma(i) = 1$, $\sigma(1) = j$. We apply the identity (b) to the elements $x_1 t_1, t_2, \ldots, t_n$, where $t_1, \ldots, t_n \in S$. Since $\sigma(1) \neq 1$, then, multiplying by $x_2$, we get

$$x_2(x_1 t_1 \ldots t_n) = x_2(t_{\sigma(1)} \ldots t_{\sigma(i-1)} x_1 t_{\sigma(i)} t_{\sigma(i+1)} \ldots t_{\sigma(n)})$$

Now, we use the identity (b) in reverse on the elements $x_2 t_{\sigma(1)}, t_{\sigma(2)}, \ldots, t_{\sigma(i-1)}, x_1 t_{\sigma(i)}, t_{\sigma(i+1)}, \ldots, t_{\sigma(n)}$ to obtain

$$x_2 x_1 t_1 \ldots t_n = x_1 t_1 \ldots t_{j-1} x_2 t_j t_{j+1} \ldots t_n$$

Then we use (b) on the elements: $t_1, \ldots, t_{j-1}, x_2 t_j, t_{j+1}, \ldots, t_n$, and we get

$$x_2 x_1 t_1 \ldots t_n = x_1 x_2 t_{\sigma(1)} \ldots t_{\sigma(i-1)} t_{\sigma(i)} \ldots t_{\sigma(n)}$$

Finally, we use (b) in reverse on $t_{\sigma(1)}, t_{\sigma(2)}, \ldots, t_{\sigma(n)}$, and so $x_2 x_1 t_1 \ldots t_n = x_1 x_2 t_1 \ldots t_n$. Thus, the assertion follows with $m = n$.

Now, consider the case when $\sigma(1) = 1$. Let $m$ be the least integer such that $\sigma(m) \neq m$. Then $uy_1 \ldots y_r = uy_{\tau(1)} \ldots y_{\tau(r)}$ for any $u \in S^{m-1}$ and any $y_1, \ldots, y_r \in S$, where $r = n - m + 1$ and $\tau \in \mathcal{S}_r$ is such that $\tau(1) \neq 1$. Thus, repeating the above arguments starting from elements $x_1 s_1, s_2, \ldots, s_r$, we get

$$
\begin{aligned}
ux_2(x_1 s_1 \ldots s_r) &= ux_2(s_{\tau(1)} \ldots s_{\tau(k-1)} x_1 s_{\tau(k)} \ldots s_{\tau(r)}) \\
&= ux_1 s_1 \ldots s_{l-1} x_2 s_l \ldots s_r \\
&= ux_1 x_2 s_{\tau(1)} \ldots s_{\tau(r)} \\
&= ux_1 x_2 s_1 \ldots s_r
\end{aligned}
$$

for any $s_1, \ldots, s_r \in S$, where $\tau(k) = 1$ and $\tau(1) = l$. Therefore, $uxyv = uyxv$ for any $u \in S^{m-1}$, $v \in S^r$, $x, y \in S$, and so this identity holds for any $u$, $v \in S^{n-1}$, $x, y \in S$.

Our next result is an immediate consequence of Lemma 5.

**Corollary 6** If $S$ is a permutative semigroup, then $K[S]^m (xy - yx) K[S]^m = 0$ for some $m \geq 1$ and any $x, y \in S$. Consequently, the commutator ideal of $K[S]$ is nilpotent.

The following extension of Theorem 2 is due to Nordahl [171].

**Theorem 7**  Let $S$ be a permutative semigroup. Then $\mathcal{J}(K[S]) = I(\rho)$, where $\rho$ is the congruence on $S$ such that

$$S/\rho = \begin{cases} (S/\mu)/\xi & \text{if } \mathrm{ch}(K) = 0 \\ (S/\mu)/\xi_p & \text{if } \mathrm{ch}(K) = p > 0 \end{cases}$$

where $\mu$ is the least commutative congruence on $S$ and $\xi$, $\xi_p$ are the least separative and $p$-separative congruences on $S/\mu$. Moreover, $\mathcal{J}(K[S])$ is a sum of nilpotent ideals of $K[S]$ and coincides with the set of nilpotent elements of $K[S]$.

*Proof.*  The congruence determined by the commutator ideal of $K[S]$ is a commutative congruence on $S$, so that it contains $\mu$. From Corollary 6 it then follows that the ideal $I(\mu)$ is nilpotent. Since $S/\mu$ is commutative, then the first assertion follows through Theorem 2. Further, since $\mathcal{J}(K[S/\mu])$ is a sum of nilpotent ideals and coincides with the set of nilpotents of $K[S/\mu]$, then the same holds for $\mathcal{J}(K[S])$.

It is not known whether, if $S$ is not necessarily commutative, $\mathcal{J}(K[S])$ always is a nil ideal even if $S$ is a group. Simple examples showing that $\mathcal{J}(K[S])$ may not be of the form $I(\rho)$ for a congruence $\rho$ on $S$ are obtained via Maschke's theorem.

Our next goal is to look for an extension of Theorem 2 to the class of arbitrary $PI$-semigroup-algebras. It turns out that the case where $S$ is cancellative will be crucial. It was settled by Okniński in [183].

**Theorem 8**  Let $S$ be a cancellative semigroup, and let $\mathrm{ch}(K) = p$ (not necessarily $> 0$). Assume that $K[S]$ is a $PI$-algebra. Then $S$ has a (two-sided) group of fractions $G$ such that $K[G]$ is a $PI$-algebra. Moreover,

(i)  $\mathcal{B}(K[S]) = \mathcal{J}(K[S]) = \mathcal{J}(K[G]) \cap K[S]$, and $\mathcal{J}(K[S]) = 0$ if $p = 0$, or if $p > 0$ and $G$ has no normal subgroups of order divisible by $p$.

(ii)  For any $s, t \in S$, $(s,t) \in \rho_{\mathcal{J}(K[S])}$ if and only if $s$ and $t$ are in the same coset of some normal $p$-subgroup of $G$.

*Proof.*  The first assertion is established in Theorem 1 in Chapter 20.

(i) Since $K[S]$ is a $PI$-algebra, then the $FC$-subgroup $\Delta(G)$ of $G$ has finite index in $G$; see Theorem 3 in Chapter 18. Hence (i) follows from Corollary 5 in Chapter 11.

(ii) Let $s, t \in S$ and $\mathrm{ch}(K) = p > 0$. By (i) $(s,t) \in \rho_{\mathcal{J}(K[S])}$ if and only if $(s,t) \in \rho_{\mathcal{J}(K[G])}$. Thus, the assertion is a direct consequence of Corollary 5 in Chapter 11.

A special case of interest arises when considering separative semigroups as defined in Chapter 6. Out aim is to get an extension of Proposition 1 and Theorem 8 to this class. While the proof might be obtained through subdirect decompositions by an extension of the method used in the proof of Theorem 2, we offer another interesting technique coming from a general result on semilattices of semigroups.

**Proposition 9** Let $S$ be a semilattice $\Omega$ of semigroups $S_\alpha$, $\alpha \in \Omega$. Assume that $\mathcal{J}(K[S_\alpha]) = 0$ for every $\alpha \in \Omega$. Then $\mathcal{J}(K[S]) = 0$.

*Proof.* Suppose that $0 \neq a \in \mathcal{J}(K[S])$. Write $a$ as $\sum_{\alpha \in \Omega} a_\alpha$, where $\mathrm{supp}(a_\alpha) \subseteq S_\alpha$. Let $\beta \in \Omega$ be maximal with respect to the natural ordering of $\Omega$ in the set $\{\alpha \in \Omega \mid a_\alpha \neq 0\}$. Define $I = \{\alpha \in \Omega \mid \alpha \not\geq \beta \text{ in } \Omega\}$. Then $I$ is an ideal of $\Omega$, with an associated ideal $S(I) = \{s \in S \mid s \in S_\alpha \text{ for some } \alpha \in I\}$ of $S$. It is clear that $T = S/S(I)$ is a semilattice $\Omega/I$ of semigroups $S_\alpha$, $\alpha \in \Omega \setminus I$, and $S_\theta = \{\theta\}$ if $I \neq \varnothing$; see Chapter 6. Therefore, $K[S_\beta]$ is an ideal in the algebra $K_0[T]$. Moreover, by the choice of $\beta$, $0 \neq \varphi(a_\beta) = \varphi(a) \in \mathcal{J}(K_0[T]) \cap K[S_\beta] \subseteq \mathcal{J}(K[S_\beta])$ where $\varphi: K[S] \to K_0[T]$. This contradicts the hypothesis and proves the result.

Let $G$ be a group generated by it subsemigroup $S$. Assume that $S$ is $p$-separative for some prime $p$ and that $G/\Delta(G)$ is a finite group (for example, this is the case if $K[S]$ is a $PI$-algebra). From Corollary 8 in Chapter 7, we know that $G = S(Z(\Delta(G)) \cap S)^{-1}$. Suppose that $g = st^{-1}$, $s \in S$, $t \in Z(\Delta(G))$, is such that $g^p = 1$. There exists $m \geq 1$ satisfying $s^m t = ts^m$ because $[G : C_G(t)] < \infty$. Then $g = st^{-1} = s^m(ts^{m-1})^{-1}$. Since $g^p = 1$ and $s^m$ centralizes $ts^{m-1}$, then $(s^m)^p = (ts^{m-1})^p$. The hypothesis on $S$ implies that $s^m = ts^{m-1}$, and, consequently, $s = t$. Then $g = 1$, which shows that $G$ has no $p$-torsion. Therefore, $\mathcal{J}(K[S]) = 0$ holds, by Corollary 5 in Chapter 11, for every field $K$ with $\mathrm{ch}(K) = p$.

**Theorem 10** Let $S$ be a separative semigroup. Then $K[S]$ is a $PI$-algebra if and only if there exists a polynomial identity that is satisfied in all algebras $K[T]$, $T$—a cancellative subsemigroup of $S$. Moreover, in this case, $\mathcal{J}(K[S]) = 0$ if $\mathrm{ch}(K) = 0$, or if $\mathrm{ch}(K) = p > 0$ and $S$ is $p$-separative.

*Proof.* From Theorem 5 in Chapter 6, we know that $S$ is a semilattice $\Omega$ of cancellative semigroups $S_\alpha$, $\alpha \in \Omega$. Every $S_\alpha$ has a group of fractions $G_\alpha$, by Theorem 1 in Chapter 20 provided that $K[S_\alpha]$ is a $PI$-algebra. Moreover, $K[G_\alpha]$ satisfies the same multilinear identities as $K[S_\alpha]$ in this case. Since $K[S]$ embeds into a subdirect product of the algebras $K[G_\alpha]$, $\alpha \in \Omega$, by Theorem 5 and Proposition 8, both in Chapter 6, then the former assertion follows. The latter is a direct consequence of Theorem 8 and the comment preceding the theorem.

We note that a description of $\mathcal{J}(K[S])$ for a semigroup $S$ being a semilattice of semigroups $S_\alpha$, $\alpha \in \Omega$, may not be given in general in terms of the components $K[S_\alpha]$ only. Namely, examples of semilattices of groups $G_\alpha$, $\alpha \in \Omega$, are known such that $\mathcal{J}(K[S]) = 0$, but $\mathcal{J}(K[G_\alpha]) \neq 0$ for $\alpha \in \Omega$. The structure of the radical appears to be dependent on the relations between the components $G_\alpha$. This observation leads to many examples illustrating the complexity of the general problem of describing $\mathcal{J}(K[S])$ for a semilattice of (semi)groups, and also for inverse semigroups [164], [217], [258].

To deal with the case of arbitrary $PI$-semigroup-algebras, we need the following extension of Corollary 18 in Chapter 5.

**Lemma 11** Let $S$ be a subsemigroup of a completely 0-simple semigroup $T = \mathfrak{M}^0(G,I,M,P)$. Then,

$$\mathcal{J}(K_0[S])^3 \subseteq \sum_{\substack{i \in I \\ m \in M}} \mathcal{J}(K_0[S_{im}])$$

where $S_{im}$ denotes the set of all elements of $S$ lying in the $i$th row and the $m$th column of $T$.

*Proof.* Let $b \in K_0[S]$ be a quasi-inverse of an element $a \in K_0[S_{im}]$ for some $i \in I$, $m \in M$. Thus, $ba = a + b = ab$. The former equality implies that supp($b$) is contained in the $m$th column of $T$. Similarly, supp($b$) lies in the $i$th row of $T$ by the latter equality. This means that $b \in K_0[S_{im}]$, which yields $\mathcal{J}(K_0[S]) \cap K_0[S_{im}] \subseteq \mathcal{J}(K_0[S_{im}])$. Now, if $c \in \mathcal{J}(K_0[S])$, and $x = (g,i,n)$, $y = (h,j,m) \in S$, then $xcy \in \mathcal{J}(K_0[S]) \cap K_0[S_{im}] \subseteq \mathcal{J}(K_0[S_{im}])$. Hence,

$$K_0[S]\mathcal{J}(K_0[S])K_0[S] \subseteq \sum_{\substack{i \in I \\ m \in M}} \mathcal{J}(K_0[S_{im}])$$

and the result follows.

Recall that if $I$ is an ideal of $K[S]$, then by $\phi_I$ we denote the natural homomorphism $K[S] \to K[S]/I$. Moreover, for every congruence $\rho$ on $S$, $\phi_\rho$ is the associated homomorphism $K[S] \to K[S/\rho]$.

Now we are ready to state our first general result, obtained by Okniński in [183].

**Theorem 12** Assume that $K[S]$ satisfies a polynomial identity. Then $\mathcal{J}(K[S]) = \mathcal{B}(K[S])$.

*Proof.* In view of the fact that every prime $PI$-algebra has no nonzero nilideals (see Theorem 1 in Chapter 18), it is enough to show that $\mathcal{J}(K[S])$ is a nilideal. The algebra $K[S]/\mathcal{B}(K[S])$ is a subdirect product of prime algebras $A_i$, $i \in \mathcal{A}$, and it follows from Theorem 1 in Chapter 18 that each $A_i$ embeds into the matrix algebra $M_n(D_i)$ for some division algebra $D_i$ and a fixed integer $n$. Let $a \in \mathcal{J}(K[S])$, $\phi_{\mathcal{B}(K[S])}(a) = (a_i)_{i \in \mathcal{A}}$. Then $\varphi_{\rho_{M_i}}(a) \in \mathcal{J}(K[S/\rho_{M_i}])$, where $M_i$ is the kernel of the given epimorphism $\phi_{M_i} : K[S] \to A_i$. Let $\sigma_i : K[S/\rho_{M_i}] \to A_i$ be the homomorphism such that $\sigma_i \phi_{\rho_{M_i}} = \phi_{M_i}$. It is sufficient to show that $\phi_{\rho_{M_i}}(a)$ is a nilpotent element for all $i \in \mathcal{A}$. In fact, then $a_i = \sigma_i \phi_{\rho_{M_i}}(a)$ is also nilpotent; hence, the fact that any nilpotent in $M_n(D_i)$ has the nilpotency index at most $n$ (Proposition 13 in Chapter 2) implies that $a_i^n = 0$. Thus, $\phi_{\mathcal{B}(K[S])}(a)^n = 0$, which means that $a^n \in \mathcal{B}(K[S])$. This implies that $\mathcal{J}(K[S])$ is a nilideal.

In view of the above, since $S/\rho_{M_i}$ embeds into $A_i$, we may assume that $S \subseteq M_n(D)$ for some division ring $D$. Moreover, by Corollary 9 in Chapter 4, we may assume that $0 \in S$ adjoining it to $S$ if necessary. Let $M_n(D) = I_n \supset \cdots \supset I_1 \supset I_0 = 0$ be the chain of ideals of the multiplicative semigroup $M_n(D)$ defined in Chapter 1. Suppose first that $S$ embeds into some $I_j/I_{j-1}, j \in \{1, \ldots, n\}$. Since, for any cancellative subsemigroup $T$ of $S$, $K[T]$ is a $PI$-algebra, then $\mathcal{J}(K[T])$ is a nilideal by Theorem 8. Therefore, Lemma 11 implies that $\mathcal{J}(K_0[S])^3$ has a basis consisting of nilpotents and, since $K[S]$ is a $PI$-algebra, then [239], Theorem 4.2.8, shows that $\mathcal{J}(K_0[S])$ is a nilideal. Thus, $\mathcal{J}(K[S])$ is a nilideal, too.

Now, let $S \subseteq M_n(D)$ be an arbitrary semigroup such that $K[S]$ is a $PI$- algebra. Put $S_j = I_j \cap S$ for $j = 0, \ldots, n$, and choose the least integer $k$ such that $S = S_k$. Now, $S_k/S_{k-1} \subseteq I_k/I_{k-1}$, and we may identify the algebras $K[S_k]/K[S_{k-1}]$ and $K_0[S_k/S_{k-1}]$ by Lemma 7 in Chapter 4. Since $a \in \mathcal{J}(K[S])$,

$$\phi_{K[S_{k-1}]}(a) \in \mathcal{J}(K_0[S_k/S_{k-1}]) \subseteq K_0[I_k/I_{k-1}]$$

Hence, from the preceding part of the proof, it follows that

$$\phi_{K[S_{k-1}]}(a^r) = (\phi_{K[S_{k-1}]}(a))^r = 0 \qquad \text{for some } r \geq 1$$

This means that $a^r \in K[I_{k-1}]$, and so

$$a^r \in \mathcal{J}(K[S]) \cap K[S_{k-1}] \subseteq \mathcal{J}(K[S_{k-1}])$$

If $S_{k-1} = \theta$, then $a^r = 0$. Otherwise, we may repeat the above procedure with regard to the element $a^r$ and the semigroup $S_{k-1}$ in the place of $a$ and $S_k = S$. The nilpotency of the element $a$ will be thus established after at most $k$ such steps. This completes the proof of the theorem.

Observe that, in contrast to the group ring case, there is no obvious way of reducing the proof of the above result to the case where the semigroup is finitely generated. Therefore, one cannot simply use the Nullstellensatz for $PI$-algebras ([17] or Theorem 1 in Chapter 18), even if $S$ is commutative, to prove Theorem 12.

**Corollary 13** Let $L$ be a field with $ch(L) = ch(K)$. If $K[S]$ is a $PI$-algebra, then the congruences $\rho_{\mathcal{J}(L[S])}$, $\rho_{\mathcal{J}(K[S])}$ coincide. Moreover, if $L \subseteq K$, then $\mathcal{J}(K[S]) = K \cdot \mathcal{J}(L[S])$.

*Proof.* Assume first that $F$ is the prime subfield of $K$. From Theorem 12 it follows that $\mathcal{J}(F[S]) \subseteq \mathcal{J}(K[S])$. Since $F$ is a perfect field and $K[S] \cong K \otimes_F F[S]$, then the equality $\mathcal{J}(K[S]) = K \cdot \mathcal{J}(F[S])$ follows from a general result on the radical of tensor products; see [203], Theorem 7.3.8. Clearly, $s - t \in \mathcal{J}(K[S])$, $s, t \in S$, if and only if $s - t \in \mathcal{J}(F[S])$. Similarly, $\mathcal{J}(L[S]) = L \cdot \mathcal{J}(F[S])$, and the result follows.

The above motivates studying $S$ through its two parts: the congruences $\rho_{\mathcal{J}(K[S])}$ for prime fields $K$, and the "radical-free" parts $S/\rho_{\mathcal{J}(K[S])}$.

Our next aim is to extend the description of $\mathcal{J}(K[S])$ for commutative $S$ given in Theorem 2. This will be given in terms of some cancellative semigroups arising from $S$. Thus, the case of cancellative semigroups handled by Theorem 8 is crucial for characterizing $\mathcal{J}(K[S])$ for $PI$-algebras $K[S]$.

If $\sim$ is a congruence in a semigroup $T$ and if $R$ is a subsemigroup of $T$, then we write $R/\sim$ for the image of $R$ under the natural homomorphism $T \to T/\sim$.

**Theorem 14** Assume that $K[S]$ is a $PI$-algebra, and let $a \in K[S]$. Then the following conditions are equivalent:

(1) $a \in \mathcal{J}(K[S])$.
(2) For any $h, g \in S^1$ and, for any congruence $\sim$ in $S$, we have $\phi_\sim(gah) \in \mathcal{J}(K[(gSh)/\sim])$ if $(gSh)/\sim$ is a 0-cancellative semigroup.

*Proof.* If $h, g \in S^1$, then $g\mathcal{J}(K[S])h \subseteq K[gSh] \cap \mathcal{J}(K[S]) \subseteq \mathcal{J}(K[gSh])$ since $\mathcal{J}(K[S])$ is a nilideal by Theorem 12. Thus, the implication $(1) \Rightarrow (2)$ follows.

By the Kaplansky theorem, to prove the converse, it is enough to show that $\phi(a) = 0$ for any epimorphism of $K$-algebras $\phi : K[S] \to M_n(D)$, where $D$ is a division algebra, $n \geq 1$. As in the proof of Theorem 12, define semigroup ideals $I_j$ of $M_n(D)$, and put $S_j = \phi(S) \cap I_j$, $j = 0, \ldots, n$. Let $i$ be the least nonzero integer such that $S_i \neq 0, \emptyset$.

Take any $\bar{g}$, $\bar{h} \in S_i$, and let $g$, $h \in S$ be such that $\phi(g) = \bar{g}$, $\phi(h) = \bar{h}$. Then $\bar{g}\phi(S)\bar{h} \subseteq S_i$, the latter being embedded into the completely 0-simple semigroup $I_i/I_{i-1}$. If $\bar{h}\bar{g} = 0$, then $(\bar{g}\phi(a)\bar{h})^2 = 0$. If $\bar{h}\bar{g} \neq 0$, then $\bar{g}\phi(S)\bar{h}$ lies in a subgroup (with zero adjoined) of $I_i/I_{i-1}$. Therefore, by condition (2), $\bar{g}\phi_\sim(a)\bar{h} = \phi_\sim(gah) \in \mathcal{J}(K[(gSh)/\sim])$, where $\sim = \rho_{\ker \phi}$. From Theorem 12 it then follows that $\bar{g}\phi_\sim(a)\bar{h}$ is nilpotent, so that its image $\bar{g}\phi(a)\bar{h}$ is nilpotent. This shows that the set $Y = \{\bar{g}\phi(a)\bar{h} \mid \bar{g}, \bar{h} \in S_i\} \subseteq M_n(D)$ consists of nilpotents. Since $S_i$ is an ideal of $\phi(S)$, then the $K$-space $V$ generated by $Y$ is an ideal of $M_n(D)$. From [239], Theorem 4.2.8, it follows that $V$ is a nil ideal, and so $V = 0$. Consequently, $M_n(D)\phi(a)M_n(D) = 0$, which shows that $\phi(a) = 0$. This completes the proof.

The above description may be improved under some additional hypothesis on the semigroup $S$.

**Theorem 15**   Let $K[S]$ be a $PI$-algebra. Assume that any irreducible linear representation of $S$ over a field of characteristic $ch(K)$ has a nonzero, completely 0-simple ideal. Then the following conditions are equivalent for an element $a \in K[S]$:

(1)   $a \in \mathcal{J}(K[S])$.
(2)   For any congruence $\sim$ on $S$ and any $g$, $h \in S^1$, we have $\phi_\sim(gah) \in \mathcal{J}(K[(gSh)/\sim])$ if $(gSh)/\sim$ is a group possibly with zero.

*Proof.*   Observe first that any epimorphism of $K$-algebras $K[S] \to M_n(D)$, $D$—a division algebra, $n \geq 1$, yields an epimorphism $L[S] \to M_n(D) \otimes_K L \to M_n(D) \otimes_{Z(D)} L \cong M_r(L)$, where $r \geq n$ and $L$ is a splitting field of the finite dimensional $Z(D)$-algebra $D$; see Proposition 2 in Chapter 18. Thus, the hypothesis on the linear representations of $S$ is equivalent to the analogous hypothesis on the skew linear representations of $S$. Now, with the notation of the proof of Theorem 14, let $R$ be a completely 0-simple ideal of $\phi(S)$. If $\bar{g}$, $\bar{h} \in R$ satisfy $\bar{h}\bar{g} \neq 0$, then $\bar{g}R\bar{h} \subseteq \bar{g}\phi(S)\bar{h} \subseteq R$, and it follows that $\bar{g}\phi(S)\bar{h} = \bar{g}R\bar{h}$ is a maximal subgroup of $R$ (with zero if it is in $R$). Thus, the proof may be completed as in Theorem 14.

Observe that, in view of Theorem 10 in Chapter 3 and Theorem 13 in Chapter 20, the hypothesis on $S$ in the above theorem is satisfied whenever $S$ is a weakly periodic semigroup.

The following description of the congruence $\rho_{\mathcal{J}(K[S])}$ is directly derived from Theorems 8 and 14.

**Corollary 16**   Let $K[S]$ be a $PI$-algebra. Then, for any $x$, $y \in S$, $(x,y) \in \rho_{\mathcal{J}(K[S])}$ if and only if, for every congruence $\sim$ on $S$ and for any $g$, $h \in S^1$

such that $(gSh)/\sim$ is 0-cancellative, $gxh$, $gyh$ lie in the same coset of a normal ch$(K)$-subgroup of the group of fractions of $((gSh)/\sim) \setminus \{\theta\}$.

## Comments on Chapter 21

The nilradical of commutative semigroup rings $R[S]$ with arbitrary coefficients $R$ was described by Parker and Gilmer in [200]; see [64], §9. Jespers and Puczyłowski [102] found a characterization of $\mathcal{J}(R[S])$ for commutative and cancellative $S$, and not necessarily commutative coefficients $R$. This answered affirmatively a question of Krempa and Sierpińska [129] and extended Amitsur's result on the radical of polynomial rings. Starting from this and Munn's result (our Theorem 2), a series of papers by Munn [160], [161], Okniński and Wauters [195], and Jespers [97] approached a general solution. A description of $\mathcal{B}(R[S])$, and of the locally nilpotent radical of $R[S]$, were found in [195]. In a similar way, based on the result of Jespers et al. in the cancellative case [101], the Brown–McCoy radical of $R[S]$ was characterized. The strongly prime radical of $R[S]$ was described in [96]. The history of the problem on $\mathcal{J}(R[S])$, up to recent results in [97], may be found in a survey paper of Jespers and Wauters [103]. We also mention some further results of Jespers [98] and Kelarev [108], [110], extending the knowledge on the nature of $\mathcal{J}(R[S])$ for commutative semigroups $S$.

Semigroups $S$ satisfying the identity $uxyv = uyxv$ for all $u, v, x, y \in S$ are called medial. Thus, Lemma 5 asserts that some power of a permutative semigroup is medial. The statement of Corollary 6 was obtained, under some more general hypothesis, by Putcha and Yaqub in [229]. In [171], Nordahl finds an intrinsic characterization of the congruence $\rho$ considered in Theorem 7.

Proposition 9 was first established by Teply et al. [258]. It is related to the results on the connections between the semilattice decompositions and the subdirect product decompositions discussed in Chapter 6.

The problem of describing the radicals of semigroup rings of semilattices of semigroups seems to be very complex. Since the basic case occurs when all the components are groups, this is related to the corresponding problem for the class of inverse semigroups. The first result in this line is due to Domanov [47], who showed that $\mathcal{J}(K[S]) = 0$ whenever $\mathcal{J}(K[G]) = 0$ for all maximal subgroups $G$ of $S$. The converse is not true, as proved by Teply et al. [258]. A variety of striking counterexamples was then constructed by Ponizovskii [217] and Munn [164] ($S$ is even bisimple here); see also Teply [257]. Following the results from [258] on the semisimplicity of some special types of semigroup algebras of semilattices of semigroups, Ponizovskii settled the case of so-called strong semilattices in [218]. The Jacobson radical of the algebra $K[S]$ for a band (=semigroup consisting of idempotents) $S$ was characterized by Munn in [166]. More generally, necessary and sufficient conditions for an abstract radical $\mathcal{S}(R)$ of a band-graded ring $R$ to be determined by the radicals of the components of $R$ were found by Kelarev in [110]. In a series of papers, Munn extended the knowledge of the Jacobson and nil radicals of inverse semigroup rings [158], [163], [164], [165]; see also [246] and the survey paper [162]. Recently, in [165], Munn showed that, for inverse semigroups $S$ whose semilattice of idempotents satisfy certain finiteness conditions introduced in [258], the ring theoretic properties of $K[S]$ are, in fact, determined by those of the group rings $K[G]$ of the maximal subgroups $G$ of $S$. A further discussion of Munn's result has been given by Ponizovskii in [221a].

A general theory of radicals of semigroup rings originated from a paper of Krempa [122], and it was developed by Puczyłowski [224] and Kelarev [111]. We refer to [185] for a survey of the main topics studied in the theory of classical and abstract radicals of group and semigroup rings.

Recently, Okniński and Putcha [192] showed that $\mathcal{J}(K[T])$ is nilpotent for every subsemigroup $T$ of the full linear monoid $S = M_n(F)$ over a field $F$ provided that $\mathrm{ch}(K) = 0$. Moreover, $\mathcal{J}(K[T]) = 0$ for a class of naturally arising finite semigroups, called semigroups of Lie type.

# 22
# Prime PI-Algebras

In this chapter, we discuss the structure of prime semigroup algebras satisfying polynomial identities. General prime $PI$-algebras form a particularly nice class of $PI$-algebras, with a deep structure theorem obtained by Posner; see Theorem 1 in Chapter 18. Our approach is an intermediate step toward the general problem of characterizing the classical Krull dimension of $K[S]$ considered in Chapter 23.

If $G$ is a group, from Theorem 3 in Chapter 18 and Connell's theorem, it follows that the group algebra $K[G]$ is a prime $PI$-algebra if and only if $\Delta(G)$ has finite index in $G$, and $\Delta(G)$ is torsion-free. Thus, $\Delta(G)$ is an abelian torsion-free subgroup of finite index in $G$. Moreover, from Theorems 19 in Chapter 7 and 1 in Chapter 20, it follows that, if $S$ is a subsemigroup of $G$ with $SS^{-1} = G$, then $K[S]$ is a prime $PI$-algebra, and every prime $PI$-algebra of a cancellative semigroup $S$ arises in this way. This fact is the basis for constructing the following typical class of semigroups yielding prime $PI$-semigroup-algebras.

**Example 1**   Let $T$ be a cancellative semigroup with a group of fractions $G$ such that $K[G]$ is a prime $PI$-algebra. Let $S = \mathfrak{M}(T,n,n,P)$ be a semigroup of matrix type over $T$, with the sandwich matrix $P$ that is not a zero divisor in $M_n(K[T])$. Clearly, $P$ is not a zero divisor in $M_n(K[G])$ in this case. We claim that $K[S]$ is a prime $PI$-algebra. In fact, $K[S] \subseteq K[\mathfrak{M}(G,n,n,P)]$, the latter satisfying a polynomial identity by Proposition 6 in Chapter 20. If $I$ is a nonzero ideal of $K[S]$ and $s \in S$, then $K[S_{11}]IK[S_{11}] \subseteq I \cap K[S_{11}]$, where $S_{11} = \{(t,1,1) \mid t \in T\}$. Since $SS_{11}$ intersects all $\mathcal{H}$-classes of $\mathfrak{M}(G,I,M,P)S_{11}$, then the left annihilator of $K[SS_{11}]$ in $K[S]$ is zero; see Lemma 2 in Chapter 5. Similarly, the right annihilator of $K[S_{11}S]$ is zero. Thus, $K[S_{11}]IK[S_{11}] \neq 0$, and so $K[S_{11}]$ intersects $I$ nontrivially. But $K[S_{11}] \cong K[T]$ is prime. Consequently, $K[S]$ is prime.

Let $R$ be a prime $PI$-algebra. Since $RZ(R)^{-1}$ is a classical ring of quotients of $R$ (see Theorem 1 in Chapter 18), then a right ideal $J$ of $R$ is essential if and only if it contains an element $a$ that is not a left (and so, two-sided) zero divisor in $R$; see [149], 3.4 and 3.5. Thus, $a(bz^{-1}) = 1$ for some $b \in R$, $z \in Z(R)$, so that $J$ contains a nonzero central element of $R$. In particular, $J$ contains a nonzero two-sided ideal of $R$. We will need the following extension of this fact.

**Lemma 2**   Assume that $K[S]$ is a prime $PI$-algebra, and let $I$ be a right ideal of $S$. If $K[I]$ is an essential right ideal of $K[S]$, then $I$ contains a (two-sided) ideal of $S$.

*Proof.*   By the foregoing remark, we have $K[I] \cap Z(K[S]) \neq 0$. Let $0 \neq z \in K[I] \cap Z(K[S])$, and let $m$ be the minimal integer such that there exist $r \in S$, $\alpha_i \in K$, $s_i \in S$ with $0 \neq \alpha_1 s_1 + \cdots + \alpha_m s_m = zr \in zS$. If $t \in S$, then $t(\alpha_1 s_1 + \cdots + \alpha_m s_m) = tzr = ztr \in zS$. Suppose that $ts_1 = ts_i$ for some $i \neq 1$. Then $tzr = z(tr) = 0$ by the choice of $m$ because $|\operatorname{supp}(tzr)| < |\operatorname{supp}(zr)|$. Since $K[S]$ is prime, then the elements of $Z(K[S])$ are not zero divisors in $K[S]$. This implies that $0 = tr \in S$, a contradiction. Therefore, $ts_1 \neq ts_i$ for all $i \neq 1$ and, consequently, $ts_1 \in \operatorname{supp}(tzr)$. Then $ts_1 \in I$ because $tzr \in zS \subseteq K[I]$. Since $t \in S$ is arbitrary, $Ss_1 \subseteq I$, and so $Ss_1S \subseteq I$ as desired.

**Lemma 3**   Let $S$ be a subsemigroup of a completely simple semigroup $T = \mathfrak{M}(G,I,M,P)$. Assume that, for every maximal subgroup $H$ of $T$, the semigroup $S \cap H$ has a group of fractions. If $T_{im} = \{g \in G \mid (g,i,m) \in S\}$,

$S_{im} = \{(g,i,m) \mid g \in G\} \cap S, i \in I, m \in M$, then, for every $j \in I, n \in M$, $T_{im}T_{im}^{-1} = T_{jn}T_{jn}^{-1}$ is a group isomorphic to $S_{im}S_{im}^{-1}$, $p_{mi} \in T_{im}T_{im}^{-1}$, and $S_{im}S_{im}^{-1} = \{(x,i,m) \mid x \in T_{im}T_{im}^{-1}\}$.

*Proof.* Note that $S_{im}$ is the intersection of $S$ with a maximal subgroup of $T$. We use $\circ$ for the multiplication in $G$ to distinguish it from that in $T$. It is clear that $T_{im} \circ p_{mj} \circ T_{jm} \subseteq T_{im}$. Therefore,

(a) $\qquad p_{mj} \circ T_{jm} \subseteq T_{im}^{-1} \circ T_{im} = (p_{mi} \circ T_{im})^{-1} \circ (p_{mi} \circ T_{im})$

Since the mapping $S_{im} \to p_{mi} \circ T_{im}$ defined by $(g,i,m) \to p_{mi} \circ g$ is a semigroup isomorphism, then the right side of (a) is a group isomorphic to the group of fractions of $S_{im}$. Now, for every $j \in I$, we get

$$T_{jm}^{-1} \circ T_{jm} = (p_{mj} \circ T_{jm})^{-1} \circ (p_{mj} \circ T_{jm}) \subseteq (T_{im}^{-1} \circ T_{im})^{-1} \circ (T_{im}^{-1} \circ T_{im})$$
$$= T_{im}^{-1} \circ T_{im}$$

Since $i, j \in I, m \in M$ are arbitrary, then, similarly, $T_{jm}^{-1} \circ T_{jm} \supseteq T_{im}^{-1} \circ T_{im}$, which implies that $T_{jm}^{-1} \circ T_{jm} = T_{im}^{-1} \circ T_{im}$. This establishes the first assertion with respect to two maximal subgroups of $T$ lying in the same row of $T$ because $T_{im} \circ T_{im}^{-1} = T_{im}^{-1} \circ T_{im}$ by the hypothesis on $S$. A similar argument applied to any $n \in M$ and starting from the inclusion $T_{im} \circ p_{mi} \circ T_{in} \subseteq T_{in}$ may be used to show that $T_{im}^{-1} \circ T_{im} = T_{in}^{-1} \circ T_{in}$. It is clear that $p_{mi} \in T_{im} \circ T_{im}^{-1}$. Finally, for $s, u \in T_{im}$, we have

$$(s,i,m)(u,i,m)^{-1} = (s,i,m)(p_{mi}^{-1}u^{-1}p_{mi}^{-1},i,m) = (su^{-1}p_{mi}^{-1},i,m)$$

so that $S_{im}S_{im}^{-1} = \{(x,i,m) \mid x \in T_{im} \circ T_{im}^{-1}\}$.

A modification of the above proof leads to a related result on certain linear semigroups.

**Lemma 4** Let $S$ be an irreducible subsemigroup of a matrix algebra $M_n(D), n \geq 1$, over a division algebra $D$ such that all nonzero matrices in $S$ have the same rank. Assume that, for every maximal subgroup $H \neq \{0\}$ of $M_n(D)$, $S \cap H$ has a group of fractions if it is nonempty. Then all groups $(S \cap H)(S \cap H)^{-1}$ are isomorphic. If, additionally, $S$ has no zero element, then $S$ is contained in the completely simple semigroup that is the union of such groups.

*Proof.* We know that $S$ embeds into a completely 0-simple principal factor $J = \mathfrak{M}^0(G,I,M,P)$ of the multiplicative semigroup of $M_n(D)$; Theorem 6 in Chapter 1. Let $H_1, H_2$ be two (nonzero) maximal subgroups of $J$ nontrivially intersecting $S$. Choose $s \in S \cap H_1, t \in S \cap H_2$. We claim that $usz \subseteq S \cap H_2$

for some $u, z \in S$. If this is not the case, then $(tS)s(St) = 0$ because $tSt \subseteq H_2 \cup \{0\}$. Thus, $tM_n(D)t = 0$ because $S$ spans $M_n(D)$ as a $Z(D)$-space. This contradiction proves our claim. Now $(us)s(sz) \subseteq S \cap H_2$, and $x = us$ is in the row of $J$ in which $t$ lies and in the column of $J$ in which $s$ lies. Similarly, the $\mathcal{H}$-class of $J$ determined by the intersection $sJ \cap Jt$ contains $y = sz$, and so it intersects $S$ nontrivially. Let $T_k = \{g \in G \mid (g, i_k, m_k) \in S \cap H_k\}$ for the appropriate $i_k \in I$, $m_k \in M$, $k = 1,2$. Put $i = i_1, j = i_2, m = m_1, n = m_2$. Then $p_{mi}, p_{nj}$ are nonzero and

(b)         $x \circ p_{mi} \circ T_1 \circ p_{mi} \circ y \subseteq T_2$      $y \circ p_{nj} \circ T_2 \circ p_{nj} \circ x \subseteq T_1$

where $\circ$ stands for the product in $G$. As in the proof of Lemma 3, $T_1 \circ T_1^{-1}$ is a group and $p_{mi} \in T_1 \circ T_1^{-1} = T_1^{-1} \circ T_1$. Therefore,

$$x \circ T_1 \circ T_1^{-1} \circ x^{-1} = x \circ p_{mi} \circ T_1 \circ T_1^{-1} \circ p_{mi}^{-1} \circ x^{-1}$$
$$= (x \circ p_{mi} \circ T_1 \circ p_{mi} \circ y) \circ (x \circ p_{mi} \circ T_1 \circ p_{mi} \circ y)^{-1} \subseteq T_2 \circ T_2^{-1}$$

Using the second inclusion in (b), we also show that

$$x^{-1} \circ T_2^{-1} \circ T_2 \circ x \subseteq T_1^{-1} \circ T_1$$

It follows that $T_1 \circ T_1^{-1} = T_2 \circ T_2^{-1}$. Consequently, $(S \cap H_1)(S \cap H_1)^{-1} \cong (S \cap H_2)(S \cap H_2)^{-1}$ as desired.

Assume now that $S$ has no zero element. Then $S$ does not intersect nonidempotent $\mathcal{H}$-classes of $M_n(D)$, and so $\bigcup \{H \mid H$ is an $\mathcal{H}$-class of $M_n(D)$ such that $S \cap H \neq \varnothing\}$ is a completely simple semigroup.

We note that semigroups $S$ described in Lemma 3 need not be of the form $S \cong \mathfrak{M}(U, I, M, P)$ for a cancellative semigroup $U$. For example, take $T = \mathfrak{M}(U, 1, 2, P)$ for a monoid $U$ (with a group of fractions) that is not a group, and let $p_{11} = p_{21} = 1$. If $z \in U$ is such that $Uz \neq U$, then $S = \{(x, 1, 1) \mid x \in U\} \cup \{(xz, 1, 2) \mid x \in U\}$ is not of this type, but $UU^{-1} = Uz(Uz)^{-1}$.

Now we prove a general result describing prime $PI$-semigroup-algebras and showing the crucial role of the class of semigroups considered in Example 1. We will exploit one more general fact. If $J$ is an ideal in a prime algebra $R$ and if $J$ satisfies a polynomial identity, then the elements of $Z(J)$ are not zero divisors in $R$, and $JZ(J)^{-1}$ is an ideal in $RZ(J)^{-1}$. By Theorem 1 in Chapter 18, the former has an identity, so that the primeness of $RZ(J)^{-1}$ (see Lemma 14 in Chapter 7) implies that $R \subseteq RZ(J)^{-1} = JZ(J)^{-1}$ is also a $PI$-algebra.

The following result was obtained by Zelmanov in [276].

**Theorem 5** Let $S$ be a semigroup. Then the following conditions are equivalent:

(i) $K[S]$ is a prime $PI$-algebra.

(ii) $S \subseteq M_m(L)$ is an irreducible linear semigroup over a field $L \supseteq K$ such that $S$ is linearly independent over $K$.

(iii) There exists an ideal $\bar{S}$ of $S$ such that $r_{K[S]}(\bar{S}) = 0$, and $\bar{S}$ embeds into a completely simple semigroup $U = \mathfrak{M}(H,n,n,P)$ over a group $H$ with $K[U]$ prime $PI$, satisfying $S_{ij}^{-1} S_{ij} = U_{ij}$ or every $i, j \in \{1,2,\ldots,n\}$, where $U_{ij} = \{(g,i,j) \in U \mid g \in G\}$ and $S_{ij} = U_{ij} \cap \bar{S}$.

(iv) There exists a subsemigroup $R$ of $S$ isomorphic to a semigroup of matrix type $\mathfrak{M}(T,n,n,Q)$ over a cancellative semigroup $T$ such that $K[R]$ is a prime $PI$-algebra and $R$ contains an ideal $J$ of $S$ with $r_{K[S]}(J) = 0$.

*Proof.* Assume first that $K[S]$ is a prime $PI$-algebra. From Theorem 1 in Chapter 18, it follows that $Z(K[S])^{-1}K[S] = M_n(D)$ for some division algebra $D \supseteq K$, $m \geq 1$. Therefore,

$$Z(K[S])^{-1}K[S] \otimes_{Z(D)} L \cong M_m(D) \otimes_{Z(D)} L \cong M_r(L)$$

where $L$ is a splitting field for $D$ over $Z(D)$, $r \geq 1$. Since

$$K \subseteq Z(K[S])^{-1}Z(K[S]) \subseteq Z(D) \subseteq L$$

it follows that $S$ spans $M_r(L)$ as an $L$-space. Thus, (ii) holds.

Since $K[S]$ is prime, then $S$ has no zero element. Therefore, the set $\bar{S} \subseteq S$ consisting of elements of minimal rank as matrices in $M_m(D)$ is an ideal of $S$. From Lemmas 4 and 3, we know that $\bar{S}$ embeds into a completely simple semigroup $U = \mathfrak{M}(H,I,M,P)$ and that $S_{ij}^{-1} S_{ij} = U_{ij}$ for all $i \in I$, $j \in M$. (Note that all $S_{ij}$ have groups of fractions by Theorem 1 in Chapter 20.)

We know that the ideal $K[\bar{S}]$ of $K[S]$ contains a nonzero central element $a$. If $a = \sum_{l=1}^{m} \lambda_l u_l$, $\lambda_l \in K$, $u_l \in \bar{S}$, then the fact that $as = sa \neq 0$ for any $s \in S$ implies that $s$ lies in a row of $\mathfrak{M}(H,I,M,P)$ in which one of the elements $u_l s$ lies. Hence, $I$ is a finite set. A similar argument with respect to the columns of $\mathfrak{M}(H,I,M,P)$ shows that $M$ is a finite set. We can then assume that $I = \{1,\ldots,k\}$ and $M = \{1,\ldots,n\}$ for some $k, n \geq 1$.

Suppose that $k \leq n$. Choose elements $s_l \in S_{ll}$ for $l = 1, \ldots, k$. We claim that

(c) $\qquad a = s_1 + \cdots + s_k$ is not a right zero divisor in $K[\bar{S}]$

If $xa = 0$, $x \in K[\bar{S}]$, then $xs_l = 0$ for every $l$ because the elements $xs_l$ have supports contained in distinct columns of $\mathfrak{M}(H,k,n,P)$. Thus, $x \circ P \circ s_l = 0$ for $l = 1, \ldots, k$. Since $x \circ P$ is a $k \times k$ matrix, it then follows that $x \circ P = 0$ and, hence, $xK[\bar{S}] = 0$. Consequently, $x = 0$ since $K[S]$ is prime, which establishes our claim. As noted before Lemma 2, the fact that $a$ is not a right zero divisor in $K[\bar{S}]$ implies that the left ideal $K[\bar{S}]a$ is essential in $K[\bar{S}]$ and also in $K[S]$. This shows that $\bar{S}$ cannot have columns other than those in which the elements $s_1, \ldots, s_k$ lie. It follows that $k = n$. A similar argument shows that the equality $k = n$ may also be derived if we suppose that $k \geq n$. Moreover, the primeness of $K[\bar{S}]$ implies that $P$ is not a zero divisor in $M_n(K[H])$.

Let $A$ be a nonzero ideal in $K[U]$. Since $P$ is not a zero divisor in $M_n(K[H])$, then, as in Example 1, we show that $A \cap K[\bar{S}] \supseteq A \cap K[S_{11}] \neq 0$. Therefore, the primeness of $K[U]$ follows from the fact that $K[\bar{S}]$ is prime. Clearly, $r_{K[S]}(\bar{S}) = 0$. This yields (iii).

Let $S_{(l)} = \bigcup_{j=1}^{n} S_{lj}$, $S^{(l)} = \bigcup_{j=1}^{n} S_{jl}$, $l = 1, \ldots, n$. Then $T = S_{(1)}S^{(1)}$ is contained in the cancellative semigroup $S_{11}$. Choose elements $r_{1j} \in S^{(1)}S_{(j)}$, $z_{i1} \in S^{(i)}S_{(1)}$. Then $R = \bigcup_{i=1}^{n} \bigcup_{j=1}^{n} z_{i1}Tr_{1j}$ is a subsemigroup of $\bar{S}$. Define a matrix $Q = (q_{ij})$ by $q_{ij} = r_{1j}z_{i1}$. Then $q_{ij} \in T$, and we claim that $R$ is isomorphic to the semigroup of matrix type $\mathfrak{M}(T,n,n,Q)$. It is clear that $\phi((s,i,j)) = z_{i1}sr_{1j}$ defines a homomorphism that maps $\mathfrak{M}(T,n,n,Q)$ onto $R$. If $z_{i1}sr_{1j} = z_{p1}ur_{1l}$ for some $s, u \in T$ and $i, p, j, l \in \{1, \ldots, n\}$, then $i = p$ and $j = l$ since, otherwise, these elements would lie in distinct rows or columns of $\bar{S}$. Thus, $z_{i1}(s - u)r_{1j} = 0$, and so $Tz_{i1}(s - u)r_{1j}T = 0$. But $Tz_{i1}, r_{1j}T \subseteq T$, and $T$ is a cancellative semigroup. Therefore, $s = u$, which shows that $\phi$ is an isomorphism.

If we show that $R$ contains an ideal $J$ of $S$, then the primeness of $K[S]$ implies that $r_{K[S]}(J) = 0$, and $K[R]$ intersects nontrivially every nonzero ideal of $K[S]$. This establishes (iv). Let $I_1 = \bigcup_{i=1}^{n} z_{i1}S_{(1)}$, $I_2 = \bigcup_{j=1}^{n} S^{(1)}r_{1j}$. Then $I_1, I_2$ are right, respectively left, ideals of $S$. Put $b = z_{11}r_{11} + z_{21}r_{12} + \cdots + z_{n1}r_{1n}$. As in (c), we show that $K[I_1]$ is an essential right ideal of $K[S]$ because $b \in K[I_1]$. From Lemma 2 it follows that $I_1$ contains a two-sided ideal $J_1$ of $S$. A similar argument shows that there exists an ideal $J_2$ of $S$ such that $J_2 \subseteq I_2$. Then $J = J_1J_2 \subseteq I_1I_2 = \bigcup_{i=1}^{n} \bigcup_{j=1}^{n} z_{i1}S_{(1)}S^{(1)}r_{1j} = R$ is an ideal of $S$. This completes the proof of (iv).

Assume now that (iii) or (iv) hold. Since $J$ is an ideal of $R$, then $K[J]$ is a prime $PI$-algebra. As in Example 1, we also show that $K[\bar{S}]$ is a prime $PI$-algebra. In view of the annihilator hypotheses, $K[S]$ is prime in any of these cases. As noted before the theorem, $K[S]$ must then be a $PI$-algebra. Thus (i) holds.

If (ii) holds, and $A, B$ are ideals of $K[S] \subseteq M_m(L) = LK[S]$, $m \geq 1$, with $AB = 0$, then $(LA)(LB) = 0$, and $LA, LB$ are ideals in $M_m(L)$. Thus, $A = 0$ or $B = 0$, which shows that $K[S]$ is prime. Since $K[S]$ satisfies a polynomial identity by Theorem 1 of Chapter 18, this establishes (i) and completes the proof.

## Comments on Chapter 22

The presentation in this chapter is a slight modification of Zelmanov's results [276]. For material related to the situation described in Lemma 3, we refer to the fragments of Petrich's book [209] concerning so-called matrix compositions. General orders in completely 0-simple semigroups have recently been extensively studied in a series of papers of Fountain, Gould, and Petrich, see [59a], [71].

# 23

# Dimensions

In this chapter, we discuss the dimensions of semigroup algebras. We will be concerned with the classical Krull dimension, the Gelfand–Kirillov dimension, and the Krull dimension in the sense of [70], denoted by $\operatorname{cl} K \dim$, $GK \dim$, $K \dim$, respectively. For the definitions and basic properties of these dimensions, we refer to [70], [121], [149], and [169]. While the general problem of computing a given dimension in terms of the underlying semigroup has not been settled for group algebras (except $GK \dim$—see Chapter 8), we restrict ourselves to the class of semigroup algebras satisfying polynomial identities. As usual, when considering the Krull dimensions, we deal with algebras with unity. Therefore, for the sake of simplicity, monoid algebras are mainly considered in this chapter.

For any semigroup $S$, we define the rank of $S$ and denote it by $rk(S)$, as $\sup\{n \in \mathbb{N} \mid S$ has a free commutative subsemigroup on $n$ free generators$\}$.

It turns out that this notion coincides with that defined by the torsion-free rank when restricted to commutative cancellative semigroups.

**Proposition 1** Let $S$ be a commutative cancellative semigroup. Then $rk(S) = rk(SS^{-1})$ is the torsion-free rank of the group $SS^{-1}$. Specifically, if $\langle s_1 t^{-1}, \ldots, s_n t^{-1} \rangle$ is a free commutative subsemigroup of rank $n$ in $S^{-1}$ for some $s_1, \ldots, s_n, t \in S$, then either $\langle s_1, \ldots, s_n \rangle$ or $\langle s_1 t, \ldots, s_n t \rangle$ is a free commutative subsemigroup of rank $n$ in $S$.

*Proof.* It is straightforward that, for every free commutative semigroup $T$ with free generators $t_1, \ldots, t_n$, $TT^{-1}$ is a free abelian group with free generators $t_1, \ldots, t_n$. Therefore, $rk(S)$ does not exceed the torsion-free rank of $SS^{-1}$, the latter being less or equal to $rk(SS^{-1})$. Thus, assume that $M$ is a free commutative subsemigroup of $SS^{-1}$ with free generators $u_1, \ldots, u_n$. We can write $u_i = s_i t^{-1}$ for some $s_1, \ldots, s_n, t \in S$ and, as above, $u_1, \ldots, u_n$ freely generate the group $UU^{-1}$. Let deg denote the total degree in $UU^{-1}$ with respect to generators $u_1, \ldots, u_n$.

Assume first that $\deg(t) \neq -1$. We claim that $s_1, \ldots, s_n$ freely generate a subsemigroup in $S$. If not, then there exist $s_{i_1}, \ldots, s_{i_k}$ and $s_{j_1}, \ldots, s_{j_m}$ with $1 \leq i_p, j_r \leq n$ such that $s_{i_1} \ldots s_{i_k} = s_{j_1} \ldots s_{j_m}$ and $\{s_{i_1}, \ldots, s_{i_k}\} \cap \{s_{j_1}, \ldots, s_{j_m}\} = \emptyset$. This implies that $t^k u_{i_1} \ldots u_{i_k} = t^m u_{j_1} \ldots u_{j_m}$ and $\{u_{i_1}, \ldots, u_{i_k}\} \cap \{u_{j_1}, \ldots, u_{j_m}\} = \emptyset$. Computing degrees, we get $k \cdot \deg(t) + k = m \cdot \deg(t) + m$. Since $\deg(t) \neq -1$, then $k = m$. Consequently, $u_{i_1} \ldots u_{i_k} = u_{j_1} \ldots u_{j_m}$, which contradicts the fact that $u_1, \ldots, u_n$ are free generators of $UU^{-1}$ and establishes our claim.

Assume now that $\deg(t) = -1$. Then $\deg(t^2) = -2$ and $u_1 = (s_1 t) t^{-2}$, $\ldots, u_n = (s_n t) t^{-2}$. By the foregoing, $s_1 t, \ldots, s_n t$ freely generate a subsemigroup in $S$.

It follows that $rk(SS^{-1}) \leq rk(S)$, which completes the proof.

The following extension of the above observation will be crucial for applications of the rank in the class of $PI$-semigroup-algebras.

**Proposition 2** Let $S$ be a cancellative semigroup with the permutational property. Assume that $G$ is a group of fractions of $S$, $H$ a subgroup of finite index in $G$, and $F \subseteq H$ a finite normal subgroup of $G$ such that $[H, H] \subseteq F$. Then,

(i) $rk(S) = rk(G) = rk(H/F) = rk((S \cap H)/\sim_F)$, where $\sim_F$ is the congruence determined by $F$.

(ii) If $I$ is a right (left) ideal of $S$, then $rk(I) = rk(S)$.

*Proof.* From Theorem 8 in Chapter 19, we know that suitable groups $G$, $H$, $F$ exist. Since $H$ is a group of fractions of $S \cap H$ by Lemma 5 in Chapter 7, then $H/F$ is a group of fractions of $(S \cap H)/\sim_F$. Therefore, $rk((S \cap H)/\sim_F) = rk(H/F)$ by Proposition 1.

It is clear that $rk(S) \leq rk(G)$.

If $A$ is a torsion-free abelian subgroup of $G$, then $A \cap H$ has finite index in $A$, and $(A \cap H) \cap F = \{1\}$. Therefore, $rk(A \cap H) = rk(A)$ by Proposition 1, and $A \cap H$ embeds into $H/F$. Hence, $rk(G) \leq rk(H/F)$. It remains to show that $rk((S \cap H)/\sim_F) \leq rk(S)$. Let $T$ be a finitely generated free commutative subsemigroup of $(S \cap H)/\sim_F$, and let $U$ be a finitely generated subsemigroup of $S \cap H$ mapping onto $T$ under the natural homomorphism $S \cap H \to (S \cap H)/\sim_F$. From Corollary 13 in Chapter 19, we know that $UU^{-1}$ has an abelian subgroup $B$ of finite index. Then $B/(B \cap F)$ has finite index in $TT^{-1}$. Therefore, by Proposition 1, $rk(T) = rk(TT^{-1}) = rk(B/(B \cap F)) \leq rk(B)$. The latter coincides with $rk(B \cap U)$ because $B$ is the group of fractions of $B \cap U$ by Lemma 5 in Chapter 7. It follows that $rk((S \cap H)/\sim_F) \leq rk(S)$ because $B \cap U \subseteq S$. This completes the proof of (i).

(ii) By Lemma 6 in Chapter 7, $II^{-1} = SS^{-1}$, that is, the groups of fractions of $I$ and $S$ coincide. Now, the assertion follows from (i).

We note that the obvious fact that the rank of a commutative semigroup does not grow under homomorphisms was used in the above proof. This is certainly no longer true for arbitrary semigroups; for example, $rk(X) = 1$ for every free (noncommutative) semigroup $X$. This is the main obstacle when we try to use the rank within classes of semigroups far from commutative.

Let $X$ be a free commutative monoid on $n$ letters. Then $K[X]$ is the polynomial algebra in $n$ indeterminates, and it is well known that $\text{cl}\,K \dim K[X] = rk(X) = n = GK \dim K[X]$. The former equality was extended to the class of semigroup algebras of cancellative commutative monoids by Arnold and Gilmer; see [64], Theorems 17.1 and 21.4. Our first aim is to show that the same holds in the class of $PI$-semigroup-algebras of cancellative semigroups. We start with the following well-known auxiliary result.

**Lemma 3** Assume that $G$ is a group with an abelian normal subgroup $H$ of finite index. Then $\text{cl}\,K \dim K[G] \geq \text{cl}\,K \dim K[H]$.

*Proof.* Observe that the group $G/H$ acts on $K[H]$ by conjugation. For any ideal $R$ of $K[H]$, define $\bar{R} = \bigcap_{g \in G/H} R^g$. Let $P \subsetneqq Q$ be prime ideals of $K[H]$. Suppose that $\bar{P} = \bar{Q}$. Since $P$ is prime and $G/H$ is finite, then $Q^g \subseteq P$ for some $g \in G/H$. Now, $Q \supseteq P \supseteq Q^g$, and so $Q \supseteq Q^g \supseteq Q^{g^2} \supseteq \cdots \supseteq Q^{g^r} = Q$, where $r = |G/H|$. This implies that $P = Q$, a contradiction. Therefore, $\bar{P} \subsetneq \bar{Q}$. Clearly, $\bar{P}$, $\bar{Q}$ are semiprime ideals of $K[H]$. Since $\bar{P}$ is $G/H$-invariant, then $\bar{P}K[G]$ is an ideal of $K[G]$. From Corollary 17 in Chapter 4, we know that $\bar{P}K[G] \cap K[H] = \bar{P}$. Hence, by Zorn's lemma, there exists an ideal $P_1$ of $K[G]$ that is maximal with respect to the property $P_1 \cap K[H] = \bar{P}$. If $J_1 J_2 \subseteq P_1$ for some ideals $J_1$, $J_2$ of $K[G]$ with $J_1, J_2 \supseteq P_1$, then $(J_1 \cap K[H])(J_2 \cap K[H]) \subseteq P_1 \cap K[H] = \bar{P} \subseteq P$. The primeness of $P$ implies that $J_i \cap K[H] \subseteq P$ for some $i \in \{1, 2\}$. Then $\bar{P} \subseteq J_i \cap K[H] = \overline{J_i \cap K[H]} \subseteq \bar{P}$ for some $i$. The choice of $P_1$ implies that $P = J_1$ or $P = J_2$, showing that $P_1$ is a prime ideal of $K[G]$. Let $J = ((\bar{Q}/\bar{P})(K[G]/P_1)) \cap K[H]/\bar{P}$. If $q \in \bar{Q}$, $g \in G$, then $(qg)^r = q_1 q_2 \cdots q_r g^r \in q_1 q_2 \cdots q_r H$ for some $G$-conjugates $q_i$ of $q$, so that $q_1, \ldots, q_r \in \bar{Q}$. Thus, if $J \neq \bar{Q}/\bar{P}$, then $J/(\bar{Q}/\bar{P})$ has a $K$-basis consisting of nilpotents (being the image of suitable elements $qg$ under the natural homomorphism $K[G] \rightarrow K[G]/P_1$). Since $K[G]$ is a $PI$-algebra by Theorem 3 in Chapter 18, then, from [239], Theorem 4.2.8, it follows that the algebra $J/(\bar{Q}/\bar{P})$ is not semiprime. This contradicts the fact that $\bar{Q}$ is a semiprime ideal of $K[H]$ and shows that $J = \bar{Q}/\bar{P}$. By taking complete inverse images, we then get $(\bar{Q}K[G] + P_1) \cap K[H] = \bar{Q}$. Therefore, by Zorn's lemma, there exists an ideal $Q_1$ of $K[G]$, which is maximal with respect to the properties $Q_1 \cap K[H] = \bar{Q}$, $Q_1 \supseteq P_1$. As above, we can check that $Q_1$ is a prime ideal of $K[G]$. Moreover, $P_1 \subsetneq Q_1$ since $\bar{P} \subsetneq \bar{Q}$. Proceeding this way with respect to an ascending chain of prime ideals of $K[H]$, we obtain a chain of prime ideals of $K[G]$ of the same length. This establishes the assertion.

Lemma 3 is a very special case of general results on the classical Krull dimension of crossed products of finite groups obtained in [138], as well as those on finite normalizing extensions; see [149].

We are now ready for the first main result of this chapter.

**Theorem 4** Let $S$ be a cancellative monoid such that $K[S]$ is a $PI$-algebra. Then,

$$\mathrm{cl}K \dim K[S] = GK \dim K[S] = rk(S)$$

*Proof.* From Theorems 1 in Chapter 20 and 3 in Chapter 18, it follows that $S$ has a group of fractions $G$ with a normal subgroup $H$ of finite index such that the commutator subgroup $H'$ is a finite $p$-group, where $p = \mathrm{ch}(K)$ (possibly $p = 0$). If $T$ is a finitely generated subsemigroup of $S$, then the group of fractions $F$ of $T$ has an abelian subgroup $Z$

of finite index by Corollary 13 in Chapter 19. Therefore, the results
of Chapter 8 yield $GK \dim K[T] = GK \dim K[F] = GK \dim K[Z]$, the
latter equal to $rk(Z)$ by Proposition 1. Hence, Proposition 2 implies that
$GK \dim K[T] = rk(Z) = rk(F) = rk(T)$. Consequently, $GK \dim K[S] =$
$\sup_T GK \dim K[T] = \sup_T rk(T) = rk(S)$, where $T$ runs over all finitely
generated subsemigroups of $S$.

Since prime $PI$-algebras are Goldie (see Theorem 1 in Chapter
18), then, from [121], Corollary 3.16, it follows that $GK \dim K[S] \geq$
$cl K \dim K[S]$. Moreover, Lemmas 15 and 13 in Chapter 7 yield

$$cl K \dim K[S] \geq cl K \dim K[G].$$

Thus, in view of Lemma 3, we get $cl K \dim K[S] \geq cl K \dim K[H/H']$. Since
$rk(H/H') = rk(S)$ by Proposition 2, and $cl K \dim K[H/H'] = rk(H/H')$ by
[64], Theorem 17.1, then $cl K \dim K[S] \geq rk(S)$. This completes the proof.

If $G$ is a finitely generated group, that is not nilpotent-by-finite, then
$GK \dim K[G] = \infty$; see Chapter 8. However, we can have $rk(G) = 1$ in this
case. For example, if $G$ is a free non-abelian group, then every subgroup
of $G$ also is free, so that $rk(G) = 1$ follows. Thus, the hypothesis on the
$PI$-property of $K[S]$ is essential in Theorem 4.

If $S$ is a cancellative monoid with zero adjoined, then, from Corollary 9
in Chapter 4, it follows that $cl K \dim K[S] = cl K \dim K_0[S]$, and $K_0[S]$ may
be identified with the semigroup algebra of the cancellative semigroup
$S \setminus \{\theta\}$. Thus, the assertion of Theorem 4 also holds for this class of
semigroups.

To state the main result of this chapter, extending Theorem 4, we
need some more preparatory results on the general properties of the rank.
Clearly, the ranks of all homomorphic images $S/\rho_P$ of $S$ determined by
prime ideals $P$ of $K[S]$ are of basic importance.

**Lemma 5** Let $S$ be a subsemigroup of the multiplicative semigroup of
$M_n(L)$ for a field $L$ and some $n \geq 1$. If $S$ has the permutational property,
then $rk(S) = rk(Z)$, where $Z$ is a subsemigroup of $S$ lying in a subgroup
of some principal factor $J_r/J_{r-1}$, $r \in \{1,\ldots,n\}$, of $M_n(L)$.

*Proof.* Let $T$ be a free commutative subsemigroup of $S$, and let $j$ be
the least integer with $T \cap J_j \neq \emptyset$. Since $T \cap J_j$ is an ideal of $T$, then,
from Proposition 2, it follows that $rk(S) = rk((S \cap J_r)/(S \cap J_{r-1}))$ for some
$r \in \{1,\ldots,n\}$. Let $J = J_r/J_{r-1}$ and $\bar{S} = (S \cap J_r)/(S \cap J_{r-1})$. If $U$ is a free
commutative subsemigroup of $J$, then it is easy to see that $U$ lies in a

subgroup of $J$; see Chapter 1. It then follows that $rk(S) = \sup\{rk(\bar{S} \cap G)\}$, where the supremum is taken over all maximal subgroups of $J$ intersecting $\bar{S}$.

Let $I$ be a nonzero principal right (or left) ideal of $J$. The union of all nonempty intersections $\bar{S} \cap G$, $G \neq \{\theta\}$—a maximal subgroup of $I$, is a subsemigroup of $\bar{S}$ contained in a completely simple subsemigroup of $J$ (which is the union of the corresponding maximal subgroups of $I$). From Lemma 3 in Chapter 22, it follows that all groups of fractions $(\bar{S} \cap G)(\bar{S} \cap G)^{-1}$ (existing by Theorem 8 in Chapter 19) are isomorphic.

Suppose now that $\bar{S} \cap G_1, \ldots, \bar{S} \cap G_k$, $k = \binom{n}{r}+1$, are nonempty for some maximal subgroups $G_i \circ J_r \setminus J_{r-1}$ and that all groups $H_i = (\bar{S} \cap G_i)(\bar{S} \cap G_i)^{-1}$ are pairwise nonisomorphic. Choose $x_i \in \bar{S} \cap G_i$. If $y \in \bar{S}$ and $i \neq j$, then the element $x_i y x_j$ cannot be in a maximal subgroup of $J_r \setminus J_{r-1}$ because otherwise, by the preceding paragraph, the groups $H_i$, $H_j$ would be isomorphic. It follows that $(x_i y x_j)^2 = \theta$. Then, also, $(e_i z e_j)^2 = \theta$ for the identities $e_i$, $e_j$ of $H_i$, $H_j$ and every $z \in \langle e_1, \ldots, e_k \rangle$. This contradicts Corollary 7 in Chapter 3 and shows that, up to isomorphism, there are finitely many groups of fractions $(\bar{S} \cap G)(\bar{S} \cap G)^{-1}$ where $G$ is a maximal subgroup of $J$. Since $rk(S) = \sup_G\{rk(\bar{S} \cap G)\}$, Proposition 2 implies that $rk(S) = rk((\bar{S} \cap G)(\bar{S} \cap G)^{-1}) = rk(\bar{S} \cap G)$ for some $G$ of this type.

**Lemma 6**  For every semigroup $S$, we have $rk(S) \leq \sup\{rk(S/\rho_P)\} = \sup\{rk(S/\rho)\}$, where $P$ runs over the set of prime ideals of $K[S]$ and $\rho$ runs over the set of all congruences on $S$.

*Proof.*  Let $T$ be a free commutative subsemigroup of $S$. Define $P$ as an ideal of $K[S]$ that is maximal with respect to the property $P \cap K[T] = 0$. It is straightforward that $P$ is a prime ideal of $K[S]$, and $T$ embeds into $S/\rho_P$. Thus, $rk(T) \leq rk(S/\rho_P)$, and the desired inequality follows. Clearly, we also have $\sup_P\{rk(S/\rho_P)\} \leq \sup_\rho\{rk(S/\rho)\}$. On the other hand, for every congruence $\rho$ on $S$, the first paragraph of the proof yields $rk(S/\rho) \leq \sup_Q\{rk((S/\rho)/\rho_Q)\}$, where $Q$ runs over the set of all prime ideals of $K[S/\rho]$. Since the inverse image $P$ of $Q$ in $K[S]$ is a prime ideal of $K[S]$, and $S/\rho_P \cong (S/\rho)/\rho_Q$, then we also get $\sup_\rho\{rk(S/\rho)\} \leq \sup_P\{rk(S/\rho_P)\}$.

Observe that, in view of Lemma 6, $\sup_P\{rk(S/\rho_P)\}$ is not dependent on the choice of the coefficient field $K$. We will denote this invariant of $S$ by $Rk(S)$. It appears that $Rk(S)$ is entirely determined by the congruence $\rho_{\mathcal{B}(K[S])}$.

**Proposition 7**  Let $S$ be a subsemigroup of the algebra $\prod_{\alpha \in A} M_{n_\alpha}(D_\alpha)$ for some division algebras $D_\alpha$ and integers $n_\alpha$, $\alpha \in A$, such that $n_\alpha \leq N$

for some $N \geq 1$. Assume that $S$ satisfies the permutational property. Then $rk(S) = Rk(S)$.

*Proof.* Let $\varphi : S \to V$ be an onto homomorphism of semigroups, and let $T \subseteq V$ be a free commutative semigroup. It is enough to show that $rk(S) \geq rk(T)$. Thus, we can assume that $S = \langle s_1, \ldots, s_n \rangle$, where $\varphi(s_i) = t_i$, $i = 1, \ldots, n$, is the free generator set of $T$. We define elements $r_1^{(i)}, \ldots, r_n^{(i)}, i = 0, 1, \ldots, N + 1$, and $u_1, u_2, \ldots, u_{N+2}$ of $S$ inductively as follows:

$$r_j^{(0)} = s_j \qquad j = 1, \ldots, n, \qquad u_1 = s_1$$
$$r_j^{(i)} = u_i^N r_j^{(i-1)} u_i^N \qquad j = 1, \ldots, n, \qquad u_{i+1} = r_1^{(i)} \ldots r_n^{(i)} \text{ for } i \geq 1$$

Let $U = \langle r_1^{(N+1)}, \ldots, r_n^{(N+1)} \rangle$. From the construction of $r_j^{(i)}$ it follows that

$$\varphi(r_1^{(N+1)}) = t_1^{\gamma+1} t_2^{\beta} \ldots t_n^{\beta}$$
$$\varphi(r_k^{(N+1)}) = t_1^{\gamma} t_2^{\beta} \ldots t_{k-1}^{\beta} t_k^{\beta+1} t_{k+1}^{\beta} \ldots t_n^{\beta}$$

for some positive integers $\gamma$, $\beta$. Since the $n \times n$ $\mathbb{Z}$-matrix

$$\begin{pmatrix} \gamma+1 & \beta & & \cdots & & \beta \\ \gamma & \beta+1 & & \cdots & & \beta \\ \gamma & \beta & \beta+1 & & & \beta \\ \vdots & \vdots & & \ddots & & \vdots \\ & & & & \beta+1 & \\ \gamma & \beta & & & \beta & \beta+1 \end{pmatrix}$$

is nonsingular, then it follows easily that the elements of $T$ corresponding to two different (commuting) monomials in $r_1^{(N+1)}, \ldots, r_n^{(N+1)}$ are different. In other words, $\varphi(U)$ is a free commutative semigroup of rank $n$.

It is enough to show that $U$ is cancellative because, from Theorem 4 (in view of Corollary 13 in Chapter 19), it then follows that $rk(U) \geq rk(\varphi(U)) = n$, and so $rk(S) \geq rk(U) \geq rk(T)$ as desired.

Fix some $\alpha \in A$. For every $s \in S$, let $s(\alpha)$ denote the projection of $s$ into $M_{n_\alpha}(D_\alpha)$, and let $S(\alpha) = \{s(\alpha) \mid s \in S\}$. It is enough to show that every $U(\alpha)$, $\alpha \in A$, is a cancellative semigroup because then $U \subseteq \prod_{\alpha \in A} U(\alpha)$ also is cancellative.

Clearly, $\rho(u_1(\alpha)) \geq \rho(u_2(\alpha)) \geq \cdots \geq \rho(u_{N+2}(\alpha))$, where $\rho$ stands for the rank of matrices in the corresponding algebra $M_{n_\alpha}(D_\alpha)$, so that the inequality $n_\alpha \leq N$ implies that $\rho(u_i(\alpha)) = \rho(u_{i+1}(\alpha))$ for some $i < N + 2$. From the choice of the $u_j$, $r_j^{(i)}$, it then follows that $\rho(r_j^{(i)}(\alpha)) = \rho(u_i(\alpha)) = \rho(u_i^N(\alpha))$. We use the fact that $\rho(u_i^N(\alpha)) = \rho(u_i^{2N}(\alpha))$ and so $u_i^N(\alpha)$ lies in a maximal subgroup $H$ of $M_{n_\alpha}(D_\alpha)$; see Corollary 7 in Chapter 1. Since

$r_j^{(i)}(\alpha) = u_i^N(\alpha)r_j^{(i-1)}(\alpha)u_i^N(\alpha)$, then it follows that all $r_j^{(i)}(\alpha), j = 1, \ldots, n$, lie in $H$, too. Therefore $\left\langle r_1^{(i)}, \ldots, r_n^{(i)} \right\rangle \subseteq H$ is a cancellative semigroup. Finally, $U(\alpha) = \left\langle r_1^{(N+1)}(\alpha), \ldots, r_n^{(N+1)}(\alpha) \right\rangle \subseteq \left\langle r_1^{(i)}, \ldots, r_n^{(i)} \right\rangle$ also is cancellative. This completes the proof.

**Corollary 8**  Assume that $K[S]$ is a $PI$-algebra. Then $rk(S/\rho_{\mathcal{B}(K[S])}) = Rk(S)$.

*Proof.*  Since $K[S]/\mathcal{B}(K[S])$ satisfies the hypothesis of Proposition 7, and $S/\rho_{\mathcal{B}(K[S])}$ embeds into this algebra by Lemma 5 in Chapter 4, then the equality $rk(S/\rho_{\mathcal{B}(K[S])}) = Rk(S/\rho_{\mathcal{B}(K[S])})$ follows. Now, the assertion is a consequence of the fact that $Rk(S) = Rk(S/\rho_{\mathcal{B}(K[S])})$ by Lemma 6.

Our next aim is to connect the rank with the Gelfand–Kirillov dimension of the prime homomorphic images of $K[S]$.

**Lemma 9**  Let $R$ be a prime $PI$-algebra with a classical quotient ring $Q$. If $a \in R$ is an element such that $aQa = eQe$ for some nonzero idempotent $e \in Q$, then $GK \dim aRa = GK \dim R$.

*Proof.*  Since $Q$ is a central localization of $R$ (see Theorem 1 in Chapter 18), then every nonzero element $x$ of $aQa$ may be written as $abz^{-1}a$ for some $b \in R$, $z \in Z(R)$. Now, $a$ is invertible in $aQa = eQe$, so that $x = aba^3c$, where $c$ is the inverse of $aza$ in $aQa$. Thus, $aQa$ is a ring of right quotients of $aRa$. From [121], Proposition 5.5, it then follows that $GK \dim aRa = GK \dim aQa = GK \dim Q = GK \dim R$ because $aQa \cong M_r(D)$, $Q \cong M_n(D)$ for some division algebra $D$ and some integers $n \geq r \geq 1$.

**Proposition 10**  Assume that $P$ is a prime ideal of a $PI$-algebra $K[S]$. Then $GK \dim K[S]/P \leq rk(S/\rho_P)$.

*Proof.*  Let $Q$ be a ring of quotients of $K[S]/P$. By Proposition 8 in Chapter 1, there exists an element $s \in S/\rho_P \subseteq K[S]/P \subseteq Q$ such that $s(S/\rho_P)s$ is a 0-cancellative semigroup, and $sQs = eQe$ for some $0 \neq e = e^2 \in Q$. Since $s(K[S]/P)s$ is a homomorphic image of $K[s(S/\rho_P)s]$, then, by Theorem 4, we get $GK \dim s(K[S]/P)s \leq GK \dim K[s(S/\rho_P)s] = rk(s(S/\rho_P)s) \leq rk(S/\rho_P)$. Thus, the assertion follows from Lemma 9.

**Proposition 11**  Let $S$ be a monoid such that $K[S]$ is a $PI$-algebra. If $P$ is a prime ideal of $K[S]$, then,

$$\operatorname{cl}K \dim K[S]/P \leq GK \dim K[S]/P \leq rk(S/\rho_P)$$
$$\leq \operatorname{cl}K \dim K[S/\rho_P] \leq GK \dim K[S/\rho_P]$$

*Proof.* Since $K[S]/P$ is a Goldie ring by Theorem 1 in Chapter 18, then, from [121], Corollary 3.16, we know that $\operatorname{cl}K \dim K[S]/P \le GK \dim K[S]/P$. The same argument with respect to prime ideals of $K[S/\rho_P]$ yields

$$\operatorname{cl}K \dim K[S/\rho_P] \le GK \dim K[S/\rho_P].$$

Thus, in view of Proposition 10, it is enough to show that $rk(S/\rho_P) \le \operatorname{cl}K \dim K[S/\rho_P]$. We know that $S/\rho_P \subseteq K[S]/P \subseteq M_n(L)$ for some $n \ge 1$ and a field $L$. Then, by Lemma 5, $rk(S/\rho_P) = rk(Z)$ for a subsemigroup $Z$ of $S/\rho_P$ contained in a subgroup $G$ of a principal factor $J_i/J_{i-1}$ of the multiplicative semigroup of $M_n(L)$. Since $K[(S/\rho_P) \cap G]$ is a $PI$-algebra, then $(S/\rho_P) \cap G$ has a group of fractions $H$ by Theorem 1 in Chapter 20, and $H$ embeds into $G$. Thus, from Lemma 21 in Chapter 7, it follows that

$$\operatorname{cl}K \dim K[H] \le \operatorname{cl}K \dim K[(S/\rho_P)/((S/\rho_P) \cap J_{i-1})] \le \operatorname{cl}K \dim K[S/\rho_P]$$

On the other hand,

$$\operatorname{cl}K \dim K[H] = \operatorname{cl}K \dim K[((S/\rho_P) \cap G)^1]$$
$$= rk((S/\rho_P) \cap G) \ge rk(Z) = rk(S/\rho_P)$$

by Theorem 4 and Proposition 2. Then $rk(S/\rho_P) \le \operatorname{cl}K \dim K[S/\rho_P]$ follows.

We now state the main result of this chapter, extending the assertion of Theorem 4.

**Theorem 12** Let $S$ be a monoid such that $K[S]$ is a $PI$-algebra. Then,

$$rk(S) \le \operatorname{cl}K \dim K[S] = \sup_P \{GK \dim K[S]/P\} = Rk(S)$$

where the supremum is taken over all prime ideals of $K[S]$.

*Proof.* Since, for every prime ideal $P$ of $K[S]$, we have natural homomorphisms $K[S] \to K[S/\rho_P] \to K[S]/P$, then,

$$\operatorname{cl}K \dim K[S] = \sup_P \{\operatorname{cl}K \dim K[S]/P\} = \sup_P \{\operatorname{cl}K \dim K[S/\rho_P]\}$$

From Proposition 11 we know that

$$\operatorname{cl}K \dim K[S] = \sup_P \{GK \dim K[S]/P\} = Rk(S).$$

The remaining inequality $rk(S) \le Rk(S)$ was established in Lemma 6.

An immediate question arising from Theorem 12 is: When does $rk(S)$ equal $\operatorname{cl}K \dim K[S]$? Proposition 7 provides examples of semigroups of this type. Another important case is settled below.

**Corollary 13** The following conditions are equivalent for a $PI$-algebra $K[S]$:

(i)  $\operatorname{cl} K \dim K[S] = 0$.

(ii)  $Rk(S) = 0$.

(iii)  $S$ is a periodic semigroup.

*Proof.* If $S$ is periodic, then $rk(S/\rho) = 0$ for every congruence $\rho$ on $S$. Thus, $Rk(S) = 0$. The assertion now follows from Theorem 12.

Let $X$ be a free monoid on two letters $x$, $y$. In [121], [128], examples of $PI$-algebras of the form $K[X/I]$ for a suitable ideal $I$ of $X$, which have an arbitrary $\geq 2$ Gelfand–Kirillov dimension are given. From Theorem 11 in Chapter 24, we know that $GK \dim K[X/I]/\mathcal{B}(K[X/I]) \leq 1$ and, from the construction of $I$, it follows that $rk(X/I) = 1$. Thus, $GK \dim K[X/I] > GK \dim K[X/I]/\mathcal{B}(K[X/I]) = \operatorname{cl} K \dim K[X/I] = 1$ in this case. This shows that, in general, one cannot expect the equality $GK \dim K[S] = \operatorname{cl} K \dim K[S]$, as in the case where $S$ is cancellative. The simplest examples of the above type are obtained by defining $I$ as the $n$th power of the ideal of $X = \langle x, y \rangle^1$ generated by $y$. Then we get $GK \dim K[X/I] = n$; see [121], Example 5.11.

Two more special cases in which the equality holds are given below.

**Theorem 14**  Assume that any of the following holds for a monoid $S$:

(i)  $K[S]$ is a noetherian $PI$-algebra.

(ii)  $S = T^1$ for a permutative semigroup $T$.

Then $\operatorname{cl} K \dim K[S] = GK \dim K[S] = Rk(S)$. Moreover, if (i) holds, then $Rk(S)$ is finite and, if (ii) holds, then $Rk(S) = rk(S)$.

*Proof.* (i) $S$ is finitely generated by Theorem 14 in Chapter 19. From [121], Theorem 10.15, we know that $GK \dim K[S] = \operatorname{cl} K \dim K[S] < \infty$ in this case. The remaining equality follows from Theorem 12.

(ii) Assume now that $T$ is a permutative semigroup. Let $\varphi : T \to V$ be a homomorphism onto a semigroup $V$ containing a free commutative subsemigroup $U$ of rank $n$. We will show that $T$ has a free commutative subsemigroup of rank $n$. Let $m$ be chosen as in Lemma 5 in Chapter 21. Then $\varphi(T^m) \cap U$ is an ideal in $U$ and, by Proposition 2, $rk(\varphi(T^m) \cap U) = rk(U)$. Hence, we can replace $U$ be a free commutative subsemigroup $U'$ of $\varphi(T^m) \cap U$ and replace $T$ be a finitely generated semigroup $\langle s_1, \ldots, s_n \rangle \subseteq T^m$ such that $t_1 = \varphi(s_1)$, ..., $t_n = \varphi(s_n)$ are generators of $U$. This allows us to assume, in view of Lemma 5 in Chapter 21, that the identity $uxyv = uyxv$ is satisfied in $T$. This implies that, for every $j \geq 2$, elements $u$, $v$, $x_1$, ..., $x_j \in T$, and every permutation $\sigma$, we have $ux_1 \ldots x_j v = ux_{\sigma(1)} \ldots x_{\sigma(j)} v$. Hence, every $s_i T s_i$ is a commutative semigroup.

Then $\varphi(s_iTs_i) = t_i^2U$ is an ideal in $U$ and so, by Proposition 2, we get $rk(\varphi(s_iTs_i)) = rk(U)$. Since $s_iTs_i$ is commutative, then, clearly, $rk(s_iTs_i) \geq rk(\varphi(s_iTs_i))$. It follows that $rk(T) \geq rk(s_iTs_i) \geq rk(U)$. This shows that $rk(T) \geq Rk(T)$.

In view of Theorem 12, it remains to prove that $GK \dim K[S] \leq rk(T)$. Since the Gelfand–Kirillov dimension and the rank are determined by their values on finitely generated subsemigroups of $T$, then again we can assume that $T = \langle s_1,\ldots,s_n \rangle$ is finitely generated. For $m \geq 1$, $S \setminus T^m$ is a finite set, which implies that $K[S]$ is a finitely generated module over the monoid algebra $K[(T^m)^1]$. Hence, $GK \dim K[S] = GK \dim K[T^m]$ by [121], Proposition 5.5. As above, we can assume that the identity $uxyv = uyxv$ is satisfied in $T$. Let $r > 1$, and let $w \in T$ be a word of length $2rn + 2$ in $s_1,\ldots,s_n$. Then $w = s_ivs_j$ for $v \in S$, $i,j \in \{1,\ldots,n\}$, and some $s_k$ occurs at least $2r$ times in $v$. Using the above identity, we can write $w = s_iw_1\ldots w_rs_j$, where every $w_l$ is in $s_kTs_k$ and has length $2n$ in $s_1, \ldots, s_n$. Let $W = \{s_1,\ldots,s_n\}$, $W_l = \{s_lvs_l \mid v$ has length at most $2n - 2$ in $s_1, \ldots, s_n\}$ for $l = 1, \ldots, n$, and let $d(r)$, $d_l(r)$ denote the growth functions determined by these subsets. It then follows that

$$d(r) \leq d(2rn + 2) \leq n^2 \sum_{l=1}^{n} d_l(r)$$

Therefore, $GK \dim K[T] \leq \max_l \{GK \dim K[s_lTs_l]\}$. Since each $s_lTs_l$ is commutative, then (i) (applied to finitely generated subsemigroups of $s_lTs_l$) implies that $rk(s_lTs_l) = GK \dim K[s_lTs_l]$. Hence, $GK \dim K[T] \leq \max_l \{rk(s_lTs_l)\} \leq rk(T)$, which completes the proof.

We note that any permutative monoid $S$ must be commutative. Thus, to handle a more general case, we had to artificially adjoin an identity to $T$ in condition (ii) of the above theorem.

Let us comment on the semigroups—finitely generated permutative semigroups—arising in the above proof. Though they are very close to commutative, the corresponding semigroup algebras are not noetherian in general. In fact, let $S$ be a free semigroup in the variety determined by the identity $uxyv = uyxv$, and assume that $s_1, \ldots, s_n$ is a minimal generating set for $S$. Then, for all $m > k$, and $s \neq t$, $s,t \in \{s_1,\ldots,s_n\}$, $st^ms \notin st^ksS^1$ because from the defining identity it follows that $st^ms$ is uniquely presented in the generators $s_1, \ldots, s_n$. Therefore, $S$ has no a.c.c. on right ideals, and $K[S]$ is not right noetherian.

**Remark 15**  It is not known whether we always have $rk(S) = Rk(S)$ provided that $K[S]$ is a $PI$-algebra. As in the proof of Theorem 14, we may restrict ourselves to finitely generated semigroups $S = \langle s_1, \ldots, s_n \rangle$ and a mapping $\varphi : S \to T$ onto a free commutative $T = \langle t_1, \ldots, t_n \rangle$ such that $\varphi(s_i) = t_i$, $i = 1, \ldots, n$. Moreover, in view of Proposition 7, we can assume that the kernel of the corresponding homomorphism $\bar{\varphi} : K[S] \to K[T]$ lies in $\mathcal{B}(K[S])$. (Note that $K[S] \to K[T]$ always factors through $K[S/\rho_{\mathcal{B}(K[S])}]$.) But $\mathcal{B}(K[S])$ must be nilpotent; see Theorem 1 in Chapter 18. Therefore, if $X = \langle x_1, \ldots, x_n \rangle$ is a free (noncommutative) semigroup and $I$ is the ideal of the free algebra $K[X]$ determined by the identity $\prod_{i=1}^{n}(y_i z_i - z_i y_i) = 0$, then the homomorphism $K[X] \to K[T]$ may be factorized as follows:

$$K[X] \longrightarrow K[X]/I \longrightarrow K[S] \longrightarrow K[T]$$

where $n$ is the nilpotency index of $\mathcal{B}(K[S])$, and each $x_i$ maps onto $t_i$. The question is whether in this case $S$ always has a free commutative subsemigroup of rank $n$.

**Remark 16**  If $S$ is commutative, then, for any prime ideal $P$ of $K[S]$, $S/\rho_P$ embeds into a domain $K[S]/P$, so that it is a 0-cancellative semigroup. Therefore, while $rk(S) = Rk(S)$, when $rk(S)$ is computed it is enough to compute the ranks of the cancellative semigroups $(S/\rho_P) \setminus \{\theta\}$, i.e., $rk(S) = \sup_P \{rk((S/\rho_P) \setminus \{\theta\})\}$. This was established by purely semigroup theoretical methods in [190].

Our last aim in this chapter is to give some conditions on the existence of the Krull dimension in the sense of [70]. By the Krull dimension we mean the right Krull dimension. It is known that if the group algebra $K[G]$ has Krull dimension, then $G$ has a.c.c. on normal subgroups and on finite subgroups [275], [199]. We will need a consequence of these results, which is stated with a proof.

**Proposition 17**  Let $G$ be a group with the property $\mathfrak{P}$. If $K[G]$ has Krull dimension, then $G$ is finitely generated and abelian-by-finite.

*Proof.*  From Theorem 8 in Chapter 19, we know that $G$ has a normal subgroup $H$ with $[G : H] < \infty$, $|H'| < \infty$. Since $K[H]$ is a direct summand of the left $K[H]$-module $K[G]$, then $K \dim K[H]$ exists; see Corollary 17 in Chapter 4. Hence, $K \dim K[H/H']$ also exists, and it is enough to show that $H/H'$ is finitely generated; then $G$ is finitely generated, and the assertion follows from Corollary 13 in Chapter 19. Thus, we may assume that $G$ is abelian. Let $T$ be the torsion subgroup of $G$. We first show that $T$ is finite. Let $T = \oplus_p T_p$, $T_p$—a $p$-group. If $\mathrm{ch}(K) = q > 0$, then, since

$K[T_q]$ has Krull dimension and $\mathcal{J}(K[T_q])$ is nil by [203], Lemma 8.1.8, then $\mathcal{J}(K[T_q])$ is nilpotent [149], Corollary 6.3.8. Therefore, by [203], Corollary 8.1.14, $T_q$ must be finite. This shows that, passing to the algebra $K[T/T_q]$ if $ch(K) = q > 0$, we may assume that $T$ has no $ch(K)$-torsion. Suppose that $H_0 \subsetneq H_1 \subsetneq H_2 \subsetneq \cdots$ is a chain of finite subgroups of $T$. Put $x_i = \sum_{g \in H_i} g$. If $i < j$, let $\{s_1, \ldots, s_m\}$ be a right transversal for $H_i$ in $H_j$. Since $x_i(s_1 + \ldots + s_m) = x_j$, then $x_j K[T] \subseteq x_i K[T]$. Moreover, $x_i K[T] = x_j K[T] + (mx_i - x_j)K[T]$ (note that $M \mid |H_j|$ is invertible in $K$). We show that this sum is direct. Suppose that $x_j r_1 = (mx_i - x_j)r_2$ for some $r_1, r_2 \in K[T]$. Let $\{t_1, \ldots, t_m\}$ be a left transversal for $H_i$ in $H_j$. Then,

$$
\begin{aligned}
x_j r_1 &= x_j m r_1 m^{-1} = (t_1 + \cdots + t_m) x_j r_1 m^{-1} \\
&= (t_1 + \cdots + t_m)(mx_i - x_j) r_2 m^{-1} \\
&= (mx_j - mx_j) r_2 m^{-1} = 0
\end{aligned}
$$

proving our claim. Let $r$ be the Goldie dimension of $K[T]$ ($r < \infty$ by [149], Lemma 6.2.6), and put $m_i = [H_i : H_{i-1}]$. Then,

$$
\begin{aligned}
x_0 K[T] &= x_1 K[T] \oplus (m_1 x_0 - x_1)K[T] \\
&= x_2 K[T] \oplus (m_2 x_1 - x_2)K[T] \oplus (m_1 x_0 - x_1)K[T] \\
&= x_r K[T] \oplus (m_r x_{r-1} - x_r)K[T] \oplus \cdots \oplus (m_1 x_0 - x_1)K[T]
\end{aligned}
$$

is a direct sum of $r + 1$ nonzero ideals, a contradiction. This shows that $T$ is finite. Now, passing to the algebra $K[G/T]$, we may assume that $G$ is torsion-free. If there exists a finitely generated subgroup $F$ of $G$ with $G/F$ being torsion, then, as above, $G/F$ is finite and $G$ is finitely generated as desired. Otherwise, $G$ has a free abelian subgroup $E$ of infinite rank. Clearly, $E$ has an infinite torsion homomorphic image, which again contradicts the first paragraph of the proof. This completes the proof.

**Proposition 18** Assume that $S$ is a monoid such that $K[S]$ is a $PI$-algebra with Krull dimension. Then,

(1) $Rk(S) < \infty$.
(2) $K \dim K[S] = clK \dim K[S] = Rk(S)$.
(3) If $rk(S) = 0$, then $S$ is finite and $Rk(S) = 0$.

*Proof.* Assume first that $S = S/\rho_P$ for a prime ideal $P$ of $K[S]$. Then, as in the proof of Proposition 11, $S$ is a subsemigroup of some $M_n(L)$, where $L$ is a field, $n \geq 1$, and $rk(S) = rk(G \cap S)$, where $G$ is a subgroup of a principal factor of $M_n(L)$. Since $K \dim K[(G \cap S)(G \cap S)^{-1}]$ exists by Lemma 21, in Chapter 7, then from Proposition 17 it follows that

$(G \cap S)(G \cap S)^{-1}$ is a finitely generated group. Hence, Proposition 2 yields $rk(S) = rk(G \cap S) = rk((G \cap S)(G \cap S)^{-1}) < \infty$.

Now, let $S$ be an arbitrary monoid satisfying the hypothesis of the proposition. From [169], Lemma 5.5.1, it follows that $K \dim K[S] = K \dim K[S]/P$ for a prime ideal $P$ of $K[S]$. It is known that, for any ring $R$ with Krull dimension, $K \dim R \geq \operatorname{cl} K \dim R$, and the equality holds if $R$ is a $PI$-algebra and the latter dimension is finite; see [169], Theorems 5.3.3 and 5.5.2. From Proposition 11 it then follows that $K \dim K[S] = \operatorname{cl} K \dim K[S]/P \leq rk(S/\rho_P) < \infty$, and further $K \dim K[S] = \operatorname{cl} K \dim K[S]$. Moreover, $\operatorname{cl} K \dim K[S] = Rk(S)$ by Theorem 12.

Finally, if $rk(S) = 0$, then $S$ is periodic, and so $Rk(S) = 0$. Hence, $K \dim K[S] = 0$, i.e., $K[S]$ is right artinian. From Theorem 23 in Chapter 14, it then follows that $S$ must be finite.

Observe that, in view of the above result, the finiteness of $rk(S)$ is not a sufficient condition for $K \dim K[S]$ to exist. (For example, take an infinite periodic group.)

As we noticed in Proposition 17, if a group algebra $K[G]$ is a $PI$-algebra with Krull dimension, then $G$ is finitely generated and $G$ has a.c.c. on subgroups. The following example shows that, in general, the situation is much more complicated. In particular, $S$ may not satisfy a.c.c. on principal right ideals, while $K[S]$ has Krull dimension.

**Example 19** Let $S = \{\theta, 1, y^n, x_n \mid n \geq 1\}$ with the multiplication defined by the rules: $x_i x_j = \theta$ for $i, j \geq 1$, $yx_1 = x_1 y = \theta$, $yx_i = x_i y = x_{i-1}$ for $i > 1$. Plainly, $S$ is not finitely generated, and $x_1 S \subsetneq x_2 S \subsetneq \cdots$ is an ascending chain of principal ideals. The subset $I = \{x_1, x_2, \ldots\} \cup \{\theta\}$ is an ideal of $S$, and $K[S]/K[I] \cong K[\langle y, 1 \rangle]$ is the polynomial ring, i.e., it has Krull dimension 1. Hence, it has Krull dimension 1 as a $K[S]$-module. Let $0 \neq J \subseteq K_0[I]$ be an ideal of $K_0[S]$. For any $a = \sum_{i=1}^{n} \lambda_i x_i \in J$, $\lambda_n \neq 0$, we have $ay, ay^2, \ldots, ay^{n-1} \in J$. This easily implies that $x_1, \ldots, x_n \in J$. Hence, $J = K_0[\{x_1, \ldots, x_m, \theta\}]$ for some $m \geq 1$ or $J = K_0[I]$. This shows that the $K_0[S]$-module $K_0[I]$ has Krull dimension 0 and, consequently $K[I]$ is also an artinian $K[S]$-module. Thus, by [70], Lemma 1.1, $K \dim K[S] = 1$.

We conjecture that in the case where $S$ has a.c.c. on principal right ideals and $K[S]$ is a $PI$-algebra with Krull dimension, $S$ must be finitely generated and $K[S]$ is right noetherian. This would extend the result on right noetherian $PI$-semigroup-algebras established in Theorem 14 in Chapter 19. We are able to prove the conjecture in two special cases. The following auxiliary result will be useful.

**Lemma 20** Let $S$ be a subsemigroup of a group $G$. If $H$ is a subgroup of $G$ and $sS^1 = tS^1$, $s(S \cap H)^1 \subseteq t(S \cap H)^1$, for some $s, t \in S \cap H$, $s \neq t$, then $s(S \cap H)^1 = s(S \cap H) = t(S \cap H) = t(S \cap H)^1$.

*Proof.* Since $t = sz, z \in S$, then $s(S \cap H)^1 \subseteq sz(S \cap H)^1$, and the cancellativity implies that $(S \cap H)^1 \subseteq z(S \cap H)^1$. Therefore, $z \in H$, so that $z \in S \cap H$, and we get $t \in s(S \cap H)$. Thus, $t(S \cap H)^1 \subseteq s(S \cap H) \subseteq s(S \cap H)^1 \subseteq t(S \cap H)^1$, $S$ must be a monoid, and the result follows.

We state a direct consequence of the lemma and Lemma 4 in Chapter 2.

**Corollary 21** Let $R$ be a subsemigroup of a semigroup $S$ with a.c.c. on principal right ideals. Then $R$ has a.c.c. on principal right ideals in any of the following cases:

(1)  $R$ is an ideal of $S$.
(2)  $S$ is a subsemigroup of a group $G$, and $R = S \cap H \neq \varnothing$ for a subgroup $H$ of $G$.

**Lemma 22** Let $S$ be a commutative archimedean semigroup with no nonzero idempotents. If $s = sx$ for some $s, x \in S$, then $s$ is a zero element of $S$. Moreover, if $t, u \in S$ are such that $tS^1 = uS^1$, then $t = u$.

*Proof.* Let $\xi$ be the least separative congruence on $S$. Then $S/\xi$ is a cancellative semigroup [26], § 4.3, which implies that $(x, x^2) \in \xi$. From the description of $\xi$ (see Chapter 6), it follows that $x(x^2)^n = x^{2n+2}$ for some $n$. Therefore, $x$ is a periodic element. The hypothesis on $S$ implies then that $S$ is a nil semigroup, so that $s = sx = sx^2 = \cdots$ must be the zero element. If $tS^1 = uS^1$ and $t \neq u$, then $t = ty$ for some $y \in S$. This, in view of the first part of the proof, shows that $t = \theta$ and, consequently, $u = \theta$, a contradiction. This establishes the assertion.

**Proposition 23** Assume that $S$ is a monoid with a.c.c. on principal right ideals. If $K[S]$ has Krull dimension, then $S$ is finitely generated if any of the following holds:

(i)  $S$ is commutative.
(ii)  $S$ is cancellative and has the permutational property.

Moreover, in this case, $K[S]$ is noetherian, and $K \dim K[S] = rk(S) < \infty$.

*Proof.* (i) Since $K[S]$ has finite Goldie dimension (see [149], Lemma 6.2.6), then $K[S]$ has finitely many idempotents. We know that, for every $e = e^2 \in S$, $K[S] \cong K[eS] \oplus K_0[S/eS]$; see Corollary 8 in Chapter 4. Thus, it is enough to consider the case where $S$ has no idempotents other than identity and possibly zero. Let $\eta$ be the least semilattice congruence on $S$; see Chapter 6. Then $K[S/\eta]$ has Krull dimension, and the above shows that $S/\eta$ is finite. We proceed by induction on the number $m_S$ of nonzero archimedean components of $S$. If $m_S = 1$, then $S$ must be a group (possibly with zero), and so it is finitely generated by Proposition 17. Assume that $m_S > 1$. Let $L$ be the archimedean component of $S$ that maps onto an element of the semilattice $S/\eta$, which is minimal (with respect to the natural ordering of $S/\eta$) among all elements corresponding to nonzero components of $S$. By Corollary 21, $I$ has a.c.c. on principal ideals where $I = L$ if $S$ has no zero element, and $I = L \cup \{\theta\}$ otherwise.

We claim that $I$ contains finitely many ideals that are maximal among the ideals of the form $xS, x \in I$. If $z_1 S, z_2 S, \ldots$ are such ideals, then let $J$ be the union of the sets of nongenerators of all $z_i S, i \geq 1$. Since $z_i S \cap z_l S \subseteq J$ for $i \neq l$, then the algebra $K_0[S/J]$ has infinitely many independent ideals $K_0[(z_i S \cup J)/J]$. This contradicts the finiteness of the Goldie dimension of this algebra and establishes the claim.

If $z_1 S, \ldots, z_n S$ are all such ideals, then $I = \bigcup_{i=1}^{n} z_i S$, and it is easy to see that $\langle z_1, \ldots, z_n, S \setminus I \rangle$ contains $S \setminus I^2$.

Assume that $t_0 \in I^2$ is a nonzero element. We can decompose $t_0$ into products $t_0 = s_1 t_1 = s_1 s_2 t_2 = \cdots = s_1 \ldots s_m t_m = \cdots$, where $t_i = t_{i+1} s_{i+1} \in I$ and every $s_i$ is chosen from the set $\{z_1, \ldots, z_n\}$, provided that all $t_i$ are in $I^2$. Since $t_1 I^1 \subseteq t_2 I^1 \subseteq \cdots$, then there exists $k$ such that $t_k I^1 = s_k t_k I^1$. While $S$ has no nontrivial idempotents and $S \neq I$, Lemma 22 contradicts the fact that $t_k \neq \theta$. It follows that, for some $j \geq 1$, the element $t_j$ is not in $I^2$. Then $t_0 \in \langle z_1, \ldots, z_n, S \setminus I \rangle$. Since $S/I$ is a finitely generated semigroup by the induction hypothesis and $\theta \neq t_0 \in I^2$ is arbitrary, it follows that $S$ is finitely generated. Clearly, $K[S]$ is noetherian in this case.

Assume now that (ii) holds. From Theorem 8 in Chapter 19, we know that $S$ has a group of fractions $G$ that is finite-by-abelian-by-finite. Moreover, $K[G]$ has Krull dimension as a localization of $K[S]$; see Lemma 13 in Chapter 7. Then, by Proposition 17, $G$ is finitely generated, so that it has an abelian normal subgroup $Z$ of finite index; see Corollary 13 in Chapter 19. Since $K[S \cap Z]$ is a direct summand of the left $K[S \cap Z]$-module $K[S]$, then $K[S \cap Z]$ has Krull dimension. From Corollary 21, and the first part of the proof, it follows that $S \cap Z$ is finitely generated. Hence, by Lemma 1 in Chapter 11, $K[S]$ is right noetherian and, similarly, $K[S]$ is also left noetherian. Consequently, $S$ is finitely generated by Theorem 9 in Chapter 11.

Now, $K \dim K[S] = rk(S) < \infty$ follows, in both cases, from Proposition 18 and Theorem 14, Theorem 4, respectively, because, in the latter case, $K[S]$ is a $PI$-algebra by Theorem 1 in Chapter 20.

## Comments on Chapter 23

The description of the classical Krull dimension for semigroup algebras of commutative cancellative semigroups, extended by our Theorem 4, was given by Arnold and Gilmer; see [64], §21. This was generalized to arbitrary commutative semigroups by Okniński in [190], where $rk(S)$ and $Rk(S)$ were defined, and also some necessary conditions for the existence of $K \dim K[S]$ were found. We note that the proof of our Proposition 1 is taken from [102]. The Krull dimension of semigroup algebras of cancellative semigroups was also studied by Shapiro in [245]. Some other conditions on the set of prime ideals of commutative semigroup algebras were considered by Gilmer in [66] and [67]; see also [64].

It is well known that, for a finitely generated prime $PI$-algebra $R$, we have $cl K \dim R = GK \dim R =$ the transcendence degree of the field of fractions of the center of $R$ over $K$; see [149], Proposition 13.10.6. Thus, $rk(S)$ plays the role of the transcendence degree for semigroup algebras, but it can also be applied in the nonfinitely generated case.

The study of the Krull dimensions of (noncommutative) group rings was started by Smith [255] and Woods [275], from which our Proposition 17 is extracted. Some further results were given by Park in [199]. Bounds on the Krull dimensions of rings graded by polycyclic-by-finite groups were obtained using the smash product techniques by Chin and Quinn in [23]. The correspondence between prime ideals in normalizing extensions, covering the case of crossed products of finite groups, was also extensively studied. We refer to [149], Section 10, for the main results on this topic.

Proposition 23 is basically extracted from [186] and [190]. As noted before, the algebra of a finitely generated permutative (or even medial) semigroup is not noetherian in general. However, it is known that every proper subvariety $\mathcal{V}$ of the variety of medial semigroups is locally noetherian, that is, every semigroup algebra of a finitely generated member of $\mathcal{V}$ is noetherian; see [170]. Another extension of commutative semigroup algebras was considered, with respect to the homological dimension, by Dicks in [42]. These are algebras $K[S]$ where $S = X/\rho$, $X$ being a free monoid on $x_1, \ldots, x_n$, and $\rho$ being the congruence generated by a subset of the set $\{(xy,yx) \in X \times X \mid x,y \in \{x_1,\ldots,x_n\}\}$.

We note that the Gelfand–Kirillov dimension of non-$PI$-semigroup-algebras of cancellative semigroups, as well as some other specific types of semigroups, is discussed in Chapters 8 and 24.

# 24

# Monomial Algebras

Let $X$ be a free semigroup. If $I$ is an ideal of $X$, then the contracted semigroup algebra $K_0[X/I]$ is called the monomial algebra defined by the set $I$ of words of $X$. From Lemma 7 in Chapter 4, we know that $K_0[X/I]$ may be identified with the quotient $K[X]/K[I]$. Therefore, monomial algebras form a class of well-behaved graded algebras, with the grading coming from the natural grading in the free semigroup $X$. Finitely generated monomial algebras arise in a natural way in the context of the growth of associative algebras; see [121], [3]. It is known that, for every finitely generated algebra (or semigroup) $R$, there exists a free semigroup $X$ and its ideal $I$ such that the growth of $R$ is the same as that of $X/I$. The study of monomial algebras is also motivated by some problems in algebraic topology. Connections of this type and some applications of the combinatorics in free semigroups developed for monomial algebras are given in [3] and [4]. A special case of so-called path algebras of graphs and homomorphic images of these algebras is of basic importance in the theory of representations of artinian algebras; see [5], [75], [210].

Our aim in this chapter is to describe the structure of (finitely generated) monomial algebras. We also discuss the $PI$-property and the dimensions of $K_0[X/I]$. Throughout, $X$ denotes a free semigroup with free generators $x_1, \ldots, x_n$, also called letters in $X$. For any $v \in X$ by $l(v)$, we mean the length of the word $v$ in the generators $x_1, \ldots, x_n$. The relations $x_1 < x_2 < \cdots < x_n$ define an order on $X$ by $v < w$ if and only if $l(v) < l(w)$ or $l(v) = l(w)$ and $v$ precedes $w$ lexicographically; see [121]. If $0 \neq f \in K[X]$, then $\bar{f}$ denotes the dominant word (with respect to the above order) in the support of $f$. Let $J \neq 0$ be a right ideal of $K[X]$. Define $\alpha(J) = \{v \in X \mid v = \bar{f}$ for some $f \in J\}$ and $\alpha(0) = \varnothing$. Since $(\overline{fg}) = \bar{f}\bar{g}$ for any $0 \neq f, g \in K[X]$, then it is clear that $\alpha(J)$ is a right ideal of $X$. Moreover, $\alpha(J_1)\alpha(J_2) \subseteq \alpha(J_1 J_2)$ for any nonzero ideals $J_1, J_2$ of $K[X]$.

The importance of monomial algebras for the general theory of Gelfand–Kirillov dimension of arbitrary algebras comes from the fact that for a two-sided ideal $J$ of $K[X]$, $GK \dim K[X]/J = GK \dim K_0[X/\alpha(J)]$ (see [121], Section 2) or, more precisely, the growth functions of $K[X]/J$ and of the semigroup $X/\alpha(J)$ coincide. Note that the growth function $d(m)$ of every $K_0[X/I]$ is determined by the number of words in $X \setminus I$ of length not exceeding $m$.

We start with the following simple observation.

**Lemma 1** Let $J_1 \subsetneqq J_2$ be right ideals of $K[X]$ such that $J_1$ is a $K$-subspace. Then $\alpha(J_1) \subsetneqq \alpha(J_2)$.

*Proof.* Clearly, $\alpha(J_1) \subseteq \alpha(J_2)$. Suppose that the equality holds. Choose $f \in K[X]$ such that $\bar{f}$ is minimal with respect to the property $f \in J_2 \setminus J_1$. Since $\bar{f} \in \alpha(J_2) = \alpha(J_1)$, then there exists $g \in J_1$ such that $\bar{f} = \bar{g}$. Then $\overline{f - \lambda g} < \bar{f}$ for some $\lambda \in K$ and $f - \lambda g \in J_2 \setminus J_1$, contradicting the choice of $f$. ∎

We derive a direct consequence of the above fact. Here, an ideal $T$ of $X/I$ is called prime (semiprime respectively) if, for all ideals $J_1, J_2$ in $X/I$, we have $J_1 \subseteq T$ or $J_2 \subseteq T$ whenever $J_1 J_2 \subseteq T$ ($J_1 \subseteq T$ whenever $J_1^2 \subseteq T$). $(X/I)/T$ is called prime (semiprime) in this case.

**Proposition 2** Let $J$ be a $K$-algebra ideal of $K[X]$. Then $K_0[X/\alpha(J)]$ is prime (semiprime) if and only if $X/\alpha(J)$ is a prime (semiprime respectively) semigroup. Moreover, in this case, the algebra $K[X]/J$ is also prime (semiprime).

*Proof.* Let $P$, $Q$ be ideals of $K[X]$ with $P \supsetneq J$, $Q \supsetneq J$. If $PQ \subseteq J$, then $\alpha(P)\alpha(Q) \subseteq \alpha(PQ) \subseteq \alpha(J)$. Since $\alpha(P) \supsetneq \alpha(J)$, $\alpha(Q) \supsetneq \alpha(J)$ by Lemma 1, then the primeness (semiprimeness) of the ideal $\alpha(J)$ of $X$ implies the respective condition of the ideal $J$ of $K[X]$. In particular, applying this to $J' = K[\alpha(J)]$, since $\alpha(J') = \alpha(J)$, we see that $K_0[X/\alpha(J)] = K[X]/K[\alpha(J)]$ is also prime (semiprime). On the other hand, for every two ideals $I_1$, $I_2$, of $X$ with $I_1 I_2 \subseteq \alpha(J)$, $I_j \supsetneq \alpha(J)$ for $i = 1, 2$, $K[I_1]$ and $K[I_2]$ are ideals of $K[X]$ satisfying $K[I_1]K[I_2] \subseteq K[\alpha(J)]$. Thus, the converse follows.

Let $\mathcal{B}(X/I)$ denote the intersection of all prime ideals of $X/I$. As in the ring case, one may check that $\mathcal{B}(X/I)$ coincides with the (Kurosh–Amitsur) lower radical of $X/I$ determined by the class of nilpotent semigroups; see [93]. By $\mathcal{L}(X/I)$ we mean the largest locally nilpotent ideal of $X/I$.

**Proposition 3**   For every monomial algebra $K_0[X/I]$, we have

 (i)   $\mathcal{B}(K_0[X/I]) = K_0[\mathcal{B}(X/I)]$
 (ii)   $\mathcal{L}(K_0[X/I]) = K_0[\mathcal{L}(X/I)]$

*Proof.* It is clear that $K_0[J]$ is nilpotent (locally nilpotent) if $J$ is a nilpotent (locally nilpotent) ideal of $K_0[X/I]$. Therefore, $K_0[\mathcal{S}(X/I)] \subseteq \mathcal{S}(K_0[X/I])$ for $\mathcal{S} = \mathcal{B}$, $\mathcal{L}$. Thus, passing to the semigroup $(X/I)/\mathcal{S}(X/I)$, we may assume that $\mathcal{S}(X/I) = 0$. Suppose that $L$ is a nonzero nilpotent (locally nilpotent) ideal of $K_0[X/I]$, and $L'$ is the inverse image of $L$ in $K[X]$. It is easy to see that $\alpha(L')/I$ is a nonzero nilpotent (locally nilpotent) ideal of $X/I$. This contradiction shows that $\mathcal{S}(K_0[X/I]) = 0$, which establishes the assertion.

We now fix an ideal $I$ of $X$. Let $\mathcal{N}(X/I)$ denote the largest nilideal of $X/I$. For a given $\theta \neq s \in X/I$, put $T_s = \{y \in X \setminus I \mid y$ is not a subword of any $s^n$, $n \geq 1\} \cup \{\theta\}$. Then $T_s$ is an ideal of $X/I$. If $s$ is a nilpotent element, then $(X/I)/T_s$ is a finite semigroup. If $s$ is not nilpotent, then it is clear that $T_s$ is the largest ideal of $X/I$ not containing powers of $s$. Moreover, $(X/I)/T_s$ consists of all subwords of $s, s^2, \dots$ and $\theta$. Observe also that $X/X_s = (X/I)/T_s$, where $X_s = \{y \in X \mid y$ is not a subword of any $s^n$, $n \geq 1\}$. (We note that the symbols $X_s$, $T_s$, $I_s$ used throughout this chapter should not be confused with the notation for the principal factors and the ideals of nongenerators used earlier in the text.)

Recall that a word $s$ of $X$ is called primitive if $s = t$ whenever $s = t^m$ for some $t \in X$, $m \geq 1$. It is straightforward that, for any $s \in X$, there exists a primitive word $t \in X$ such that $s = t^m$, $m \geq 1$. Clearly, $X_s = X_t$ in this case, and $t$ is unique by [132], Proposition 11.5.6.

**Proposition 4** $\mathcal{N}(X/I) = \bigcap_{s \in \mathcal{A}} T_s$, where $\mathcal{A}$ is the set of non-nilpotent elements of $X/I$ that are primitive as words in $X$. Moreover, any $K_0[(X/I)/T_s]$, $s \in \mathcal{A}$, is a prime algebra of Gelfand–Kirillov dimension 1.

*Proof.* If $t \in X/I$ is any non-nilpotent element, then let $s$ be a primitive word in $X$ such that $t = s^m$ for some $m \geq 1$. Since $s \in \mathcal{A}$ and $T_s = T_t$, then $\bigcap_{s \in \mathcal{A}} T_s = \bigcap_{t \in \mathcal{B}} T_t$, where $\mathcal{B}$ is the set of all non-nilpotents of $X/I$.

If $t \notin \mathcal{N}(X/I)$, then $vtw$ is not nilpotent in $X/I$ for some words $v$, $w$ in $X^1$. The fact that $vtw \notin T_{vtw}$, and so $t \notin T_{vtw}$, shows that $\bigcap_{s \in \mathcal{A}} T_s \subseteq \mathcal{N}(X/I)$. On the other hand, if $t \notin \bigcap_{s \in \mathcal{A}} T_s$, then $t \notin T_s$ for some $s \in \mathcal{A}$. This means that $xty = s^n$ for some $n \geq 1$, $x, y \in X^1$. Therefore, $xty$ is not nilpotent, and so $t \notin \mathcal{N}(X/I)$. Thus, $\bigcap_{s \in \mathcal{A}} T_s = \mathcal{N}(X/I)$ follows.

If $z \in (X/I) \setminus T_s$, then, as above, $rzu = s^m$ for some $m \geq 1$, $r, u \in X^1$. Hence, for any nonzero ideals $J_1$, $J_2$ of $X/I$, $J_1 J_2$ contains a power of $s$, so that $J_1 J_2 \neq \theta$. Consequently, $T_s$ is a prime ideal of $X/I$. From Proposition 2 it then follows that $K_0[(X/I)/T_s]$ is a prime algebra.

Finally, let $k$ be any integer with $k \geq l(s)$. Clearly, $(X/I) \setminus T_s$ has at most $l(s)$ words of length $k$. Then the growth function of $(X/I)/T_s$ is bounded by a polynomial of degree 1, so that $GK \dim K_0[(X/I)/T_s] \leq 1$. Since $s, s^2, \ldots \notin T_s$, it follows that $GK \dim K_0[(X/I)/T_s] = 1$.

Our next result, obtained together with Proposition 4 by Okniński in [191], provides a description of all factors $X/X_s = (X/I)/T_s$ and will be crucial in the sequel.

**Proposition 5** Let $s \in X \setminus I$ be primitive as a word in $X$. Then there exists an ideal $I_s$ of $X/X_s$ such that the semigroup $(X/X_s)/I_s$ is finite of cardinality not exceeding $2 \cdot l(s)^2$ and $I_s$ is isomorphic to a matrix semigroup $\mathfrak{M}^0(\langle t \rangle, l(s), l(s), P)$, where $\langle t \rangle$ is an infinite cyclic semigroup and $P = (p_{ij})$ is the $l(s) \times l(s)$ matrix given by

$$p_{11} = 1, \quad p_{ii} = t \quad \text{for } i = 2, \ldots, l(s)$$
$$p_{ij} = \theta \quad \text{for } i \neq j$$

*Proof.* Let $s = y_1 \ldots y_m$ for some letters $y_i$ in $X$, and let $I_s$ be the subset of $X/X_s$ consisting of $\theta$ and the words in $X \setminus X_s$ having $s$ as a subword. Since any word in $X \setminus X_s$ of length at least $2m - 1$ is in $I_s$, then $(X/X_s) \setminus I_s$ has at most $2m(m - 1)$ elements. Let $z \in I_s$, $z \neq \theta$. Then $z$ may be uniquely written as $ws^n u$, where $n \geq 1$, $w$ is the empty word or equals $y_q \ldots y_m$ for some $q \in \{2, \ldots, m\}$, $u$ is the empty word or equals $y_1 \ldots y_r$ for some $r \in \{1, \ldots, m - 1\}$, and $l(w)$ is minimal with this property. Define

$\varphi(z) = (t^n, \bar{q}, \bar{r})$, where

$$\bar{q} = \begin{cases} 1 & \text{if } w \text{ is the empty word} \\ q & \text{otherwise} \end{cases}$$

$$\bar{r} = \begin{cases} 1 & \text{if } u \text{ is the empty word} \\ r+1 & \text{otherwise} \end{cases}$$

Put $\varphi(\theta) = \theta$. Clearly, $\varphi : I_s \to \mathfrak{M}^0(\langle t \rangle, m, m, P)$ is a bijective mapping. We claim that it is a semigroup homomorphism. Take $z_1, z_2 \in I_s \setminus \{\theta\}$ and, as above, let $z_1 = w_1 s^{n_1} u_1$, $z_2 = w_2 s^{n_2} u_2$. If $u_1 w_2$ is the empty word, then $z_1 z_2 = w_1 s^{n_1 + n_2} u_2$ and $\varphi(z_1)\varphi(z_2) = (t^{n_1}, \overline{q_1}, 1)(t^{n_2}, 1, \overline{r_2}) = (t^{n_1 + n_2}, \overline{q_1}, \overline{r_2}) = \varphi(z_1 z_2)$. A similar argument works if $u_1 w_2 = s$. Thus, assume that $u_1 w_2 \neq 1, s$. Since $\varphi(z_1)\varphi(z_2) = \theta$ in this case, then we have to show that $z_1 z_2 = \theta$. Suppose that $z_1 z_2 \neq \theta$, i.e. $w_1 s^{n_1} u_1 w_2 s^{n_2} u_1$ is a subword of some $s^k$, $k \geq 1$. Then $s u_1 w_2 s$ is a subword of $s^5$, and so $t s u_1 w_1 s u = s^5$ for some $t, u \in (X/X_s) \cup \{1\}$. If $t, u \in \{1, s, s^2\}$, then $u_1 w_2 \in \{1, s\}$ contradicting our assumption. Hence, $t$ or $u$, say $t$ (the other case goes similarly), is not in $\{1, s, s^2\}$. Since $ts$ is a subword in $s^5$, then $xsy = s^2$ for some nonempty words $x, y$ of $X$. Thus, $y_1 \ldots y_m = s = y_l y_{l+1} \ldots y_m y_1 \ldots y_{l-1}$ for some $l \in \{2, \ldots, m\}$. This and [132], Corollary 11.5.3, imply that the words $y_1 \ldots y_{l-1}$, $y_l \ldots y_m$ are powers of the same word, contradicting the fact that $s$ is a primitive word in $X$.

**Corollary 6** Let $s \in X$. Then $K_0[X/X_s]$ is a $PI$-algebra and $\mathcal{J}(K_0[X/X_s]) = 0$.

*Proof.* With the notation of Proposition 5, we know that $K_0[X/X_s]$ is a $PI$-algebra whenever $K_0[I_s]$ is a $PI$-algebra because the quotient algebra is finite dimensional. The latter is a direct consequence of Proposition 6 in Chapter 20. Now, by Theorem 12 in Chapter 21, $\mathcal{J}(K_0[X/X_s]) = \mathcal{B}(K_0[X/X_s])$ and, since $K_0[X/X_s]$ is prime by Proposition 4, the result follows.

**Corollary 7** $\mathcal{J}(K_0[X/I]) \subseteq K_0[\mathcal{N}(X/I)]$.

*Proof.* This follows from the fact that $K_0[\mathcal{N}(X/I)]$ is an intersection of the ideals $K_0[T_s]$ of $K_0[X/I]$, with the corresponding quotient algebras being Jacobson semisimple.

It is not known whether $\mathcal{J}(K_0[X/I])$ always is homogeneous with respect to the $X/I$-gradation on $K_0[X/I]$, that is whether $\mathcal{J}(K_0[X/I]) = K_0[T]$ for an ideal $T$ of $X/I$. If this is the case, and if $\text{ch}(K) = 0$, then, from Theorem 18 in Chapter 14, it follows that $T$ is locally finite because $K[T]$ is a local algebra. Thus, Proposition 3 implies that $\mathcal{J}(K_0[X/I]) =$

$\mathcal{L}(K_0[X/I]) = K_0[\mathcal{L}(X/I)]$, that is, $T = \mathcal{L}(X/I)$. We are able to establish only the following partial result showing that the Jacobson radical is homogeneous with respect to the gradations on $K_0[X/I]$ defined by the degrees of words of $X/I$ in the generators $x_1,\ldots,x_n$ of $X$.

**Proposition 8** $\mathcal{J}(K_0[X/I])$ is the $K$-subspace of $K_0[X/I]$ generated by all elements of $\mathcal{J}(K_0[X/I])$ of the form

(a)  $c_1 + \cdots + c_m$, where $m \geq 1, c_i$ are monomials with $\deg_x c_i = \deg_x c_j$
     for every $1 \leq i, j \leq m$, and every $x \in \{x_1,\ldots,x_n\}$.

*Proof.* Let $0 \neq a \in \mathcal{J}(K_0[X/I])$, and let $a = a_1 + \cdots + a_k$ be the decomposition of $a$ into a sum of elements $a_i$ of the form (a), with minimal $k$. If $K$ is finite, then let $L$ be a finite field extension of $K$, with the cardinality exceeding $\max \{l(w) + 1 \mid w \in \mathrm{supp}_0(a)\}$. Since $L_0[X/I] \cong L \otimes_K K_0[X/I]$ by Lemma 10 in Chapter 4 and $L : K$ is separable, then, from [203], Theorem 7.2.13, it follows that $a \in \mathcal{J}(L_0[X/I])$. If $K$ is infinite, then put $L = K$. Choose any $0 \neq \lambda \in L$, $x \in \{x_1,\ldots,x_n\}$, and define an automorphism $\varphi_\lambda$ of $L[X]$ by

$$\varphi_\lambda(x) = \lambda x \qquad \varphi_\lambda(y) = y \text{ for } y \in \{x_1,\ldots,x_n\} \setminus \{x\}$$

It is clear that $\varphi_\lambda$, $\varphi_\lambda^{-1}$ map $L[I]$ onto $L[I]$. Thus, $\varphi_\lambda$ induces an automorphism $\overline{\varphi_\lambda}$ of the algebra $L[X]/L[I] \cong L_0[X/I]$. Consequently, $\overline{\varphi_\lambda}(a) \in \mathcal{J}(L_0[X/I])$. We have $\overline{\varphi_\lambda}(a) = b_0 + \lambda b_1 + \cdots + \lambda^l b_l$, where $a = b_0 + b_1 + \cdots + b_l$ with $\deg_x w = i$ for every $w \in \mathrm{supp}_0(b_i)$, $i = 0$, 1, $\ldots$, $l$. By the assumption on the cardinality of $L$, the Vandermonde determinant used with respect to $\overline{\varphi_\lambda}(a)$ for $l + 1$ distinct nonzero elements $\lambda \in L$ yields $b_1, \ldots, b_l \in \mathcal{J}(L_0[X/I])$. Since $x \in \{x_1,\ldots,x_n\}$ is arbitrary, then the above procedure repeated with respect to all $x$ in $\{x_1,\ldots,x_n\}$ shows that $a_j \in \mathcal{J}(L_0[X/I]) \cap K_0[X/I] \subseteq \mathcal{J}(K_0[X/I])$ for every $1 \leq j \leq k$. The result follows.

Our next aim is to characterize monomial algebras satisfying polynomial identities. We first state two auxiliary results on ideals in the free semigroup $X$.

**Lemma 9** For every elements $s_1, \ldots, s_l \in X$, $l \geq 1$, $\bigcap_{i=1}^l X_{s_i}$ is a finitely generated ideal of $X$.

*Proof.* Since, for any $v \in X$, there exists a primitive word $w \in X$ such that $X_v = X_w$, then we may assume that all $s_i$ are primitive. Let $y \in \bigcap_{i=1}^l X_{s_i}$, and choose a subword $x$ of $y$ of minimal length with $x \in \bigcap_{i=1}^l X_{s_i}$. Suppose that $l(x) > r = 3 \max\{l(s_i) \mid i = 1, \ldots, l\}$. If $x = y_1 z y_2$ for some letters $y_1, y_2 \in X$ and some $z \in X$, then $y_1 z, z y_2 \notin \bigcap_{i=1}^{l} X_{s_i}$. Consequently, there exist $i, j \in \{1, \ldots, l\}$ such that $y_1 z \notin X_{s_i}$, $z y_2 \notin X_{s_j}$. For simplicity, write $s = s_i$, $t = s_j$. Since $y_1 z$ is a subword of a power of $s$, then $x = y_1(s')^m e y_2$, where $s'$ is a word coming from a cyclic permutation of the letters in $s$, $y_1$ is the last letter in $s'$, and $e$ is an initial segment of $s'$, $m \geq 1$. Similarly, $x = y_1(t')^k f y_2$, where $t'$ is a word coming from a cyclic permutation of the letters in $t$, $f y_2$ is an initial segment of $t'$ for some $f \in X$, and $k \geq 1$. Then $(s')^m e v = (s')^{m+1}$, $(t')^k f w = (t')^{k+1}$ for some $v, w \in X^1$. Moreover, $l((s')^m e) = l((t')^k f) = l(x) - 2 \geq 2 \max\{l(s), l(t)\}$ by our supposition. Therefore, [132], Lemma 11.5.5, implies that $s'$, $t'$ are powers of the same word $u$ in $X$. It then follows that $s$, $t$ are powers of some words $u'$, $u''$ obtained by a cyclic permutation of the letters in $u$. Since, $s$, $t$ are primitive, then $s = u'$, $t = u''$, which implies that $s' = u = t'$. Then $y_1$ is the last letter in $t'$. Therefore, $x = y_1(t')^k f y_2$ is a subword of a power of $t'$, and so it is a subword of a power of $t$. This means that $x \notin X_t$, contradicting our assumption. Hence, $l(x) \leq r$. Then the set $\{z \in \bigcap_{i=1}^{l} X_{s_i} \mid l(z) \leq r\}$ generates $\bigcap_{i=1}^{l} X_{s_i}$ as an ideal of $X$.

**Lemma 10** Assume that $J$ is a finitely generated nil ideal of $X/I$. If $X/I$ has the permutational property, then $J$ is nilpotent.

*Proof.* Let $J = \bigcup_{i=1}^{m} (X/I)^1 z_i (X/I)^1$ for some $z_i \in J$. We regard $X/I$ as being generated by the set $\{s_1, \ldots, s_u\}$ consisting of $z_1, \ldots, z_m$ and those generators of $X$ that are not in $I$. We choose $p \geq 1$, elements $v_1, \ldots, v_p \in X/I$, and integers $k$, $N$ for this set as in Theorem 2 in Chapter 19. Recall that then every $v_i$ is a word in $s_1, \ldots, s_u$ of length $\leq n$, where $\mathfrak{P}_n$ is satisfied in $X/I$. Moreover, every $s \in X/I$ that has length $\geq N$ (denoted by $d(s)$) as a word in $s_1, \ldots, s_u$ may be written in the form $s = v_{i_1}^{j_1} \ldots v_{i_k}^{j_k}$, where $j_l \geq 0$. Let $M$ be the largest of the nilpotency indices of those $v_i$ that are in $J$. Put $q = knM + 1$, $r = \max(N, q)$. We will show that $J^r = \theta$. Let $s \in J^r$. We know that $s = v_{i_1}^{j_1} \ldots v_{i_k}^{j_k}$ for some $j_l \geq 0$, $i_l \in \{1, \ldots, p\}$, and $d(s) \geq r \geq N$. Moreover, at least $r$ of the elements $z_1, \ldots, z_m$ (with repetitions) appear in $s$. Therefore, there exists $t$ such that $v_{i_t}^{j_t}$ contains $nM + 1$ elements of $z_1, \ldots, z_m$ (with repetitions) because $r \geq knM + 1$. Thus, $d(v_{i_t}^{j_t}) \geq nM + 1$. Also, since $d(v_{i_t}) \leq n$, then $d(v_{i_t}^{j_t}) \leq n \cdot j_t$. Consequently, $j_t \geq M$. That $v_{i_t} \in J$ (because it contains some $z$ as subwords) implies that $v_{i_t}^{j_t} = \theta$ by the choice of $M$. Hence, $s = \theta$. This completes the proof.

The following result was obtained by Okniński in [191].

**Theorem 11** Assume that $X/I$ has the permutational property. Then $K_0[X/I]$ is a $PI$-algebra. Moreover, $\mathcal{J}(K_0[X/I]) = K_0[\mathcal{N}(X/I)] = K_0[\mathcal{B}(X/I)]$ is a nilpotent ideal, and $GK \dim K_0[(X/I)/\mathcal{N}(X/I)] \leq 1$.

*Proof.* Assume that $X/I$ has the property $\mathfrak{P}_n$, $n \geq 2$. If $X/I = \mathcal{N}(X/I)$, then $X/I$ must be finite by Theorem 3 in Chapter 19, and we are done. Assume that $X/I \neq \mathcal{N}(X/I)$. Then, by Proposition 4 and Proposition 5, $K_0[(X/I)/\mathcal{N}(X/I)]$ is a subdirect product of some algebras $K_0[X/X_s]$, $s \in X$, containing ideals $J_s$ of the form $J_s = K_0[I_s] = \mathfrak{M}(K[\langle t \rangle], l(s), l(s), P_s)$ such that $K_0[X/X_s]/J_s$ has dimension $\leq 2l(s)^2$ over $K$. From the shape of $P_s$ and from Proposition 19 in Chapter 19, it follows that $I_s$ has no property $\mathfrak{P}_{2l(s)-1}$, so that $n \geq 2l(s)$. Thus $K_0[(X/I)/\mathcal{N}(X/I)]$ is a subdirect product of finitely many algebras $K_0[X/X_s]$, namely, those for which $l(s) \leq n/2$. By Lemma 9, $\mathcal{N}(X/I)$ is then a finitely generated ideal of $X/I$. Thus, Lemma 10 implies that $\mathcal{N}(X/I)$ is nilpotent. From Corollary 7 it follows that $\mathcal{J}(K_0[X/I]) = K_0[\mathcal{N}(X/I)] = K_0[\mathcal{B}(X/I)]$. Since every $K_0[X/X_s]$ is a $PI$-algebra by Corollary 6, then $K_0[(X/I)/\mathcal{N}(X/I)]$ also is a $PI$-algebra. Finally, from Proposition 4 and [121], Corollary 3.3, it follows that it is an algebra of Gelfand–Kirillov dimension 1 as a subdirect product of finitely many algebras of this type. This completes the proof.

**Remark 12** From the above proof it follows that $K_0[(X/I)/\mathcal{N}(X/I)]$ is a $PI$-algebra if and only if there is a bound on the lengths of non-nilpotent elements in $X/I$ that are primitive as words in $X$. Clearly, the latter means that there are finitely many elements of this type in $X/I$.

**Corollary 13** If $K[X/I]$ is a $PI$-algebra, then $\operatorname{cl} K \dim K[(X/I)^1] = rk(S) = Rk(S) \leq 1$, and it is equal to 0 only if $X/I$ is finite.

*Proof.* If $X/I$ is finite, then, clearly, $rk(X/I) = Rk(X/I) = 0$, and $\operatorname{cl} K \dim K_0[(X/I)^1] = 0$. Assume that $X/I$ is infinite. Then $X/I$ is not periodic by Theorem 3 in Chapter 19, so that $rk(X/I) = rk((X/I)/\mathcal{N}(X/I)) \geq 1$. Since $xy = yx$ for $x, y \in X$ implies that $x, y$ are powers of the same word in $X$ (see [132], Corollary 11.5.3), then $rk(X/I) = 1$. Now, $\operatorname{cl} K \dim K[(X/I)^1] = \operatorname{cl} K \dim K_0[(X/I)^1/\mathcal{N}(X/I)]$, and the latter algebra is semiprime by Theorem 11. Thus, the assertion follows from Theorem 12 and Corollary 8, both in Chapter 23.

**Corollary 14** If $X/I$ has the property $\mathfrak{P}$, then $GK \dim K_0[X/I]$ does not exceed the index of nilpotency of $\mathcal{N}(X/I)$.

*Proof.* It is a direct consequence of Theorem 11 and of the fact that $GK\, \dim A \leq m(GK\, \dim A/N)$ for every algebra $A$ and every ideal $N$ of $A$ such that $N^m = 0$ [121], Corollary 5.10.

Observe that, in the case described in Corollary 11, $GK\, \dim K_0[X/I]$ may happen to be equal to the nilpotency index of $X/I$. For example, the algebra $K_0[X/I]$ for the ideal $I = (X^1 y X^1)^m$, $m \geq 1$, of the free semigroup $X = \langle x, y \rangle$, is of this type; see [121], Example 5.11.

In [121], and [128], examples of monomial $PI$-algebras with a pathological behavior of the Gelfand–Kirillov dimension were given. Theorem 11 shows that the pathology must lie in $\mathcal{B}(X/I)$ in examples of this type. This cannot happen in the case where $I$ is a finitely generated ideal of $X$.

**Proposition 15** Assume that $I$ is a finitely generated ideal of $X$. Then $\mathcal{N}(X/I)$ is nilpotent, and $\mathcal{J}(K_0[X/I]) = K_0[\mathcal{B}(X/I)]$. Moreover, $K_0[X/I]$ is a $PI$-algebra if and only if there are finitely many non-nilpotent elements in $X/I$ that are primitive as words in $X/I$.

*Proof.* Assume that $I$ is generated by a subset $\{w_1, \ldots, w_n\}$ of $X$. Let $\theta \neq s \in X/I$, $s = y_1 \ldots y_m$ for some letters $y_i$ in $X$. If $st = \theta$ for some $\theta \neq t \in X/I$, then $st$ contains some $w_i$ as a subword. Hence, the right annihilator of $s$ in $X/I$ is generated as a right ideal of $X/I$ by the set $R_s = \{u \in X \setminus I \mid y_k \ldots y_m u = w_j \text{ for some } k \in \{1, \ldots, m\}, j \in \{1, \ldots, n\}\} \cup \{\theta\}$. Then, any nonzero $u \in R_s$ is a right factor of some $w_j$, and so there are a finite number of right annihilators of single elements of $X/I$. Therefore, there are finitely many right annihilator ideals in $X/I$. Similarly, $X/I$ has finitely many left annihilators and, from Proposition 13 in Chapter 2, it follows that $\mathcal{N}(X/I)$ is nilpotent. Hence, the result follows from Remark 12 and Corollary 7.

In the case where the property $\mathfrak{P}$ is imposed on $\mathcal{N}(X/I)$ only, the local finiteness of $\mathcal{N}(X/I)$, following from Theorem 3 in Chapter 19, may be strengthened. Recall that a word $s$ of $X$ is called periodic with a period $t \in X$ if $s = t^k u$ for an initial segment $u$ of $t$ and some $k \geq 1$. It is easy to see that if $sx = tsy$ for some $s, t, x, y \in X$ with $l(s) > l(t)$, then $s$ is periodic with a period $t$; see [121], Section 2.

**Proposition 16** Assume that $\mathcal{N}(X/I)$ has the permutational property. Then $\mathcal{N}(X/I) = \mathcal{B}(X/I)$.

*Proof.* Replacing $X/I$ by $(X/I)/\mathcal{B}(X/I)$, we may assume that $\mathcal{B}(X/I) = \theta$. If $\mathcal{N}(X/I)$ is commutative, then clearly $\mathcal{N}(X/I) = \mathcal{B}(X/I)$ is a union of nilpotent ideals and we are done. Suppose that $\mathcal{N}(X/I)$ is not commutative. Then there exists a maximal integer $m$ such that there exist elements $s_1$, $\ldots, s_m \in \mathcal{N}(X/I)$ with $\theta \neq s_1 \ldots s_m \neq s_{\sigma(1)} \ldots s_{\sigma(m)}$ for any $1 \neq \sigma \in \mathcal{S}_m$. (In fact, $\mathcal{N}(X/I)$ has $\mathfrak{P}_n$ for some $n > 2$; hence, any integer $m$ satisfying the above condition must satisfy $m \leq n$, and the set of integers satisfying this condition is nonempty since it contains 2.) Thus, for any $x \in \mathcal{N}(X/I)$ with $xs_1 \ldots s_m \neq \theta$, we have

(b)                  $$xs_1 \ldots s_m = s_{\sigma(1)} \ldots s_{\sigma(i-1)} x s_{\sigma(i)} \ldots s_{\sigma(m)}$$

for some $\sigma \in \mathcal{S}_m$ and some $i$ such that the above permutation of $x$, $s_1$, $\ldots$, $s_m$ is nontrivial. Observe that $i \neq 1$ since otherwise $s_1 \ldots s_m = s_{\sigma(1)} \ldots s_{\sigma(m)}$, $\sigma \neq 1$, contradicting the choice of $s_1$, $\ldots$, $s_m$. If $l(x) > l(s_{\sigma(1)} \ldots s_{\sigma(i-1)})$, then (b) implies that $x$ is periodic as a word in $X$ with period $s_{\sigma(1)} \ldots s_{\sigma(i-1)}$. However, $s_{\sigma(1)} \ldots s_{\sigma(i-1)} \in \mathcal{N}(X/I)$, and so it is nilpotent. This shows that there are a finite number of elements $x \in \mathcal{N}(X/I)$ with this property. In other words, any set $P_{\sigma,i} = \{x \in \mathcal{N}(X/I) \mid xs_1 \ldots s_m \neq \theta \text{ and (b) holds}\}$ is finite. Now, the left ideal $\mathcal{N}(X/I)s_1 \ldots s_m$ is the union of all $P_{\sigma,i}$, $\sigma \in \mathcal{S}_m$, $i > 1$, so that it is also finite. Hence, $\mathcal{N}(X/I)s_1 \ldots s_m$ is nilpotent, which implies that $(X/I)s_1 \ldots s_m$ is also nilpotent. Thus, $s_1 \ldots s_m \in \mathcal{B}(X/I) = \theta$. This contradiction completes the proof.

We briefly discuss chain conditions for monomial algebras. By the right multiplication graph of $X/I$ we mean an oriented graph with $X \setminus I$ as the set of vertices, where $(v, w)$ is an edge if and only if $vx = w$ for a letter $x \in X$, $v, w \in X \setminus I$. We say that a word $s$ is under a word $t$ (and $t$ is above $s$), $s, t \in X \setminus I$, if $tz = s$ for some $z \in X$.

**Proposition 17** The following conditions are equivalent:

(1)   $K_0[X/I]$ is right noetherian as a $K$-algebra.
(2)   $X/I$ has a.c.c. on right ideals.
(3)   The right multiplication graph of $X/I$ has finitely many splittings.

Moreover, in this case, $K_0[X/I]$ is a $PI$-algebra of Gelfand–Kirillov dimension not exceeding 1.

*Proof.* (1) $\Rightarrow$ (3) Suppose that there is an infinite subset $Z = \{z_1, z_2, \ldots\}$ of elements of $X/I$ such that $z_i s_i \neq \theta$, $z_i t_i \neq \theta$ for some letters $s_i, t_i \in X$, $s_i \neq t_i$, $i = 1, 2, \ldots$. If infinitely many elements of $Z$ are not under $z_1 s_1$, then put $u_1 = z_1 s_1$. Otherwise, infinitely many elements of $Z$ are not under $z_1 t_1$, and we put $u_1 = z_1 t_1$. Hence, there is an infinite subset $Z_1$ of elements of $Z$, the elements of which are not under and not above $u_1$. Replacing $Z$

by $Z_1$ and enumerating the elements of $Z_1$ so that $Z_1 = \{z_2, z_3, \ldots\}$, we may repeat the above procedure with respect to the element $z_2 \in Z_2$. Continuing this way, a sequence of elements $u_1, u_2, \ldots$ of $X/I$ may be chosen so that $u_i(X/I)^1 \not\supseteq u_j(X/I)^1$ for any $i \neq j$. Then $K_0[\bigcup_{i=1}^n u_i(X/I)^1]$, $n = 1, 2, \ldots$, is an increasing chain of right ideals of $K_0[X/I]$ contradicting (1).

The fact that (2) implies (1) is a direct consequence of Lemma 1.

Assume that (3) holds. Then there exists $m \geq 1$ such that the right multiplication graph of $X/I$ restricted to words of length not less than $m$ consists of a finite number $k$ of disjoint chains. It is clear that $X/I$ has a.c.c. on right ideals in this case. This proves that (1), (2), and (3) are equivalent. Moreover, for any $i \geq m$, $X/I$ has at most $k$ words of length $i$, and so $GK \dim K_0[X/I] \leq 1$.

Now, since $\mathcal{N}(X/I)$ is a nil semigroup in a right noetherian algebra, then it is nilpotent by Proposition 13 in Chapter 2. Thus, $K_0[\mathcal{N}(X/I)]$ also is nilpotent. To prove that $K_0[X/I]$ is a $PI$-algebra, we can assume that $\mathcal{N}(X/I) = \theta$. We know that the right multiplication graph of $X/I$ is a union of finitely many chains. Since every infinite chain with no splittings and with zero adjoined is a right ideal of $X/I$ and $X/I$ has no nonzero nil ideals, then every infinite subchain contains a non-nilpotent element. Thus, we may choose non-nilpotents $s_1, \ldots, s_m \in X/I$ such that

(i)   $M_i = \{t \in X/I \mid t \text{ lies below } s_i\}$ is a chain for every $i = 1, \ldots, m$.

(ii)   Every $t \in (X/I) \setminus \bigcup_{i=1}^m M_i$ lies above some $s_i$.

Then $s_i^k \in M_i$ for all $k > 1$, and so $M_i$ consists of subwords of powers of $s_i$. Hence, $M_i \cap T_{s_i} = \emptyset$. Therefore, $\bigcap_{i=1}^m T_{s_i} = \theta$. Thus, the remaining assertion follows as in Theorem 11.

A partial converse to the above is given below.

**Corollary 18**   If $K_0[X/I]$ is a $PI$-algebra, then $K_0[X/I]/\mathcal{B}(K_0[X/I])$ is noetherian as a $K$-algebra.

*Proof.*   From the proof of Theorem 11, we know that $\mathcal{N}(X/I) = T_{s_1} \cap \cdots \cap T_{s_r}$ for some $s_i \in X/I$ and $K_0[\mathcal{N}(X/I)] = \mathcal{B}(K_0[X/I])$. Then $(X/I)/\mathcal{N}(X/I)$ is a homomorphic image of $X/(X_{s_1} \cap \cdots \cap X_{s_r})$, and it is enough to show that the latter satisfies the conditions of Proposition 17. If $s \in X$, $s = y_1 \ldots y_m$ for some letters $y_j$ in $X$, $m \geq 1$, then, for any $\sigma \in \mathcal{S}_m$, put $s^\sigma = y_{\sigma(1)} \ldots y_{\sigma(m)}$, $V_s^\sigma = \{(s^\sigma)^k y_{\sigma(1)} \ldots y_{\sigma(j)} \mid k \geq 0, j = 1, \ldots, m\}$. It is clear that $X \setminus (X_{s_1} \cap \cdots \cap X_{s_r})$ is the union of all $V_{s_i}^\sigma$, $i = 1, \ldots, r$, $\sigma$—a cyclic permutation of $\mathcal{S}_{l(s_i)}$. Therefore, the right multiplication graph of

$X/(X_{s_1} \cap \cdots \cap X_{s_r})$ is a union of finitely many chains. Thus, the condition (3) of Proposition 17 is satisfied for this semigroup, and the result follows.

We note that it was shown in [254] that semiprime (unitary) algebras of Gelfand–Kirillov dimension 1 are noetherian $PI$-algebras of Krull dimension 1. Thus, the assertions of Corollaries 18 and 13 are valid in a more general setting.

In the case where $I$ is a finitely generated ideal of $X$, the result on the $PI$-property of $K_0[X/I]$ may be strengthened. This is due independently to Ufnarovskii [260] and Anick [4].

**Theorem 19** Assume that $I$ is a finitely generated ideal of $X$. Then the following conditions are equivalent:

(1)  $K_0[X/I]$ is a $PI$-algebra.
(2)  $GK \dim K_0[X/I] < \infty$.
(3)  $X/I$ has no noncommutative free subsemigroups.

*Proof.* The implication (1) $\Rightarrow$ (2) holds for any finitely generated algebras (see [121], Corollary 10.7), but it is also a special case of our Corollary 14. It is clear that (2) implies (3).

Assume that (3) holds. Let $Y = \{w \in X \setminus I \mid l(w) \geq m\} \cup \{\theta\}$, where $\{w_1, \ldots, w_n\}$ is a generating set of $I$ as an ideal of $X$ and $m = \max\{l(w_i) \mid i = 1, \ldots, n\}$. We have ideals $Y \cap \mathcal{N}(X/I) \subseteq Y$ of $X/I$, the former being nilpotent by Proposition 15, with the factor $(X/I)/Y$ being finite. If $Y \subseteq \mathcal{N}(X/I)$ then, clearly, $K_0[X/I]$ is a $PI$-algebra. Otherwise, there exists a non-nilpotent element $s \in Y$. Suppose that $s(T_s \cap Y)s \subseteq \mathcal{N}(X/I)$. Since $K_0[\mathcal{N}(X/I)]$ is a semiprime ideal of $K_0[X/I]$ by Proposition 4, then $T_s \cap Y \subseteq \mathcal{N}(X/I)$. Thus, $T_s \cap Y$ is nilpotent, $T_s/(T_s \cap Y)$ is finite, and $K_0[(X/I)/T_s]$ is a $PI$-algebra. Hence, $K_0[X/I]$ also is a $PI$-algebra in this case. Therefore, let $s(T_s \cap Y)s \not\subseteq \mathcal{N}(X/I)$, i.e., $sts \notin \mathcal{N}(X/I)$ for some $t \in T_s \cap Y$. Observe that if $vw = \theta$ for $v, w \in X/I$, then $v'w' = \theta$, where $v'$ is a right factor of $v$ with $l(v') \leq m$, and $w'$ is a left factor of $w$ with $l(w') \leq m$. Thus, the fact that $s^2 \neq \theta$ and $(sts)^2 \neq \theta$ implies that the subsemigroup $S$ of $X/I$ generated by $\{s, sts\}$ does not contain the zero element. This means that $S$ embeds into $X$. If the elements $s$, $sts$ commute, then, by [132], Corollary 11.5.3, they are powers of a word $u$ in $X$. Thus, $s$, $t$ are powers of $u$, and so $T_s = T_u = T_t$, which is impossible because $t \in T_s$. Therefore, $s^2ts$, $sts^2$ are distinct elements of $S$. Since $l(s^2ts) = l(sts^2)$, then it follows that $\langle s^2ts, sts^2 \rangle$ is a free noncommutative subsemigroup of $X/I$. This shows that (3) implies (1).

The hypothesis that $I$ is finitely generated is, in fact, essential when proving implication (3) $\Rightarrow$ (1) in the above theorem. Namely, examples of semigroups $X/I$ with $GK \dim K_0[X/I] < \infty$ and such that $K_0[X/I]$ is not a $PI$-algebra are given in [121]; see also [128].

Let $I$ be a finitely generated ideal of $X$. From Theorem 19, we know that $GK \dim K_0[X/I] < \infty$ whenever $K_0[X/I]$ is a $PI$-algebra. We will show that $GK \dim K_0[X/I]$ actually must be an integer in this case. Moreover, an algorithm to compute this dimension will be given. To this end, we will associate a finite graph with $X/I$. This graph respects the growth of $X/I$, and an easy graph theoretic criterion to determine $GK \dim X/I$ will be given. The presented ideas and results come from the paper of Ufnarovskii [260].

Let $W = \{w_1, w_2, \ldots, w_r\}$ be the minimal generating set of the ideal $I$. Let $k$ be such that $k + 1 = \max \{l(w_i) \mid i = 1, \ldots, r\}$. We can assume that $k \neq 0$ because otherwise $X/I$ is a free semigroup or $X/I = \theta$. Define an oriented graph $\Gamma = \Gamma(I)$, whose set of vertices is $V = \{v \in X \setminus I \mid l(v) = k\}$. If $s, t \in V$, then the edge $(s, t)$ with the origin in $s$ and the terminal vertex $t$ is in $\Gamma$ if there exist letters $x, y$ in $X$ such that $sx = yt \notin I$. Note that there exists a bijective correspondence between the edges in $\Gamma$ and the words of length $k + 1$ not lying in $I$. In fact, let $y_1 y_2 \ldots y_{k+1} \in X \setminus I$ for some letters $y_i$ in $X$. Then $y_1 \ldots y_k, y_2 \ldots y_{k+1} \notin I$ are vertices in $\Gamma$, and so $(y_1 \ldots y_k, y_2 \ldots y_{k+1})$ is an edge in $\Gamma$. It is also clear that whenever $sx = yt \notin I$ for some letters $x, y \in X$ and $s, t \in X$ with $l(s) = l(t) = k$, then $s = y_1 \ldots y_k$, $t = y_2 \ldots y_{k+1}$ for some letters $y_i$ in $X$.

**Example 20** Let $X = \langle x_1, x_2, \ldots, x_n \rangle$, and let $W = \{x_j x_i \mid j > i, \ i, j \in \{1, \ldots, n\}\}$. Then $k = 1$, $V = \{x_1, x_2, \ldots, x_n\}$, and the set of edges in $\Gamma$ consists of all $(x_j, x_i)$, where $j \leq i$. We present $\Gamma$ in the case where $n = 4$:

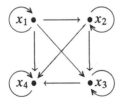

Observe that, if $R = K[x_1, \ldots, x_n]$ is the polynomial algebra in $n$ commuting indeterminates and $J$ is the kernel of the natural homomorphism $K[X] \to R$, then the associated ideal $\alpha(J)$ of $X$ is exactly the ideal generated by $W$.

We note that, in general, a finitely presented algebra $K[S]$ does not lead to a finitely generated ideal $I$ of the corresponding free semigroup $X$, so that one cannot associate graphs of the above type with arbitrary finitely presented algebras. This may happen even in the case of cancellative weakly nilpotent semigroups, as the following example shows.

**Example 21** Let $S$ be the semigroup defined by the generators $x$, $y$, $z$ subject to the relations

$$zx = xz, \quad zy = yz, \quad xy = zyx$$

It may be verified that $S$ is a cancellative semigroup with a group of fractions $G$ that is a free nilpotent group of class 2 on generators $x, y$. This, together with Theorem 3 in Chapter 8, shows that $GK \dim K[S] = GK \dim K[G]$. The latter equals 4 in view of the formula for the growth of nilpotent groups; see Chapter 8. Consider the ordering $z < x < y$ in the free semigroup $X = \langle x,y,z \rangle$. In view of the defining relations, the ideal $I$ of $X$ corresponding to $S$ (that is, $I = \alpha(J)$, where $J$ is the kernel of the natural homomorphism $K[X] \to K[S]$) contains then the elements $xz$, $yz$, $zyx$. But, also, we have $zy^nx^n = y^{n-1}zyxx^{n-1} = y^{n-1}xyx^{n-1}$ for all $n \geq 1$. Since $l(zy^nx^n) = 2n + 1 > l(y^{n-1}xyx^{n-1})$, then $zy^nx^n \in I$ for all $n \geq 1$. Suppose that $I = \bigcup_{i=1}^{r} X^1 w_i X^1$ for some $w_i \in X$, $r \geq 1$. Then one of the $w_i$'s, say $w_1$, is a subword of infinitely many $zy^nx^n$, $n \geq 1$. It follows that $w_1$ is a subword of some $zy^r$ or some $y^r x^r$. Then $zy^r$, or $y^r x^r$, respectively, lies in $I$. However, from the defining relations for $S$ it follows that this is impossible. This shows that $I$ is not a finitely generated ideal of $X$, and $K_0[X/I] = K[X]/K[I]$ is not a finitely presented algebra.

**Example 22** Let $X = \langle x,y \rangle$ and let $W = \{x^3, yxy, xyx\}$. Then $k = 2$, and $V = \{x^2, xy, yx, y^2\}$. $\Gamma$ may be presented as

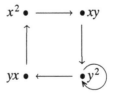

By a route of length $m \geq 1$ in $\Gamma$, we mean an alternating sequence of vertices and edges $v_0, y_1, v_1, y_2, \ldots, v_{m-1}, y_m, v_m$ such that $y_i = (v_{i-1}, v_i)$, that is, the vertex $v_{i-1}$ is the origin, and the vertex $v_i$ is the end of the edge $y_i$ for all $i = 1, \ldots, m$. We define the growth function $d_\Gamma(m)$ of the graph $\Gamma$ as the number of routes in $\Gamma$ with length not exceeding $m$, $m \geq 1$.

As in the case of semigroups and algebras, the growth of $\Gamma$ is defined as the equivalence class of the function $d_\Gamma(m)$ with respect to the equivalence relation $\leftrightarrow$ defined in the class of nondecreasing functions $f : \mathbb{N} \to R_+$ by

$f \leftrightarrow g$     if $f \prec g$ and $g \prec f$ where
$f \prec g$     if there exists positive integers $c, n$ such that $f(m) \le cg(mn)$ for every $m \ge 1$.

We define the Gelfand–Kirillov dimension of $\Gamma$ by

$$GK \dim \Gamma = \limsup_m \log_m d_\Gamma(m)$$

see [121].

Recall that a chain in $\Gamma$ is a route all the edges of which are distinct. A chain is called simple if all the vertices through which it passes are distinct. A cycle (that is, a chain the terminal vertex of which coincides with its origin) is simple if all its vertices, except the origin and the end, are distinct. The following example will be crucial.

**Example 23** Let $\Gamma$ be the graph of the form

(c)     $\xrightarrow{z_1} \left(\sigma_1\right) \xrightarrow{z_2} \left(\sigma_2\right) \xrightarrow{z_3} \cdots \xrightarrow{z_d} \left(\sigma_d\right) \xrightarrow{z_{d+1}}$

where the circles denote simple cycles $\sigma_i$, the arrows denote simple chains $z_i$ while the punctured arrows mean that the simple chains at the ends may be absent. Then $d$ is called the length of the circuit (c). We claim that $d_\Gamma(m) \leftrightarrow m^d$, that is, the growth of $\Gamma$ is polynomial and $GK \dim \Gamma = d$. Let $n$ be the number of edges of $\Gamma$, and $n_1, n_2$ the maximum and the minimum of the lengths of the cycles $\sigma_i$. Let $P(m) \subseteq (\mathbb{Z}_+)^d$, where $\mathbb{Z}_+$ is the set of non-negative integers, be the solution set of the inequality $k_1 + k_2 + \cdots + k_d \le m$ and $p(m) = |P(m)|$. With each $d$-tuple $k_1, \ldots, k_d$ in $P(m)$, we associate a route of length not exceeding $n_1 m + n$ consisting of "going around" $\sigma_1$ $k_1$ times, passing to $\sigma_2$ (via $z_2$ and possibly a part of $\sigma_1$), "going around" $\sigma_2$ $k_2$ times, passing to $\sigma_3$, etc. Thus, $d_\Gamma(n_1 m + n) \ge p(m)$, which implies that $p \prec d_\Gamma$. On the other hand, with any route of length not exceeding $n_2 m$ we can associate a sequence $l_1, l_2, \ldots, l_d$ of non-negative integers by counting the number of times the cycles $\sigma_1, \sigma_2, \ldots, \sigma_d$ occur. If two routes corresponding to $l_1, l_2, \ldots, l_d$ have the same origin and end, then these routes coincide. Therefore, $d_\Gamma(n_2 m)/k^2 \le p(m)$, where $k$ is the number of vertices in $\Gamma$. Then $d_\Gamma(m) \le d_\Gamma(n_2 m) \le k^2 p(m)$, and so it follows that $d_\Gamma(m) \leftrightarrow p(m)$, that is, the functions $d_\Gamma(m), p(m)$ are equivalent. It can be verified that $p(m) = \binom{m+d}{d}$, so that it is a polynomial of degree $d$ in $m$ and, hence, it is equivalent to $m^d$; see [121], Example 1.6. This proves the claim.

If $V = \{v_1, \ldots, v_r\}$, then the $r \times r$ matrix $B$ defined by $b_{ij}$ = the number of edges with the origin in $v_i$ and the end in $v_j$, is called the incidence matrix of $\Gamma$. It is well known, and easy to check, that the $(i,j)$th entry of $B^m$, $m > 1$, is the number of routes of length $m$ in $\Gamma$ with the origin in $v_i$ and the end in $v_j$; see [84], Theorem 13.1. Since $B$ is an $r \times r$ matrix, then $B^r = \sum_{i=0}^{r-1} \lambda_i B^i$ for some $\lambda_i \in \mathbf{Z}$. Then $B^{m+r} = \sum_{i=0}^{r-1} \lambda_i B^{m+i}$ and, adding all entries of the matrices on both sides of this equality, we see that $c_{m+r} = \sum_{i=0}^{r-1} \lambda_i c_{m+i}$, where $c_j = d_\Gamma(j) - d_\Gamma(j-1)$ for all $j > 1$ is the number of routes of length $j$. It is well known that the series $\sum_{m=1}^{\infty} c_m t^m$ is a rational function in this case and that the growth of $\Gamma$ is either exponential or polynomial; see [81], Section 3.1. It is called that Hilbert series of $\Gamma$. Thus, we derive the following consequence; see [73], [260].

**Proposition 24** The Hilbert series of the graph $\Gamma$ is a rational function, and the growth of $\Gamma$ is either exponential or polynomial.

The following result allows us to use graphs of the above type when computing the growth of $K_0[X/I]$.

**Theorem 25** The semigroup $X/I$ and the associated graph $\Gamma(I)$ have the same growth, and so $GK \dim K_0[X/I] = GK \dim \Gamma(I)$.

*Proof.* Let $d_I(m)$ denote the number of words in $X \setminus I$ of length not exceeding $m$. Let $m > k$, and let $w = y_1 y_2 \ldots y_m \in X \setminus I$, where $y_1, \ldots, y_m$ are letters in $X$. Then $y_j y_{j+1} \ldots y_{j+k-1} \in X \setminus I$ for $j = 1, \ldots, m-k+1$, and so with $m$ we can associate a route of length $m-k$ in $\Gamma$ with the successive vertices

(d)        $y_1 y_2 y_3 \ldots y_k, \ y_2 y_3 \ldots y_{k+1}, \ \ldots, \ y_{m-k+1} y_{m-k+2} \ldots y_m$

On the other hand, every route of length $m - k$ must be of the form (d) for some $w = y_1 y_2 \ldots y_m \in X \setminus I$. (In fact, if $w \in I$, then some generator $u \in W$ of the ideal $I$ is a subword of $w$. Since $l(u) \le k+1$, then one of the elements $y_j y_{j+1} \ldots y_{j+k}$ contains $u$ as a subword, and so it lies in $I$. This contradicts the fact that $y_j y_{j+1} \ldots y_{j+k-1}$ is a vertex in $\Gamma$.) It then follows that $d_I(m) = d_I(k) + d_\Gamma(m-k)$ for $m > k$. It is easy to see that the functions $d_I$, $d_\Gamma$ are equivalent in this case; see [121]. The result follows.

We now describe a criterion for determining the growth of $\Gamma$. Some preparatory graph theoretic results will be needed. A vertex of $\Gamma$ is said to be doubly cyclic if there are two distinct simple cycles passing through it. A graph with such a vertex is presented in Example 22.

**Lemma 26** Let $\Gamma$ be a finite oriented graph.

(1) If there is a route passing through any two edges of $\Gamma$ and if $v$ is a vertex not contained in any cycle, then no more than one edge goes from it or to it.

(2) Assume that $\Gamma$ has no doubly cyclic vertices. Let $\sigma$ be a simple cycle in $\Gamma$, and $y$ an edge not belonging to $\sigma$ but such that the origin (respectively end) $v$ of $y$ belongs to $\sigma$. Then, for every route $z$ beginning (respectively ending) with the edge $y$, $v$ is the only vertex in $\sigma$ contained in $z$.

(3) Assume that $y_1$, $y_2$ are edges in $\Gamma$ such that there is no route passing through $y_1$ and $y_2$. Let $\Gamma_1 = \Gamma \setminus \{y_1\}$, $\Gamma_2 = \Gamma \setminus \{y_2\}$ be the graphs obtained by removing the edge $y_1$, respectively $y_2$, from $\Gamma$. Then the functions $d_\Gamma(m)$, $d_{\Gamma_1}(m) + d_{\Gamma_2}(m)$ are equivalent, and so $GK \dim\Gamma = \sup\{GK \dim\Gamma_i \mid i = 1,2\}$.

*Proof.* (1) If there are two edges $y_1, y_2$ starting from (or ending at) $v$, then the hypothesis implies that there is a route containing the edges $y_1, y_2$, and so passing twice the vertex $v$. Then there is a route, the starting vertex and the terminal vertex of which is $v$. Such a route of minimal length is a cycle containing $v$.

(2) Assume that $v$ is the origin of $y$. Suppose that the route $z$ passes through some vertex $u$ in $\sigma$. Choosing $z$ of minimal length, we may assume that $z$ is a simple cycle or a simple chain. The former is impossible because then $v = u$ would belong to two distinct simple cycles $\sigma, z$, contradicting the hypothesis on $\Gamma$. Thus, let $z$ be a simple chain. Then $u \neq v$, and there exists a simple chain in $\sigma$ beginning at $u$ and ending at $v$. Then, again, there is a simple chain $\neq \sigma$ passing through $v$. This contradiction proves the assertion.

The proof in the case where $v$ is the end of $y$ is analogous.

(3) The hypothesis implies that for every $m$

$$d_\Gamma(m) = d_{\Gamma_1}(m) + d_{\Gamma_2}(m) - d_{\Gamma_3}(m)$$

where $\Gamma_3 = \Gamma_1 \setminus \{y_2\} = (\Gamma \setminus \{y_1\}) \setminus \{y_2\}$. Thus, $d_\Gamma(m) \leq d_{\Gamma_1}(m) + d_{\Gamma_2}(m)$. Clearly, $d_{\Gamma_i}(m) \leq d_\Gamma(m)$, $i = 1, 2$, so that $d_\Gamma \leftrightarrow d_{\Gamma_1} + d_{\Gamma_2}$, and (3) follows.

Our last result shows how circuits of the form (c) in Example 23 may be applied to determine the growth of $\Gamma = \Gamma(I)$, and so the growth of $X/I$.

**Theorem 27** The growth of $\Gamma = \Gamma(I)$ is either exponential or polynomial. Moreover,

(i)   $GK \dim \Gamma = \infty$ if and only if $\Gamma$ has a doubly cyclic vertex.

(ii)  $GK \dim \Gamma < \infty$ if and only if $\Gamma$ has no doubly cyclic vertices. In this
      case, $GK \dim \Gamma$ is the length of a maximal circuit in $\Gamma$ of the form (c).

*Proof.*  Assume first that $\Gamma$ has a doubly cyclic vertex $v$. Let $n$ be the
maximum of the lengths of the two simple cycles $\sigma_1$, $\sigma_2$ that pass through
$v$. Let

$$A_{\epsilon_i} = \begin{cases} \sigma_1 & \text{if } \epsilon_i = 1 \\ \sigma_2 & \text{if } \epsilon_i = 0 \end{cases}$$

With every sequence $\epsilon = \epsilon_1, \ldots, \epsilon_m$ with $\epsilon_i \in \{0, 1\}$ we associate a route
$A_\epsilon$, which is the composition of the routes $A_{\epsilon_1}, A_{\epsilon_2}, \ldots, A_{\epsilon_m}$. Clearly, $A_\epsilon$
is a route of length not exceeding $mn$, and $A_\epsilon \neq A'_\epsilon$ if $\epsilon \neq \epsilon'$. Therefore,
$d_\Gamma(mn) \geq 2^m$, and so $GK \dim \Gamma = \infty$ in this case. Assume now that $\Gamma$ has no
doubly cyclic vertices. If $\Gamma_1, \Gamma_2, \ldots, \Gamma_r$ are the connected components of $\Gamma$,
then, clearly, $d_\Gamma(m) = \sum_{i=1}^r d_{\Gamma_i}(m)$ because every route of length $\leq m$ lies
in one of the $\Gamma_i$. Thus, we can further assume that $\Gamma$ is a connected graph.
Moreover, Lemma 26 (3) allows us to assume that there is a route passing
through any two edges. If $\Gamma$ has no cycles, then, clearly, there is a bound
on the lengths of routes in $\Gamma$. This means that $d_\Gamma(m) = d_\Gamma(m+1) = \cdots$ for
some $m$ and, consequently, $GK \dim \Gamma = 0$. Thus, assume that $\sigma$ is a simple
cycle in $\Gamma$. We claim that no more than one edge goes from $\sigma$. In fact,
if there are two such edges, then there is a route containing them, which
contradicts Lemma 26 (2). Suppose that there is an edge coming out of a
cycle. Then, by the above, it is unique. It may be extended to a maximal
chain, all of which vertices except the first (and possibly the last) do not
belong to a cycle. If the last vertex is contained in a cycle, then we construct
the corresponding simple cycle, and we apply the same arguments to it. We
note that, by Lemma 26 (2), only one edge can go to it. If the last vertex
does not belong to any cycle, then (by the maximality) no edge begins in it.

Proceeding this way, we construct an alternating sequence of simple
cycles and simple chains. It cannot be closed because $\Gamma$ has no doubly cyclic
vertices. Therefore, we have constructed a circuit described in Example 23.
Let $\Gamma'$ be a maximal subgraph of $\Gamma$ of this form. We claim that $\Gamma' = \Gamma$. Since
$\Gamma$ is connected, then it is enough to check that if $y = (v_1, v_2)$ is an edge in
$\Gamma$ such that $v_1$ or $v_2$ is in $\Gamma'$, then $y$ is in $\Gamma'$. This has been verified above
for $y$ such that $v_1$ or $v_2$ lies in a cycle. If this is not the case, then $y$ is in $\Gamma'$
by Lemma 26 (1). Thus, $\Gamma' = \Gamma$, and the result follows from Example 23.

We note that case (ii) in Theorem 27 corresponds to the conditions in
Theorem 19, while Example 22 can be used to illustrate (i).

## Comments on Chapter 24

We refer to the paper of Anick [3] for a study of the ideals $\alpha(J)$ of $X$ associated with arbitrary ideals $J$ of the free algebra $K[X]$. Their role for the theory of growth of associative algebras was indicated in the monograph [121]. In fact, the proofs of general results on a finitely generated algebra $R$ are often obtained via a translation to the free semigroup $X$ such that $K[X]$ maps onto $R$. This is particularly the case with respect to the paper of Small et al. [254], where it is shown that every finitely generated algebra $R$ of Gelfand–Kirillov dimension 1 is $PI$, and $R/\mathcal{B}(R)$ is a finitely generated module over its center, which is noetherian. This settles a more general context for the assertions of our Theorem 11 and Corollary 18. Finitely presented monomial algebras have been studied by many authors. The first result—the rationality of the Hilbert series—is due to Govorov [73]. Our proof in Proposition 24 follows the paper of Ufnarovskii [260]. We note that it is conjectured that the same is true for more general finitely presented algebras (see [121]), and it is known for finitely generated noetherian $PI$-algebras. Anick [4] obtained an alternative proof of Theorem 19 and showed that $K_0[X/I]$ is finitely presented whenever it has finite global dimension. In [14], Borisenko, using the graph technique, extended the statement of Theorem 19 proving that finitely presented monomial algebras that satisfy polynomial identities are embeddable into matrix algebras over fields. Gateva-Ivanova and Latyshev [63] and Gateva-Ivanova [62] established algorithms for determining several algebraic properties of finitely presented algebras $K_0[X/I]$. In a recent paper [261], Ufnarovskii extends the graph technique to the case of some infinitely generated ideals $I$. This allows him to generalize the results of [4], [191], and [260]. The special case of monomial algebras that are path algebras has been extensively studied. We refer to the papers of Anick and Green [5] and Green et al. [75] for results on the projective dimensions and the connection with the representation theory of algebras. We note, in conjunction with our Proposition 3, Corollary 7, and Proposition 8, that the radicals of semigroup rings of noncommutative free semigroups with arbitrary coefficients are discussed by Puczyłowski in [223]. The nilpotency of the Jacobson radical of any finitely generated $PI$-algebra (see Theorem 11) was established by Braun in [17].

# 25

# Azumaya Algebras

In this chapter, we are concerned with an important special class of $PI$-algebras. Recall that an algebra $R$ with unity over a commutative unitary ring $A$ is called separable if $R$ is a projective left $R \otimes_A R^\circ$-module under the action given by $(\sum_i x_i \otimes y_i)a = \sum_i x_i a y_i$ for all $a, x_i \in R, y_i \in R^\circ$. Here $R^\circ$ stands for the opposite algebra of $R$. $R$ is an Azumaya algebra if it is separable over its center $Z(R)$. For the basic examples, motivations, and results on this class of algebras, we refer to [38]. We first summarize some of the fundamental facts repeatedly used in the sequel.

**Lemma 1**  Assume that $R$ is an Azumaya algebra. Then,

  (i)  $R$ is a finitely generated projective module over its center $Z(R)$, and $Z(R)$ is a direct summand of this module.
  (ii)  For every ideal $J$ of $R$ and $I$ of $Z(R)$, we have $(J \cap Z(R))R = J$, $IR \cap Z(R) = I$.

(iii)   For every ideal $J$ of $R$, $Z(R/J) = (Z(R)+J)/J$, and $R/J$ is an Azumaya
        algebra.

The following observation shows that we may expect an intrinsic characterization of Azumaya semigroup algebras in terms of $S$, and the characteristic of the base field only. This is an easy application of the basic results on tensor products, which are Azumaya algebras.

**Lemma 2**   Let $L$ be a subfield of $K$. Then $K[S]$ is an Azumaya algebra if and only if $L[S]$ is an Azumaya algebra.

*Proof.*   Since $K[S] \cong K \otimes_L L[S]$, then $Z(K[S]) \cong K \otimes_L Z(L[S])$. Hence, $K[S] \cong K \otimes_L Z(L[S]) \otimes_{Z(L[S])} L[S] \cong Z(K[S]) \otimes_{Z(L[S])} L[S]$. Now, the fact that $Z(L[S])$ is a direct summand of the $Z(L[S])$-module $Z(K[S])$ implies, in view of [38], Corollary II.1.10, that $L[S]$ is an Azumaya algebra whenever $K[S]$ is also. The converse follows from [38], Corollary II.1.7.

Our first aim is to describe Azumaya group algebras. We start with the case in which the group is finite. Clearly, any finite dimensional group algebra $K[G]$ over a field $K$ with $\text{ch}(K) = 0$ is $K$-separable, and so it is $Z(K[G])$-separable; see [38], Proposition II.1.12. To deal with the case of the prime characteristic, we need the following auxiliary result; see [203], the proof of Lemma 6.1.10.

**Lemma 3**   Let $G$ be a finite group, and let $p$ be a prime number. Then $p$ does not divide the order of the commutator subgroup $G'$ of $G$ if and only if $G$ has an abelian Sylow $p$-subgroup $P$ and a normal $p$-complement. Moreover, in this case, for every algebraically closed field $K$ with $\text{ch}(K) = p$, we have

$$K[G] \cong M_{n_1}(K[P_1]) \oplus \cdots \oplus M_{n_t}(K[P_t])$$

for some $t$, $n_i \geq 1$ and some subgroups $P_i$ of $P$.

**Proposition 4**   Let $G$ be a finite group. If $\text{ch}(K) = p > 0$, then $K[G]$ is an Azumaya algebra if and only if $|G'|$ is invertible in $K$.

*Proof.*   In view of Lemma 2, we can assume that $K$ is algebraically closed. Let $K[G]$ be an Azumaya algebra. Let $B$ be a block ideal of $K[G]$, that is, an ideal generated by a minimal nonzero central idempotent of $K[G]$. Then $B$ is an Azumaya algebra as a homomorphic image of $K[G]$. We claim that $B/\mathcal{J}(B)$ is a simple algebra. Suppose the contrary. Then $B/\mathcal{J}(B)$ has a nontrivial central idempotent $f$. Since, by Lemma 1, $Z(B/\mathcal{J}(B))$ is the image of $Z(B)$ under the natural homomorphism $\phi : B \to B/\mathcal{J}(B)$, then there exists $u \in Z(B)$ such that $\phi(u) = f$. Then $u - u^2 \in \mathcal{J}(B)$, and it is well known that there exists $e = e^2 \in Z(B)$ with $e - u \in \mathcal{J}(B)$; see [93], p. 54. This contradicts the fact that $B$ has no nontrivial central idempotents. Thus, $B/\mathcal{J}(B)$ is a simple algebra. Consequently, each block of $K[G]$ has only one simple module. By [32], Corollary 65.3, it follows that $G$ has a normal $p$-complement $N$. Thus, $G = NP$, where $P$ is a Sylow $p$-subgroup of $G$. We will show that $P$ is abelian. Observe first that $P$ is a homomorphic image of $G$, and so $K[P]$ is an Azumaya algebra. From Lemma 1, we know that $\mathcal{J}(K[P]) = IK[P]$ for an ideal $I$ in $Z(K[P])$. Moreover, $\mathcal{J}(K[P]) = \omega(K[P])$ because $P$ is a $p$-group; see [203], Lemma 8.1.17. Hence,

$$K[P] = K + \mathcal{J}(K[P]) = K + IK[P] = Z(K[P]) + IK[P]$$

and Nakayama's lemma implies that $Z(K[P]) = K[P]$. Thus, from Lemma 3 it follows that $|G'|^{-1} \in K$.

On the other hand, if the order of $G'$ is invertible in $K$, then, by Lemma 3, $K[G] \cong \oplus_{i=1}^{t} M_{n_i}(K[P_i])$ for some abelian groups $P_i$. Consequently, every $K[P_i]$, $M_{n_i}(K[P_i])$, and so $K[G]$ are Azumaya algebras.

The case of an arbitrary group algebra $K[G]$ will be attacked through the so-called local-global property for Azumaya algebras. Recall that, if $R$ is an algebra with unity that is finitely generated over $Z(R)$, then $R$ is an Azumaya algebra provided that $R/MR$ is a separable $Z(R)/M$-algebra for every maximal ideal $M$ in $Z(R)$; see [38], Theorem II.7.1. The following result is crucial when the above criterion is applied for separability.

**Lemma 5** Let $R$ be a ring with unity that is finitely generated as a module over its center. If $I \neq Z(R)$ is an ideal of $Z(R)$, then $IR \neq R$.

*Proof.* Let $M$ be a maximal ideal of $Z(R)$ containing $I$. Suppose that $MR = R$. If $Z(R)_M$ and $R_M$ denote the localizations of $Z(R)$, $R$, respectively, with respect to $Z(R) \setminus M$, then $MR_M = R_M$. Now, $R_M$ is a finitely generated module over a local ring $Z(R)_M$, and Nakayama's lemma implies that $R_M = 0$. This contradiction shows that $MR \neq R$; hence, $IR \neq R$.

To establish the first main result of this chapter, we need one more preparatory lemma. It allows a reduction to the case of finitely generated groups, where the local-global criterion for separability can be used.

**Lemma 6** Let $G$ be a group whose center has finite index. Then there exists a finitely generated subgroup $H$ of $G$ such that $G = Z(G)H$ and

$G' = H'$. Moreover, in this case, $K[G]$ is an Azumaya algebra if and only if $K[H]$ is an Azumaya algebra.

*Proof.* If $H$ is a group generated by a finite set of coset representatives of $Z(G)$ in $G$, then, clearly, $G = Z(G)H$. If $x, y \in Z(G)$ and $g, h \in G$, then $[xg, yh] = [g, h]$, which implies that $G' = H'$.

We claim that $K[G] \cong Z(K[G]) \otimes_{Z(K[H])} K[H]$. Let $T \subseteq Z(G)$ be a set of right coset representatives of $H$ in $G$. If $g \in \Delta(G)$, then $g = ht$ for some $h \in H$, $t \in T$. Therefore, $c(g) = c(h)t$, where $c(g)$, $c(h)$ denote the $G$-conjugacy class sums of the elements $g$, $h$, respectively. Since the elements $c(g)$, $g \in \Delta(G)$, form a $K$-basis of $Z(K[G])$ (see [203] or Theorem 10 in Chapter 9), then $Z(K[G]) = \sum_{t \in T} Z(K[H])t$. Consequently,

$$Z(K[G])K[H] = \sum_{t \in T} Z(K[H])tK[H] = \sum_{t \in T} tK[H] = K[G]$$

from which our claim follows. Thus, the remaining assertion can be derived as in Lemma 2 through the results on tensor products [38], Corollaries II.1.7 and II.1.10.

The following result was obtained by DeMeyer and Janusz in [39].

**Theorem 7** Let $G$ be a group. Then $K[G]$ is an Azumaya algebra if and only if the center $Z(G)$ has a finite index in $G$ and the order of the commutator subgroup $G'$ is finite and invertible in $K$.

*Proof.* Assume that $K[G]$ is an Azumaya algebra. We know that $Z(K[G]) \subseteq K[\Delta(G)]$ and that $K[G]$ is a finitely generated $Z(K[G])$-module. Thus, $K[G]$ is a finitely generated $K[\Delta(G)]$-module, and so $\Delta(G)$ has a finite index in $G$. From Lemma 1 it follows that $Z(K[G/\Delta(G)]) = K$. While $G/\Delta(G)$ is finite, this may happen only if $G/\Delta(G)$ is a trivial group, that is, $G = \Delta(G)$.

Let $T$ be a finite set of elements of $G$ such that $K[G] = \sum_{t \in T} Z(K[G])t$. Since $T \subseteq \Delta(G) = G$, then $F = \bigcap_{t \in T} C_G(t)$ has a finite index in $G$. It is clear that $F = Z(G)$, and so Schur's lemma implies that $G'$ is a finite group; see [203], Lemma 4.1.4.

From Lemma 6 we know that there exists a finitely generated subgroup $H$ of $G$ such that $H' = G'$ and $K[H]$ is an Azumaya algebra. Since, as above, $Z(H)$ has finite index in $H$, then $Z(H)$ is finitely generated. Thus, $Z(H) = B \times E$, where $E$ is a finitely generated torsion-free (abelian) group and $B$ is finite. Clearly, $E$ is a normal subgroup of $H$. Further, $H/E$ is a finite group. Since $H' \cap E = \{1\}$, $H'$ being finite, then we come to $H' \cong (H/E)'$. Now $K[H/E]$ is an Azumaya algebra as a homomorphic image of $K[H]$, and from Proposition 4 it follows that $(H/E)' \cong H' \cong G'$ has order invertible in $K$. This completes the proof of the necessity.

Assume now that $Z(G)$ has a finite index in $G$, and $|G'|$ is invertible in $K$. We consider two cases.

Case 1: $\mathrm{ch}(K) = 0$

To show that $K[G]$ is an Azumaya algebra, it is enough to construct a separability idempotent $e$ in $K[G] \otimes_{Z(K[G])} K[G]^\circ$; see [38], Proposition II.1.1. Let $T$ be a set of coset representatives of $Z(G)$ in $G$. We put

$$e = [G : Z(G)]^{-1} \sum_{t \in T} t \otimes t^{-1} \in K[G] \otimes_{Z(K[G])} K[G]^\circ$$

Clearly, $e$ maps onto the unity under the multiplication map

$$K[G] \otimes_{Z(K[G])} K[G]^\circ \longrightarrow K[G].$$

We show that $(a \otimes 1)e = (1 \otimes a)e$ for every $a \in K[G]$. It is enough to prove this for $a \in G$. Observe that, for $t \in T$, there exist $z \in Z(G)$, $s \in T$, such that $at = sz$. Then $(a \otimes 1)(t \otimes t^{-1}) = sz \otimes t^{-1} = s \otimes zt^{-1} = s \otimes s^{-1}a = (1 \otimes a)(s \otimes s^{-1})$. As $t$ runs over the set $T$, so does $s$. Therefore, $(a \otimes 1)e = (1 \otimes a)e$ follows, completing the proof in this case.

Case 2: $\mathrm{ch}(K) = p > 0$

In view of Lemma 2 and Lemma 6, we can assume that $K$ is a finite field and that $G$ is finitely generated. Since $[G : Z(G)] < \infty$, then $K[G]$ is a finitely generated $Z(K[G])$-module. From Lemma 5 it follows that, for every maximal ideal $M$ in $Z(K[G])$, we have $MK[G] \neq K[G]$. By the local-global property for separability, it is then enough to show that $K[G]/MK[G]$ is $Z(K[G])/M$-separable. Since $G$ is finitely generated and $Z(G)$ has a finite index in $G$, then, as before, $Z(G) = B \times E$, where $B$ is a finite group and $E$ is finitely generated and torsion-free. We claim that $Z(K[G])$ is a finitely generated $K$-algebra. In fact, $Z(K[G]) = \sum_{t \in T} K[E]c(t)$, where $T$ is the set of coset representatives of $E$ in $G$, and $c(t)$ denotes the sum of all $G$-conjugates of $t$. Since $K[E]$ is a finitely generated algebra and $T$ is a finite set, then $Z(K[G])$ is finitely generated. Now, the field $L = Z(K[G]) \backslash M$ is a finitely generated $K$-algebra. Consequently, $L$ is a finite field extension of $K$. Since $K$ is finite, then $L$ is finite. Let $U$ be the kernel of the restriction of the natural homomorphism $Z(K[G]) \to L$ to the group $E$. Then $U$ is a torsion-free group, and so $U \cap G' = \{1\}$. Consequently, $(G/U)' = G'$. Moreover, $U$ has a finite index in $E$, so that $G/U$ is a finite group. From Proposition 4 it then follows that $K[G/U]$ is an Azumaya algebra.

The condition $U \cap G' = \{1\}$ implies that if $x, y$ are distinct conjugates of some $t \in G$ in $G$, then $xU, yU$ are distinct conjugates of $tU$ in $G/U$. Therefore, the class sum $c(t)$ is mapped onto the class sum $c(tU)$ in $K[G/U]$ under the natural homomorphism $\phi : K[G] \to K[G/U]$. It follows that $\phi(Z(K[G])) = Z(K[G/U])$. Observe that the natural homomorphism $K[G] \to K[G]/MK[G]$ factors through $K[G/U]$, that is, it may be written as $\varphi\phi$, where $\varphi$ maps $K[G/U]$ onto $K[G]/MK[G]$. Since $K[G/U]$ is an Azumaya algebra and $MK[G] \neq K[G]$ by Lemma 5, then its image $K[G]/MK[G]$ is separable over $\varphi(Z(K[G/U])) = \varphi\phi(Z(K[G]))$. The maximality of $M$ and the fact that $\varphi\phi(M) = 0$ imply that $\varphi\phi(Z(K[G])) = Z(K[G])/M$. The result follows.

It is interesting to consider whether a separability idempotent can be given in the case of Azumaya group algebras in positive characteristics.

The next step is to examine semigroup algebras of semigroups close to groups. We first show that Azumaya semigroup algebras of cancellative semigroups are, in fact, central extensions of some Azumaya group algebras. In view of Lemma 3 in Chapter 10, we restrict our attention to cancellative monoids.

**Proposition 8** Let $S$ be a cancellative monoid with the group of units $U$. Then the following conditions are equivalent:

  (i)   $K[S]$ is a Azumaya algebra.
  (ii)  $K[S] = Z(K[S])K[U]$, and $K[U]$ is an Azumaya algebra.
  (iii) $C_{K[S]}(U) = Z(K[S])$ and $K[U]$ is an Azumaya algebra.

Moreover, in this case, $S = \Delta(S)$, $StS = SsS$ if and only if $s \in tU$, and $s^x \in sU$ for every $s,t,x \in S$.

*Proof.* It is clear that (ii) implies (iii).

If (iii) holds, then $Z(K[U]) \subseteq Z(K[S])$, and $K[S]$ is a $K[U]-Z(K[U])$ bimodule. The fact that $K[U]$ is an Azumaya algebra implies then, in view of [38], Corollary II.3.6, that $K[S] \cong K[U] \otimes_{Z(K[U])} Z(K[S])$. It follows that $K[S]$ is an Azumaya algebra as a central tensor product extension of $K[U]$; see [38], Corollary II.1.7.

Assume that (i) holds. Since $K[S]$ is a $PI$-algebra, then $S \setminus U$ is an ideal in $S$ and $K[S]/K[S \setminus U] \cong K[U]$ is an Azumaya algebra; see Lemma 12 in Chapter 4. We know that $Z(K[S])$ is mapped onto $Z(K[U])$ under this homomorphism. Hence, for every $z \in Z(K[U])$, there exists $y \in K[S \setminus U]$ such that $z + y \in Z(K[S])$. While $S$ acts by conjugation on $\Delta(S)$ (see

Proposition 2 in Chapter 9), then every conjugate class sum $c(s)$, $s \in \Delta(S)$, is either contained in $K[U]$ or it lies in $K[S \setminus U]$. Thus, the condition $z + y \in Z(K[S])$ implies, in view of the description of $Z(K[S])$ (see Chapter 9), that $z \in Z(K[S])$. Consequently, $Z(K[U]) \subseteq Z(K[S]) \subseteq K[\Delta(S)]$. Since $U$ is an $FC$-group by Theorem 7, then every element of $U$ is contained in the support of some element of $Z(K[U])$. Therefore, $U \subseteq \Delta(S)$.

While $K[S]$ is a finite module over its center, then we may write $S = \bigcup_{i=1}^{n} \Delta(S)s_i$ for some $s_i \in S$, $n \geq 1$. From Lemma 1, we also know that $s_i \in (K[Ss_iS] \cap Z(K[S]))K[S]$, so that $s_i = z_it_i$ for some $z_i \in Ss_iS \cap \Delta(S)$, $t_i \in S$. Then there exist $x_i \in \Delta(S)$ and $l \in \{1,\ldots,n\}$ such that $t_i = x_is_l$. Let $i_1 \in \{1,\ldots,n\}$. Starting with $s_{i_1}$ and applying the above to the subsequent $s_{i_j}$ we come to

$$s_{i_1} = z_{i_1}x_{i_1}s_{i_2} = z_{i_1}x_{i_1}z_{i_2}x_{i_2}s_{i_3} = z_{i_1}x_{i_1}\ldots z_{i_n}x_{i_n}s_{i_{n+1}}$$

where $x_{i_j} \in \Delta(S)$, $z_{i_j} \in Ss_{i_j}S \cap \Delta(S)$, and $s_{i_j} = z_{i_j}x_{i_j}s_{i_{j+1}}$ for $j = 1, 2, \ldots, n$. Thus, there exist $k, m \leq n + 1$, $k < m$, such that $s_{i_k} = s_{i_m}$. Then $s_{i_k} = z_{i_k}x_{i_k}\ldots z_{i_{m-1}}x_{i_{m-1}}s_{i_m}$ implies that $z_{i_k}x_{i_k}\ldots z_{i_{m-1}}x_{i_{m-1}} = 1$ because $S$ is cancellative. Consequently, $z_{i_k} \in U$. Since $Ss_{i_k}S \supseteq Sz_{i_k}S$, then $Ss_{i_k}S = S$, and so $s_{i_k} \in U$. It follows that $s_{i_1} = z_{i_1}x_{i_1}\ldots z_{i_{k-1}}x_{i_{k-1}}s_{i_k} \in \Delta(S)U \subseteq \Delta(S)$. But $i_1 \in \{1,\ldots,n\}$ was arbitrary, so that $S = \bigcup \Delta(S)s_i = \Delta(S)$.

Now Proposition 2 in Chapter 9 implies that $Ss = sS$ for every $s \in S$, and so $SsS = Ss$. If $t \in S$ is such that $StS = SsS$, then this shows that $t \in sU$. Let $a \in Z(K[SsS])$ and $x \in S$. Then, by Proposition 2 in Chapter 9, there exists $n \geq 1$ such that $t^nx = xt^n$. Thus, $axt^n = at^nx = t^nxa = xt^na = xat^n$ because $a$ centralizes $t^nx \in SsS$ and $t^n \in SsS$. But $t$ is not a zero divisor in $K[S]$, so that $ax = xa$. While $x \in S$ is arbitrary, $a \in Z(K[S])$ and $Z(K[SsS]) \subseteq Z(K[S])$. Since $K[S]$ is an Azumaya algebra, we know that $s \in (K[SsS] \cap Z(K[S]))K[S]$, so that $s \in Z(K[SsS])K[S]$. Then $s = \sum \lambda_ic_iu_i$ for some $\lambda_i \in K$, $u_i \in S$, and $SsS$-conjugacy class sums $c_i = c_{SsS}(t_i) \in K[SsS]$ of some $t_i \in SsS$ (see the description of the center in Chapter 9.) Now, $s = t_i^yu_i$ for some $i$ and some $y \in SsS$. Then, for every $x \in S$, $s^x = (t_i^y)^xu_i^x = t_i^{xy}u_i^x \in SsS$ (note that, for every $z \in S$, $t_i^z \in \text{supp}(c_i) \subseteq SsS$ because $c_i \in Z(K[SsS]) \subseteq Z(K[S])$.) Since, by Proposition 2 in Chapter 9, $v \to v^x$, $v \in S$, is an automorphism of $S$ of finite order, it follows that $s^x$ lies in the $\mathcal{J}$-class $sU$ of $s$ in $S$.

We have shown that every $\text{supp}(c_j)$ is either in the ideal $I$ of nongenerators of $SsS$, or $\text{supp}(c_j) \cap I = \varnothing$. If $u_j \notin U$, then $\text{supp}(c_ju_j) \subseteq I$. Therefore, all elements $u_j \notin U$, and all $c_j \in K[I]$, can be skipped in the above presentation of $s$ (the omitted elements yield products lying in $K[I]$, while all remaining products are in $K[SsS \setminus I]$). It follows that $s \in Z(K[S])K[U]$. This establishes (i).

Recall that a semigroup $S$ is idempotent-free if $S$ has no idempotents except possibly identity and zero elements. We conjecture that the equivalences of Proposition 8 can be extended to this class of semigroups.

Observe that, in the case described in Proposition 8, $K[S]$ is a homomorphic image of the group ring $Z(K[S])[U(S)] \cong Z(K[S]) \otimes_K K[U(S)]$, which is an Azumaya algebra as a central tensor product extension of an Azumaya algebra $K[U(S)]$. Note also that the proof of the sufficiency, that is, the implications (ii) $\Rightarrow$ (iii), (iii) $\Rightarrow$ (i), do not require assuming that $S$ is cancellative. On the other hand, it is easy to construct examples showing that this hypothesis is essential for the proof of the necessity of (ii) and (iii).

**Example 9**  Let $S^1$ be the semigroup of $n \times n$ matrix units with zero and with an identity adjoined. Since $K_0[S] \cong M_n(K)$, then $K[S] \cong K \oplus K \oplus M_n(K)$; see Corollary 27 in Chapter 5. Hence, $K[S]$ is an Azumaya algebra. Clearly, $Z(K[S])K[U(S)] = Z(K[S]) \cong K \oplus K \oplus K$.

Proposition 8 shows that Azumaya semigroup algebras of cancellative semigroups have a nice construction. Our next aim is to give an intrinsic characterization of cancellative semigroups $S$ of this type. This was established by Okniński and Van Oystaeyen in [194].

**Theorem 10**  Let $S$ be a cancellative semigroup. Then the following conditions are equivalent:

 (i)   $K[S]$ is an Azumaya algebra.
 (ii)  $S$ is a monoid with a group of fractions $G$ such that $Z(G)$ has a finite index in $G$, the order of $G'$ is invertible in $K$, and $G = C_G(g)U(S)$ for every $g \in G$.
 (iii) $S$ is a monoid such that $Z(U(S))$ has a finite index in $U(S)$, the order of $U(S)'$ is invertible in $K$, and $S = C_S(s)U(S)$ for every $s \in S$.

Moreover, in this case $G' = U(S)'$.

*Proof.*  Assume first that $K[S]$ is an Azumaya algebra. Then $S$ is a monoid by Lemma 3 in Chapter 10 and, from Theorem 1 in Chapter 20, we know that $S$ has a group of fractions $G$. Since $S = \Delta(S)$ by Proposition 8, then $U(S)$ is a normal subgroup of $G$ (note that conjugation by $g \in G$ is an automorphism of $S$ by Proposition 2 in Chapter 9). Therefore, the commutator subgroup $U(S)'$ also is a normal subgroup of $G$. From Proposition 8, we know that $K[S] = Z(K[S])K[U(S)]$. If $\rho$ denotes the congruence on $S$ determined by $U(S)'$, then $K[S/\rho] = Z(K[S/\rho])K[U(S)/U(S)']$ is a commutative algebra. Since $G/U(S)'$ is a group of fractions of $S/\rho$, then it is an abelian group. Consequently, $G' \subseteq U(S)'$. It is obvious that $U(S)' \subseteq G'$, so that $G' = U(S)'$. Then Proposition 8 and Theorem 7 imply that $|G'| = |U(S)'|$ is finite and invertible in $K$. Moreover, by Lemma 1, $K[S] = \sum_{i=1}^{n} Z(K[S])s_i$ for some $s_i \in S$. Thus, $Z(G) = \bigcap_{i=1}^{n} C_G(s_i)$. While $G'$ is finite, then $G$ is an $FC$-group, and so $[G : Z(G)] < \infty$.

Let $D = C_{K[S]}(U(S))$. Then $D = \{\sum \lambda_s z_s \mid \lambda_s \in K, z_s \in Z\}$, where $Z$ is the set of all $U(S)$-conjugacy class sums $z_s$ of elements $s \in S$. The condition $Z(K[S]) = D$ is equivalent to saying that every $z_s \in Z$ is in $Z(K[S])$, that is, for any $g \in G$, there exists $u \in U(S)$ such that $g^{-1}sg = u^{-1}su$. The latter means that $gu^{-1} \in C_G(s)$. Thus, $G = C_G(s)U(S)$ for every $s \in S$. If $g \in G$, then $gZ(G) = sZ(G)$ for some $s \in S$ since $G/Z(G)$ is finite. Then $C_G(g) = C_G(s)$, which implies that $G = C_G(s)U(S)$ for every $g \in G$. Therefore, the implication (i) $\Rightarrow$ (ii) follows through Proposition 8.

Assume that (ii) holds. If $s \in S$, then $s = tu$ for some $t \in C_G(s)$, $u \in U(S)$. Thus, $t = su^{-1} \in S$, and so $S = C_S(s)U(S)$. Moreover, since $U(S)' \subseteq G'$, then the conditions on $U(S)$ in (iii) follow from the respective conditions on $G$ assumed in (ii).

Assume that (iii) holds. We claim that $C_{K[S]}(U(S)) = Z(K[S])$. Let $s \in S$ be such that $V = \{v^{-1}sv \mid v \in U(S)\}$ is a finite set. If $t \in S$, then $t = wu$ for some $w \in C_S(s)$, $u \in U(S)$. Thus, $st = swu = wsu = wuu^{-1}su = tu^{-1}su$. Consequently, $st \in tV$ and, similarly, $Vt \subseteq tV$. Since $S$ is cancellative, then $|Vt| = |V| = |tV|$, and so $Vt = tV$. It follows that $zt = tz$, where $z = \sum_{x \in V} x$. Since $t \in S$ is arbitrary, then $z \in Z(K[S])$. Furthermore, $C_{K[S]}(U(S))$ is the $K$-subspace generated by elements of this type, and so $C_{K[S]}(U(S)) = Z(K[S])$, as claimed. Now, $K[S]$ is an Azumaya algebra by Proposition 8. This completes the proof of the theorem.

Observe that, for $S$, $G$ as in Theorem 10, from the condition (ii), it follows that every $G$-conjugate of an element $s \in S$ is, in fact, a $U(S)$-conjugate of $s$. We also note that in this case, $G = SZ(S)^{-1}$ by Corollary 16 in Chapter 9, which shows that $K[G]$ is a central localization of $K[S]$.

**Remark 11** Assume that $K[S]$ is an Azumaya algebra and that $S$ is cancellative. Since $S = \Delta(S)$, then, from Proposition 2 in Chapter 9, it follows that, for any $s \in S$, the rule $u \to s^{-1}us$, $u \in U(S)$, defines an automorphism $\varphi_s$ of $U(S)$. If all $\varphi_s$, $s \in S$, are inner automorphisms, then, for every $s \in S$, there exists $v \in U(S)$ such that $sv^{-1} \in C_S(U(S))$. Then $S = C_S(U(S))U(S)$. Since $C_S(U(S)) = C_{K[S]}(U(S)) \cap S = Z(K[S]) \cap S = Z(S)$, then $S = Z(S)U(S)$ in this case.

Our next aim is to construct examples of Azumaya semigroup algebras of cancellative semigroups that are nontrivial in the sense that $S \neq Z(S)U(S)$, that is, they do not arise as an extension of an Azumaya group algebra $K[U]$ by a central semigroup $Z$.

**Example 12** Let $U$ be a group with a noninner automorphism $\varphi_U$ of order $p$, where $p$ is a prime number, preserving the conjugacy classes of $U$. Assume further that $[U : Z(U)] < \infty$. Examples of this type, where $U$ is finite, are given in [265]. Let $T$ be a nontrivial commutative monoid with $U(T) = \{1\}$, and let $Z$ denote the group of fractions of $T$. Put $H = U \times Z$. If $g \neq 1$ in $T$, then we may construct a group extension $G$ of $H$ such that $G/H$ is a cyclic group of order $p$ and also satisfying $G = H \cup sH \cup \cdots \cup s^{p-1}H$ for some $s \in G$ with $s^p = g$, $hs = s\varphi(h)$ for $h \in H$, where $\varphi$ is the automorphism of $H$, that extends $\varphi_U$ and acts trivially on $Z$. It is clear that $Z \subseteq Z(G)$ and also that $Z(U) \subseteq Z(G)$ by the choice of $\varphi_U$. Suppose that $s^r u \in Z(G)$ for some $u \in U$, $1 \leq r < p$. Then $s^r uv = vs^r u$ for every $v \in U$ and, therefore, $\varphi_U^r$ is an inner automorphism of $U$. Since $\varphi_U$ is an automorphism of order $p > r$, then $\varphi_U$ is inner, a contradiction. Hence, $Z(G) \subseteq H$, and so $Z(G) = Z(U)Z$.

We claim that $C_G(x)U = G$ for every $x \in G$. It suffices to check this for $x$ of the form $x = s^r u$ with $0 \leq r < p$, $u \in U$. If $r \geq 1$, then $x \in C_G(x) \setminus H$. Since $H \subseteq C_G(x)U$ and $G/H$ has order $p$, then $C_G(x)U = G$. If $r = 0$, then $s^{-1}xs = v^{-1}xv$ for some $v \in U$ in view of the choice of $\varphi_U$. Hence, $sv^{-1} \in C_G(x)$ and $s \in C_G(x)U$. Since $C_G(x)U \supseteq H$, then $C_G(x)U = G$ in this case, too.

Now, define $S$ as the subsemigroup of $G$ generated by $U \cup T \cup \{s\}$. Then $S = T \langle s \rangle U$, and $TT^{-1} = Z$ yields $SS^{-1} = S^{-1}S = G$. Clearly, $U \subseteq U(S)$. On the other hand, assume that $(ts^r u)(t's^k u') = 1$ for some $t, t' \in T$, $u, u' \in U$, $0 \leq r, k < p$. Then $tt's^{r+k}\varphi^k(u)u' = 1$, so that $s^{k+r} \in H$. Consequently, $s^{r+k}(= 1$ or $g)$ is in $Z$, and we get $tt's^{r+k} = 1$. The fact that $U(T) = \{1\}$ implies that $t = 1$ and $r = k = 0$. This shows that $U(S) = U$.

Next, consider a field $K$ such that $|U'|$ is nonzero in $K$. Then, by Theorem 10, $K[S]$ is an Azumaya algebra. However, $U(S)Z(S) = UT \neq S$.

We close this chapter with some observations on Azumaya semigroup algebras of arbitrary semigroups.

**Proposition 13** Let $K[S]$ be an Azumaya algebra. Then every 0-simple principal factor $S_t$, $t \in S$, of $S$ is of the form $\mathfrak{M}^0(G, n, n, P)$, where $G$ is a group such that $K[G]$ is an Azumaya algebra, and the matrix $P$ is invertible in $M_n(K[G])$. Moreover, in this case, $K[StS]$ is a ring direct summand of $K[S]$.

*Proof.* Assume that $t \in S$ is an element determining a 0-simple principal factor $S_t$ of $S$. Let $J_t = StS$. Since $K[S]$ satisfies a polynomial identity, then, from Theorem 13 in Chapter 20, it follows that $S_t$ is completely 0-simple. Then $S_t^2 = S_t$, and so $J_t^2 = J_t$. From Lemma 1, we then know that $K[J_t] = (K[J_t] \cap Z(K[S]))K[S]$ and that $K[J_t] \cap Z(K[S])$ is an idempotent ideal of $Z(K[S])$. Let $t = \sum_{i=1}^{m} a_i t_i$ for some $a_i \in K[J_t] \cap Z(K[S])$, $t_i \in S$. Then,

$$K[J_t] = K[S] \left( \sum_{i=1}^{m} a_i t_i \right) K[S] \subseteq \sum_{i=1}^{m} a_i K[S] \subseteq K[J_t]$$

and so

$$K[J_t] \cap Z(K[S]) = \left( \sum_{i=1}^{m} a_i K[S] \right) \cap Z(K[S]) = \sum_{i=1}^{m} a_i Z(K[S])$$

since $Z(K[S])$ is a direct summand of the $Z(K[S])$-module $K[S]$; see Lemma 1. Thus, $K[J_t] \cap Z(K[S])$ is a finitely generated idempotent ideal of $Z(K[S])$. It is known that a finitely generated idempotent ideal of a commutative ring is generated by an idempotent; see [29], p. 402, Exercise 6. Therefore, $K[J_t] \cap Z(K[S]) = eZ(K[S])$ for some idempotent $e \in Z(K[S])$. Consequently, $K[J_t] = eZ(K[S])K[S] = eK[S]$. Hence, $K[J_t]$ has an identity $e$, and so it is a ring direct summand of $K[S]$. Then $K_0[J_t]$ is an Azumaya algebra as a homomorphic image of $K[S]$. From Corollary 26 in Chapter 5, it follows that $S_t$ is of the desired form and $K_0[S_t] \cong M_n(K[G])$. Since $M_n(K[G]) \cong K[G] \otimes_{Z(K[G])} M_n(Z(K[G]))$, then $K[G]$ is an Azumaya algebra; see [38], Corollary II.1.9.

As a consequence, we show that, within the class of finite semisimple semigroups, all Azumaya semigroup algebras are, in fact, determined by the class of Azumaya group algebras described in Theorem 7. The following result was first proved by DeMeyer and Hardy in [37].

**Corollary 14** Let $S$ be a finite semisimple semigroup. Then the following conditions are equivalent:

(i)  $K[S]$ is an Azumaya algebra.

(ii)   $K[S] \cong \oplus_{i=1}^n M_{n_i}(K[G_i])$ for some $m$, $n_i \geq 1$, and some subgroups $G_i$ of $S$ such that every $K[G_i]$ is an Azumaya algebra.

*Proof.* The implication (ii) $\Rightarrow$ (i) is clear. (i) $\Rightarrow$ (ii) follows from Proposition 13 by an induction argument on the number of principal factors of $S$, similar to that used in Corollary 27 in Chapter 5.

An essentially stronger $K$-separability condition was characterized for arbitrary semigroup rings by Cheng in [20]; see also [37].

**Corollary 15**   Let $S$ be an arbitrary semigroup. Then the following conditions are equivalent:

(i)   $K[S]$ is a separable $K$-algebra.
(ii)  $K[S] \cong \oplus_{i=1}^n M_{n_i}(K[G_i])$ for some $m$, $n_i \geq 1$, and some finite subgroups $G_i$ of $S$ with orders invertible in $K$.
(iii) $S$ is finite and $\mathcal{J}(K[S]) = 0$.

*Proof.*   It is well known that (i) is equivalent to the fact that $K[S]$ is a finite dimensional $K$-algebra and $\mathcal{J}(L[S]) = 0$ for every field $L \supseteq K$; see [38], Section II.2. The equivalence of (ii) and (iii) was established in Theorem 24 of Chapter 14. Finally, from (ii) and the Maschke theorem, it follows that $\mathcal{J}(L[S]) = 0$ for every field extension $L$ of $K$. Hence, (ii) implies (i).

Our last aim in this chapter is to show how the general problem of describing Azumaya semigroup algebras may sometimes be reduced to the case of "almost idempotent-free" semigroups, that is, semigroups with only one nonzero $\mathcal{J}$-class containing idempotents.

**Lemma 16**   Let $S$ be a finitely generated semigroup. If $K[S]$ is an Azumaya algebra, then $S$ has finitely many idempotents.

*Proof.*   Assume that $S = \langle s_1, \dots, s_n \rangle$. We know that $K[S] = \sum_{i=1}^m Z(K[S])t_i$ for some $m \geq 1$, $t_i \in S$. If $s_i = \sum_j z_{ij} t_i$ for $z_{ij} \in Z(K[S])$, and $t_k t_l = \sum_j z_{klj} t_j$ for $z_{klj} \in Z(K[S])$, then $K[S] = \sum_{i=1}^m B t_i$, where $B$ is the central subalgebra of $K[S]$ generated by all $z_{ij}$, $z_{klj}$ and the identity of $K[S]$. Consequently, $B$ is noetherian, and so $K[S]$ is noetherian as a finitely generated $B$-module. It follows that $K[S]$ has finitely many central idempotents.

Let $e = e^2 \in S$. From Proposition 13, we know that $K[SeS]$ has an identity $f_e$ and that the corresponding principal factor $S_e$ has finitely many idempotents. Let $\varphi : E(S) \to Z(K[S])$ be the mapping defined by $\varphi(e) = f_e$ for $e \in E(S)$. Since $\varphi^{-1}(e)$ consists of idempotents from $S_e$, then it is finite. Therefore, the fact that $Z(K[S])$ has finitely many idempotents implies that $E(S)$ is finite.

**Corollary 17** The following conditions are equivalent for a finitely generated semigroup $S$:

(i) $K[S]$ is an Azumaya algebra.

(ii) $K[S] \cong \oplus_{i=1}^{n} K_0[S_i]$, where $S_i$ are semigroups with only one nonzero $\mathcal{J}$-class containing idempotents, and such that $K_0[S_i]$ are Azumaya algebras.

*Proof.* The implication (ii) $\Rightarrow$ (i) is clear. The converse follows through Lemma 16 by an induction (using Proposition 13) on the number of $\mathcal{J}$-classes containing idempotents.

Finally, let us recall that, for every Azumaya algebra $K[S]$, we know that $\sum_{i=1}^{n} Z(K[S])t_i = K[S]$ for some $t_i \in S$, so that $K[S]$ is a homomorphic image of the semigroup ring $Z(K[S])[T]$, where $T = \langle t_1, \ldots, t_n \rangle$ is a finitely generated semigroup. This relates the general problem of describing Azumaya semigroup algebras to the case considered in Corollary 17.

## Comments on Chapter 25

Azumaya algebras form a class of very well behaved $PI$-algebras that allow development of the Brauer group classification of commutative rings; see [38]. They also prove to be very useful as an efficient tool in the study of general algebras satisfying polynomial identities. We refer to [17] for an important application of this type.

$Z$-separability of the integral semigroup ring $Z[S]$ of an arbitrary finite semigroup $S$ was first studied by Shapiro in [244]. Then Cheng [20] obtained a description of $R$-separable rings $R[S]$ for an arbitrary commutative coefficient ring $R$. DeMeyer and Hardy independently considered this problem in [37]. An extension of these results to the class of so-called excellent extensions $R[S] \supseteq R$ for an arbitrary unitary ring $R$ was given by Okniński in [187].

In a sequence of papers, Van Oystaeyen studied group-graded Azumaya algebras; see [262] and [263]. In particular, he established some sufficient conditions on an algebra strongly graded by a group to be an Azumaya algebra. This provides another proof of the sufficiency in Theorem 7.

# Part V

# Problems

In this part, we state some of the most important open problems connected with the material covered in the book. Many of them are extracted from the text for the reader's convenience. We start with the questions concerning the finiteness conditions.

**Problem 1.** Is $S$ locally finite whenever $K[S]$ is semilocal?

The answer is not known for group algebras of positive characteristics, but this case would be sufficient for proving the semigroup algebra question. Moreover, this would yield a description of semilocal semigroup algebras; see Chapter 14. A similar question may be stated, and also the problem starts in the group algebra of positive characteristic case, with "semilocal" replaced by "algebraic $K$-algebra."

**Problem 2.** Is $S$ locally finite whenever $K[S]$ is a von Neumann regular algebra?

This seems to be the main obstacle when one is looking for a characterization of the regularity condition for arbitrary $K[S]$. One tends to believe that such a description (in terms of $S$ and ch($K$)) should exist. This may be rephrased as follows.

**Problem 3.**   Assume that $L$ is a field with ch($L$) = ch($K$). Is $K[S]$ regular if and only if $L[S]$ is regular?

Another question is whether the regularity is a "local" property of $K[S]$. This is particularly interesting since we have a description of regular algebras of 0-simple semigroups; see Chapter 15. Thus, we ask the question in Problem 4.

**Problem 4.**   Is $K[S]$ regular if and only if all the principal factor algebras $K[S_t], t \in S$, are regular, provided that $S$ is a locally finite semigroup?

An example of a periodic inverse semigroup $S$ such that $K[S]$ is not regular but all principal factor algebras are regular has been given in [112a].

While the answer to Problem 2 is affirmative for $PI$-algebras, the following special case seems promising.

**Problem 5.**   Characterize regular semigroup algebras satisfying polynomial identities.

It may seem very difficult to find a characterization of self-injective semigroup algebras $K[S]$ in terms of $S$. However, the following question is challenging.

**Problem 6.**   Does $K[S]$–right self-injective imply that $S$ is finite?

The next group of our problems is concerned with the effort to characterize the noetherian property for $K[S]$. The general problem seems very difficult, and it has not been solved for group algebras. We start with the following question.

**Problem 7.**   Is $S$ finitely generated whenever $K[S]$ is right noetherian?

One could conjecture that $S$ is finitely generated whenever it has a.c.c. on right congruences. Note that, in view of the examples of infinite noetherian groups constructed by Olshanskii [196], it is rather unlikely that a.c.c. on right congruences in $S$ implies that $K[S]$ is right noetherian. However, under some additional hypotheses, there should be a strong connection between the lattices $\mathcal{R}(S)$ and $\mathcal{R}(K[S])$ of right congruences on $S$ and right ideals of $K[S]$. We state a particular question of this nature, which makes use of the fact that the Krull dimension may be defined, and works really well, in the class of modular lattices; see [169].

**Problem 8.** Examine connections between $K \dim \mathcal{R}(S)$ and $K \dim K[S]$ (if they are defined) for $PI$-algebras $K[S]$.

We state two other general problems on noetherian algebras.

**Problem 9.** Is $\mathcal{J}(K[S])$ nilpotent whenever $K[S]$ is right noetherian?

This is true for $PI$-algebras because of Theorem 12 in Chapter 21 and the fact that the prime radical of a noetherian ring is nilpotent. Moreover, there are no known examples of semigroup algebras with non-nil Jacobson radicals.

**Problem 10.** Is the idempotent set $E(S/\rho_{\mathcal{J}(K[S])})$ finite for any right noetherian algebra $K[S]$?

We know that $S$ has finitely many idempotent $\mathcal{R}$-classes, all of which lie in completely 0-simple principal factors of $S$.

Much more information may be expected in the case of noetherian $PI$-algebras $K[S]$.

**Problem 11.** Characterize right noetherian monoid algebras satisfying polynomial identities.

One might conjecture that (at least for a large class of semigroups) a monoid algebra $K[S]$ is right noetherian $PI$ if and only if

(a)  $S$ is finitely generated, $S$ has the permutational property, and $S$ has a.c.c. on right ideals.

Perhaps (a) should be strengthened by replacing a.c.c. on right ideals by a.c.c. on right congruences. We know that the listed conditions are necessary. Moreover, the assertion is true for the class of cancellative semigroups and the class of inverse semigroups. If $S$ is regular (or even weakly periodic), with a.c.c. on right congruences, then, by [88], $S$ has finitely many right ideals. Moreover, from Proposition 13 in Chapter 4, it follows that any subgroup $G$ of $S$ has a.c.c. on right congruences, so that it is finitely generated. Therefore, $G$ is abelian-by-finite (see Theorem 15 in Chapter 19), and also $S$ has a chain of ideals $S = S_n \supseteq S_{n-1} \cdots \supseteq S_1$ with completely 0-simple factors. From the results of Chapter 20, it follows that $K[S]$ is a $PI$-algebra. It seems that the same should be true if we assume (a). To show this, it would be sufficient to prove that $S$ has d.c.c. on principal ideals or, equivalently, d.c.c. on idempotents; see [83]. However, it is not clear when a semigroup algebra of this type is right noetherian.

Since prime homomorphic images of a right noetherian algebra, as in the $PI$-case, are Goldie, then one can use the techniques developed in Chapter 23 to study the dimensions of right noetherian algebras $K[S]$ through a reduction to the cancellative case. Since the non-$PI$ cancellative case seems difficult to handle, here we state only the following.

**Problem 12.** Do we have $rk(S) = Rk(S) = \mathrm{cl}\,K\,\dim K[S]$ for every $PI$-algebra $K[S]$?

The following result has recently been obtained by Ananin [1a]: if $R$ is a right noetherian $PI$-algebra that is finitely generated, then $R$ embeds into some $M_n(A)$, where $A$ is a commutative algebra, $n \geq 1$. The hypothesis is satisfied for right noetherian $PI$-algebras $R = K[S]$. Does it follow that $rk(S) = Rk(S)$ in this case?

**Problem 13.** Study the prime spectrum of a noetherian $PI$-algebra $K[S]$, trying to generalize some of the commutative case results; see [64] and [66]. In particular, study relations between chains of prime ideals $P$ in $K[S]$ and the chains of congruences $\rho = \rho_P$ on $S$ that are determined by such ideals.

**Problem 14.** Assume that $S$ has a.c.c. on principal right ideals and that $K\,\dim K[S]$ exists. Is $K[S]$ then a finitely generated right noetherian algebra? (at least if $K[S]$ is a $PI$-algebra)

**Problem 15.** Develop the representation theory (= theory of irreducible modules) of algebras $K[S]$ of a subsemigroup $S$ of a polycyclic-by-finite group.

This seems to be a promising, and important, object to study in view of the role of the (noetherian) group algebras of groups of this type [204]. One might start with the case in which $S$ has a.c.c. on right ideals, that is, $K[S]$ is right noetherian; see Chapter 11.

**Problem 16.** Does there exist a finitely presented semigroup $S$ that is periodic but infinite?

Note that such an $S$ would lead to a non-finitely presented associated semigroup $X/I$; see Theorem 19 in Chapter 24, and Theorem 3 in Chapter 19.

**Problem 17.** Is the Hilbert series of a right noetherian (finitely generated, finitely presented) algebra $K[S]$ a rational function?

**Problem 18.** Can semigroups of polynomial growth be characterized by means of some generalized "almost nilpotency" condition as in the cancellative case?

We note that Malcev's nilpotency certainly is too restrictive here because it excludes unions of nilpotent groups that are not semilattices of these groups; see [131].

**Problem 19.** Do we have $GK \dim K[S] = \lim \log_m d_S(m)$ for every finitely generated cancellative semigroup $S$?

Our next group of problems concerns algebras close to prime, or even domains. As noted in Comments to Chapter 9, from the results on the tensor products of algebras, as in the group algebra case, one may expect an intrinsic characterization of primeness and semiprimeness of $K[S]$ in terms of $ch(K)$ and $S$. Beyond the results for cancellative semigroups (see Chapters 7 and 9) and $PI$-algebras (see Part IV), we note that Munn handled this problem for a class of inverse semigroups [165]. The following, most general, statement may seem too difficult at this stage.

**Problem 20.** Characterize prime and semiprime algebras $K[S]$.

However, at least the cancellative case should be settled.

**Problem 21.** Assume that $S$ is cancellative and $\Delta(S) \neq \emptyset$. Is $K[S]$ prime (semiprime respectively) if and only if $K[\Delta(S)]$ is also?

We note that the latter condition may be characterized in terms of $\Delta(S)$ and $ch(K)$; see Chapter 9.

**Problem 22.** Study the connections between the properties of $\Delta(S)$, $\hat{\Delta}(S)$, and $\Delta(G) \cap S$ for a cancellative semigroup $S$ with a group of right fractions $G$.

**Problem 23.** Is $\mathcal{J}(K[S]) = 0$ whenever $S$ is a u.p. semigroup?

This is the case for semigroups with groups of fractions; see Chapter 10. Clearly, one could also extend one of the main problems on group algebras by asking whether $\mathcal{J}(K[S]) = 0$ for fields $K$ with $ch(K) = 0$ and cancellative $S$.

We continue with two more questions on the Jacobson radical.

**Problem 24.** Describe the Jacobson radical of an arbitrary monomial algebra $K_0[X/I]$. Is it equal to $K_0[T]$ for some ideal $T$ of $X/I$? Is then $T = \mathcal{L}(X/I)$?; see Chapter 24.

**Problem 25.** Describe the congruences $\rho_{\mathcal{J}(K[S])}$ on $S$ coming from different fields $K$.

In the group algebra case, $\rho_{\mathcal{J}(K[S])}$ always comes from a normal $\mathrm{ch}(K)$-subgroup. In general, relations between the congruences corresponding to different fields $K$ would also be of interest; see Chapter 14 for the semilocal case.

**Problem 26.** Find necessary and sufficient conditions on $K[S]$ to be an Azumaya algebra.

It seems that the major steps in solving the above problem should be concerned with the three questions below.

**Problem 27.** Do we always have $|E(S) \setminus Z(E(S))| < \infty$ for an Azumaya algebra $K[S]$?

In other words, we ask whether $S$ has only finitely many principal factors with more than one nonzero idempotent; see Proposition 13 in Chapter 25.

**Problem 28.** Characterize Azumaya algebras $K[S]$ of almost idempotent-free semigroups.

**Problem 29.** Assume that $S$ is finite. Is an Azumaya algebra $K[S]$ always isomorphic to a finite direct product of matrix algebras over semigroup algebras of idempotent-free monoids?

**Problem 30.** Characterize the permutational property for important classes of semigroups, for example, 0-simple semigroups and their subsemigroups, or inverse semigroups.

In particular, is it true that, for $S$-inverse, $S$ has $\mathfrak{P}$ if and only if there is a bound on the number of idempotents in the principal factors of $S$, and there is $n \geq 2$ such that all subgroups of $S$ have $\mathfrak{P}_n$?

**Problem 31.** If $S$, $T$ are semigroups with the permutational property, does $S \times T$ have this property?

Note that this is the case with respect to the $PI$-property because, by Regev's theorem, the tensor product $R_1 \otimes_K R_2$ of algebras is $PI$ if and only if both $R_1$, $R_2$ are $PI$; see [239], Theorem 6.1.1.

**Problem 32.** Characterize $PI$-algebras $K[S]$ of a completely 0-simple semigroup $S$ in the case where $\mathrm{ch}(K) > 0$.

More generally, one might ask for a description of the $PI$-property for semisimple semigroups. In particular, we ask the question in Problem 33.

**Problem 33.** Assume that $S$ is a regular semigroup. Find conditions under which $K[S]$ is $PI$ whenever all principal factor algebras $K[S_t]$, $t \in S$, satisfy the same polynomial identity?

**Problem 34.** Study irreducible representations (primitive modules) of semigroups with the permutational property.

One of the important auxiliary results used in the text characterizes $\pi$-regular linear semigroups; see Theorem 10 in Chapter 3. The only known proof of this theorem employs the exterior power technique. For applications to the representations of some, other than $PI$, classes of semigroup algebras, for example, noetherian, it seems important to answer the following question.

**Problem 35.** Is the description of (strongly) $\pi$-regular linear semigroups given in Theorem 10 in Chapter 3 valid for skew linear semigroups $S \subseteq M_n(D)$ (where $D$ is a division algebra) of this type?

**Problem 36.** Study the structure and properties of semigroup algebras of linear semigroups.

The theory of linear representations of semigroups has been developed by many authors. We refer to the papers of McAlister [148], Lallement and Petrich [133], and Ponizovskii [215a] and the monograph of Clifford and Preston [26] for the main results and bibliography. Recently, decisive results on the irreducible representations of certain naturally arising linear semigroups have been obtained by Okniński and Putcha in the characteristic zero case [192], [193] and by Harris and Kuhn in the modular case [85].

**Problem 37.** Characterize monoid algebras that are semifirs, or hereditary.

Some new conditions on a monoid $M$ for the algebra $K[M]$ to be a 2-fir have been recently found by Cedo in [18a]. An example of a monoid $M$, that satisfies all the known necessary conditions for $K[M]$ to be a semifir, but that is not a directed union of free monoids, (see Chapter 17), is there also constructed.

---

**ADDED IN PROOF:** Problem 12 has recently been answered affirmatively by the author.

# References

1. Amitsur, S. A., "Algebras over infinite fields," *Proc. Amer. Math. Soc.* **7** (1956), 35–47.

1a. Ananin, A. Z., "An intriguing story about representable algebras," Ring Theory, pp. 31–38, Weizmann Science Press of Israel, Jerusalem, 1989.

2. Anderson, T., Divinsky, N., and Suliński, A., "Lower radical properties of associative and alternative rings," *J. London Math. Soc.* **41** (1966), 417–424.

3. Anick, D. J., "Non-commutative graded algebras and their Hilbert series," *J. Algebra* **78** (1982), 120–140.

4. Anick, D. J., "On monomial algebras of finite global dimension," *Trans. Amer. Math. Soc.* **291** (1985), 291–310.

5. Anick, D. J. and Green, E. L., "On the homology of path algebras," *Comm. Algebra* **15** (1987), 309–341.

6.  Azumaya, G., "Strongly $\pi$-regular rings," *J. Fac. Sci. Hokkaido Univ.* **13** (1954), 34–39.

7.  Bass, H., *The degree of polynomial growth of finitely generated nilpotent groups*, Proc. London Math. Soc. **25**(4), (1972), 603–614.

8.  Bautista, R. R., Gabriel, P., Roiter, A. V., and Salmeron, L., "Representation —finite algebras and multiplicative bases," *Invent. Math.* **81** (1985), 217–285.

9.  Beattie, M., "A generalization of the smash product of a graded ring," *J. Pure Appl. Algebra* **52** (1988), 219–226.

10. Bianchi, M., Brandl, R., and Gillio Berta Mauri, A., "On the 4-permutational property," *Arch. Math.* **48** (1987), 281–285.

11. Blyth, R. D., "Rewriting products of group elements—I," *J. Algebra* **116** (1988), 506–521.

12. Blyth, R. D., "Rewriting products of group elements—II," *J. Algebra* **119** (1988), 246–259.

13. Boffa, M. and Bryant, R. M., "Les groupes linéaires vérifiant une identité monoidale," *C. R. Acad. Sci. Paris* **308**, Ser. I (1989), 127–128.

14. Borisenko, V. V., "On matrix representations of finitely presented algebras defined by a finite set of words," *Vest. Mosk. Univ.* **4** (1985), 75–77. (In Russian.)

15. Bourbaki, N., *Algebre*, Chapitre III, Hermann, Paris, 1948.

16. Bourbaki, N., *Theories Spectrales*, Hermann, Paris, 1967.

16a. Bovdi, A. A., *Group Rings*, Uzhgorod, 1974 (In Russian).

17. Braun, A., "The nilpotency of the radical in a finitely generated *PI* ring," *J. Algebra* **89** (1984), 375–396.

18. Cedo, F., "A question of Cohn on semifir monoid rings," *Comm. Algebra* **16**(6) (1988), 1187–1189.

18a. Cedo, F. On semifir monoid rings, Publications Mathematiques **33** (1989), 123–132.

19. Chekanu, N., "Semisimple locally finite algebras," *Studies in the Theory of Rings, Algebras and Modules*, Mat. Issled. **76** (1984), 172–179. (In Russian.)

20. Cheng, C. C., "Separable semigroup algebras," *J. Pure Appl. Algebra*, **33** (1984), 151–158.

21. Cheng, C. and Wong, R. W., "Hereditary monoid rings," *Amer. J. Math.* **14** (1982), 935–942.

22. Chick, H. L. and Gardner, B. J., "The preservation of some ring properties by semilattice sums," *Comm. Algebra* **15** (1987), 1017–1038.

23. Chin, W. and Quinn, D., "Rings graded by polycyclic-by-finite groups," *Proc. Amer. Math. Soc.* **102** (1988), 235–241.

24. Chouinard, L. G., Hardy, B. R., and Shores, T. S., "Arithmetical and semihereditary semigroup rings," *Comm. Algebra* **8** (1980), 1593–1652.

25. Cliff, G. H., "Zero divisors and idempotents in group rings," *Canad. J. Math.* **32** (1980), 596–602.

26. Clifford, A. H. and Preston, G. B., *The Algebraic Theory of Semigroups*, American Mathematical Society, Providence, Vol. 1, 1961; Vol. 2, 1967.

27. Cohen, M. and Montgomery, S., "Group graded rings, smash products and group actions," *Trans. Amer. Math. Soc.* **282** (1984), 237–258; Addendum: ibid., **300** (1987), 810–811.

28. Cohn, P. M., "Some remarks on the invariant basis property," *Topology* **5** (1966), 215–228.

29. Cohn, P. M., *Algebra II*, Wiley, New York, 1977.

30. Cohn, P. M., *Skew Field Constructions*, London Math. Soc. Lecture Notes **27**, Cambridge Univ. Press, Cambridge, UK, 1977.

31. Cohn, P. M., *Free Rings and Their Relations*, Academic Press, London, 1985.

32. Curtis, C. W. and Reiner, I., *Representation Theory of Finite Groups and Associative Algebras*, Interscience, London, 1962.

33. Curzio, M., Longobardi, P., and Maj, M., "Su di un problema combinatorio in teoria dei gruppi," *Atti Acc. Lincei Rend.*, Vol. VIII, **74** (1983), 136–142.

34. Curzio, M., Longobardi, P., Maj, M., and Robinson, D. J. S., "A permutational property of groups," *Arch. Math.* **44** (1985), 385–389.

35. Dauns, J., "Generalized semigroup rings," Algebra Carbondale 1980, pp. 235–254, Lect. Notes in Math. **848**, Springer-Verlag, New York, 1981.

36. Dauns, J., "Centers of semigroup rings and conjugacy classes," *S. Forum* **38** (1989), 355–364.

37. DeMeyer, F. and Hardy, D., "Semigroup rings which are separable algebras," Ring Theory Antwerp, 1985, pp. 51–59; Lect. Notes in Math., **1197**, Springer-Verlag, Berlin, 1986.

38. DeMeyer, F. and Ingraham, E., *Separable Algebras Over Commutative Rings*, Lect. Notes in Math. **181**, Springer-Verlag, Berlin, 1971.

39. DeMeyer, F. and Janusz, G., "Group rings which are Azumaya algebras," *Trans. Amer. Math. Soc.* **279** (1983), 389–395.

40. Dicks, W., "Hereditary group rings," *J. London Math. Soc.* **20** (1979), 27–38.

41.  Dicks, W., *Groups, Trees and Projective Modules*, Lect. Notes in Math., **790**, Springer-Verlag, Berlin, 1980.

42.  Dicks, W., "An exact sequence for rings of polynomials in partly commuting indeterminates," *J. Pure Appl. Algebra* **22** (1981), 215–228.

43.  Dicks, W. and Menal, P., "The group rings that are semifirs," *J. London Math. Soc.* **19** (1979), 288–290.

44.  Dicks, W. and Schofield, A. H., "On semihereditary rings," *Comm. Algebra* **16** (1988), 1243–1274.

45.  Divinsky, N., *Rings and Radicals*, Univ. of Toronto Press, Toronto, 1965.

46.  Domanov, O. I., "On identities of semigroup algebras of completely 0-simple semigroups," *Mat. Zametki* **18** (1975), 203–212. (In Russian.)

47.  Domanov, O. I., "The semisimplicity and identities of semigroup algebras of inverse semigroups," *Rings and Modules*, Mat. Issled. **38** (1976), 123–127. (In Russian.)

48.  Domanov, O. I., "Identities of semigroup rings of 0-simple semigroups," *Sib. Math. J.* **17** (1976), 1406–1407. (In Russian.)

49.  Domanov, O. I., "Perfect semigroup rings," *Sib. Math. J.* **18** (1977), 294–303. (In Russian.)

50.  Drazin, M. P., "Pseudo-inverses in associative rings and semigroups," *Amer. Math. Monthly* **7** (1958), 506–514.

51.  Drazin, M. P., "Maschke's theorem for semigroups," *J. Algebra* **72** (1981), 269–278.

52.  Drazin, M. P., "Structure matrices of algebras," *J. Algebra* **87** (1984), 247–260.

53.  Dubreil, P., "Contribution a la theorie des demi-groupes," Mem. Acad. Sci. Inst. France (2)63, No. 3 (1941), 1–52.

53a. Engel, A. B. and Makar-Limanov, L., "On group rings which are not Ore domains," *S. Forum* **39**, No. 2 (1989), 249–254.

54.  Faith, C., *Lectures on Injective Modules and Quotient Rings*, Lect. Notes in Math. **49**, Springer-Verlag, Berlin, 1967.

55.  Faith, C., *Algebra II, Ring Theory*, Springer-Verlag, New York, 1976.

56.  Faith, C., "When self-invective rings are QF: a report on a problem," preprint.

57.  Farkas, D. R. and Snider, R. L., "$K_0$ and Noetherian group rings," *J. Algebra* **42** (1976), 192–198.

58.  Finkelstein, M. Ya. and Kozhukhov, I. B., "Perfect and semiprimary semigroup rings," *Izv. Vyss. Uceb. Zaved. Mat.* (1987), No. 12, 45–48. (In Russian)

59.  Formanek, E., "A problem of Herstein on group rings," *Can. Math. Bull.* **17** (1974), 201–202.

59a. Fountain, J. and Petrich, M., "Completely 0-simple semigroups of quotients, III," *Math. Proc. Camb. Phil. Soc.* **105** (1989), 263–275.

60.  Garzon, M. and Zalcstein, Y., "On permutation properties in groups and semigroups," *S. Forum* **35** (1987), 337–351.

61.  Garzon, M. and Zalcstein, Y., "Linear semigroups with permutation properties," *S. Forum* **35** (1987), 369–371.

62.  Gateva-Ivanova, T., "Algorithmic determination of the Jacobson radical of monomial algebras," preprint.

63.  Gateva-Ivanova, T. and Latyshev, V. N. "On recognizable properties of associative algebras," *J. Symbolic Computation*, 6 (1988), nos. 2–3, 371–388.

64.  Gilmer, R., *Commutative Semigroup Rings*, Univ. Chicago Press, Chicago, 1984.

65.  Gilmer, R., "Chain conditions in commutative semigroup rings," *J. Algebra* **103** (1986), 592–599.

66.  Gilmer, R., "Conditions concerning prime and maximal ideals of commutative monoid rings," *Group and Semigroup Rings*, North Holland, New York, 1986, pp. 19–26.

67.  Gilmer, R., "Commutative monoid rings with finite maximal or prime spectrum," *Group and Semigroup Rings*, North Holland, New York, 1986, pp. 27–34.

68.  Gilmer, R. and Teply, M. L., "Idempotents of commutative semigroup rings," *Houston J. Math.* **3** (1977), 369–385.

69.  Goodearl, K. R., *Von Neumann Regular Rings*, Pitman, London, 1979.

70.  Gordon, R. and Robson, J. C., "Krull dimension," *Mem. Amer. Math. Soc.* **133**, 1973.

71.  Gould, V. A. R., "Orders in semigroups," Contributions to General Algebra **5**, pp. 163–169, Hölder-Pichler-Tempsky, Vienna, 1987.

72.  Goursaud, J. and Valette, J., "Anneaux de groupe hereditaires et semihereditaires," *J. Algebra* **34** (1975), 205–212.

73.  Govorov, V. E., "Graded algebras," *Math. Notes* **12** (1972), 552–556.

74.  Grassman, H., "On irreducible and principal indecomposable representations of finite semigroups," *Math. Nachr.* **81** (1978), 269–277.

75.  Green, E. L., Kirkman, E., and Kuzmanovich, J., "Finitistic dimensions of finite dimensional monomial algebras," preprint.

76. Grigorchuk, R. I., "Degrees of growth of finitely generated groups and theory of invariant means," *Izv. Akad. Nauk SSSR*, Ser. Math. **48**(5) (1984), 939–985. (In Russian.)

77. Grigorchuk, R. I., "On the degrees of growth of $p$-groups and torsion-free groups," *Mat. Sb.* **126**(2) (1985), 194–214. (In Russian.)

78. Grigorchuk, R. I., "Cancellative semigroups of power growth," *Mat. Zametki* **43** (1988), 305–319. (In Russian.)

79. Groenewald, N. J., "A note on the units of semigroup rings," *Comm. Algebra* **9**(17) (1981), 1681–1691.

80. Gromov, M., "Groups of polynomial growth and expanding maps," *Publ. Math. IHES* **53**(1) (1981), 53–73.

81. Hall. M., *Combinatorial Theory*, Wiley, New York, 1967.

82. Hall. T. E., "The radical of the algebra of any finite semigroup over any field," *J. Austral. Math. Soc.* **11** (1970), 350–352.

83. Hall, T. E., "On the natural ordering of $\mathcal{J}$-classes and of idempotents in a regular semigroup," *Glasgow Math. J.* **11** (1970), 167–168.

84. Harary, F., *Graph Theory*, Addison-Wesley, Reading, Mass., 1969.

85. Harris, J. C. and Kuhn, N. J., "Stable decompositions of classifying spaces of finite abelian $p$-groups," Math. Proc. Camb. Phil. Soc. **103** (1988), 427–449.

86. Herstein, I. N., *Noncommutative Rings*, Carus Math. Monographs 15, Mathematical Association of America, 1968.

87. Hofmann, K. and Skryago, A. M., "Finite dimensional continuous representations of compact regular semigroups," *S. Forum* **28** (1984), 199–234.

88. Hotzel, E., "On semigroups with maximal conditions," *S. Forum* **11** (1975/76), 337–362.

89. Hotzel, E., "On finiteness conditions in semigroups," *J. Algebra* **60** (1979), 352–370.

90. Howie, J. M., *An Introduction to Semigroup Theory*, London Math. Soc. Monographs, Academic Press, London, 1976.

91. Irving, R. S., "An algebra whose simple modules are of arbitrary finite degree," *Ring Theory*, Antwerp, 1978, pp. 87–96; Marcel Dekker, New York, 1979.

92. Irving, R. S. and Small, L. W., "The Goldie conditions for algebras with bounded growth," *Bull. London Math. Soc.* **15** (1983), 596–600.

93. Jacobson, N., *Structure of Rings*, American Mathematical Society, Providence, 1968.

94.  Jacobson, N., *PI-Algebras, An Introduction*, Lect. Notes in Math. **441**, Springer-Verlag, New York, 1979.

95.  Janeski, J. and Weissglass, J., "Regularity of semilattice sums of rings," *Proc. Amer. Math. Soc.* **39** (1973), 479–482.

96.  Jespers, E., "On radicals of graded rings and applications to semigroup rings," *Comm. Algebra* **13** (1985), 2457–2472.

97.  Jespers, E., "The Jacobson radical of semigroup rings of commutative semigroups," *J. Algebra* **109** (1987), 266–280.

98.  Jespers, E., "When is the Jacobson radical of a semigroup ring of a commutative semigroup homogeneous?" *J. Algebra* **109** (1987), 549–560.

99.  Jespers, E., "Chain conditions and semigroup graded rings," *J. Austral. Math. Soc.* **45** (1988), 372–380.

100. Jespers, E., Krempa, J. and Puczyłowski, E. R., "On radicals of graded rings," *Comm. Algebra* **10** (1982), 1849–1854.

101. Jespers, E., Krempa, J., and Wauters, P., "The Brown–McCoy radical of semigroup rings of commutative cancellative semigroups," *Glasgow J. Math.* **26** (1985), 107–113.

102. Jespers, E. and Puczyłowski, E. R., "The Jacobson radical of semigroup rings of commutative cancellative semigroups," *Comm. Algebra* **12** (1984), 1115–1123.

103. Jespers, E. and Wauters, P., "A description of the Jacobson radical of semigroup rings of commutative semigroups," *Group and Semigroup Rings*, North-Holland, Amsterdam, 1986, 43–89.

104. Jespers, E. and Wauters, P., "Almost Krull rings," *J. Austral. Math. Soc.* (Ser. A) **41** (1986), 275–285.

105. Jespers, E. and Wauters, P., "When is a semigroup ring of a commutative semigroup local or semilocal?" *J. Algebra* **108** (1987), 188–194.

106. Justin, J. and Pirillo, G., "Comments on the permutation property for semigroups," S. Forum 39 (1989), no. 1, 109–112.

107. Karpilovsky, G., *Commutative Group Algebras*, Marcel Dekker, New York, 1983.

108. Kelarev, A. V., "Radicals of semigroup algebras of commutative semigroups," XIX Vsessojuzn. Algebr. Conf., Abstracts, Lvov, 1987, p. 122. (In Russian.)

109. Kelarev, A. V., "A simple matrix semigroup which is not completely simple," *S. Forum* **37** (1988), 123–125.

110. Kelarev, A. V., "Hereditary radicals and semilattices of associative algebras," to appear.

111. Kelarev, A. V., "Radicals and semigroup rings," *Mat. Issled.* **105**, Modules Algebras, Topol., 1988, pp. 81–92. (In Russian.)

112. Kelarev, A. V. and Volkov, M. V., "Varieties and bands of associative rings," Izv. Vyss. Ucebn. Zaved. **30** (1986), 16–23. (In Russian.)

112a. Kelarev, A. V., "On regular semigroup rings," S. Forum 40, no. 1, (1990), 113–114.

113. Kertesz, A., "Systems of equations over modules," *Acta Sci. Math.* **18** (1957), 207–234.

114. Kertesz, A., *Lectures on Artinian Rings*, Akademiae Kiado, Budapest, 1987.

115. Kljushin, A. V. and Kozhukhov, B. I., "On algebraically compact semigroup rings," *S. Forum* **39** (1989), 215–248.

116. Kozhukhov, I. B., "Chain semigroup rings," *Uspekhi Mat. Nauk* XXIX 1 (175), (1974), 169–170. (In Russian.)

117. Kozhukhov, I. B., "On semigroups with minimal or maximal condition on left congruences," *S. Forum* **21** (1980), 337–350.

118. Kozhukhov, I. B., "Self-injective semigroup rings of inverse semigroups," *Izv. Vyss. Uceb. Zaved.* **2** (1981), 46–51. (In Russian.)

119. Kozhukhov, I. B., "Free left ideal semigroup rings," *Algebra i Logika* **21** (1982), 37–59. (In Russian.)

120. Kozhukhov, I. B., "Semigroups with minimal condition and artinian semigroup rings," to appear.

121. Krause, G. R. and Lenagan, T. H., *Growth of Algebras and Gelfand–Kirillov dimension*, Pitman, London, 1985.

122. Krempa, J., "Radicals of semigroup rings," *Fund. Math.* **85** (1974), 57–71.

123. Krempa, J., "On semigroup rings," *Bull. Acad. Polon. Sci.* **25** (1977), 225–231.

124. Krempa, J., "Special elements in semigroup rings," *Bull Acad. Polon. Sci.* **28** (1980), 17–23.

125. Krempa, J., "On semisimplicity of tensor products," *Ring Theory*, Antwerp, 1978, pp. 105–122; Marcel Dekker, New York, 1979.

126. Krempa, J., "Isomorphisms of group rings," Banach Center Publications, Vol. **9**, PWN, Warsaw, 1982, pp. 233–255.

127. Krempa, J. and Okniński, J., "Semilocal, semiperfect and perfect tensor products," *Bull. Acad. Polon. Sci.* **28** (1980), 249–256.

128. Krempa, J. and Okniński, J., "Gelfand–Kirillov dimensions of a tensor product," *Math. Z.* **194** (1987), 487–494.

129. Krempa, J. and Sierpińska, A., "The Jacobson radical of certain group and semigroup rings," *Bull. Acad. Polon. Sci.* **26** (1978), 963–967.

130. Kuzmanovich, J. and Teply, M. L., "Semihereditary monoid rings," *Houston J. Math.* **10** (1984), 525–534.

131. Lallement, G., "On nilpotency in semigroups," *Pacific J. Math.* **42** (1972), 693–700.

132. Lallement, G., *Semigroups and Combinatorial Applications*, Wiley-Interscience, New York, 1979.

133. Lallement, G. and Petrich, M., "Irreducible matrix representations of finite semigroups," *Trans. Amer. Math. Soc.* **139** (1969), 393–412.

134. Lawrence, J., "Semilocal group rings and tensor products," *Mich. Math. J.* **22** (1975), 309–313.

135. Lawrence, J., "A countable self-injective ring is quasi-Frobenius," *Proc. Amer. Math. Soc.* **65** (1977), 217–220.

136. Lawrence, J. and Woods, S. M., "Semilocal group rings in characteristic zero," *Proc. Amer. Math. Soc.* **60** (1976), 8–10.

137. Longobardi, P. and Maj, M., "Sui gruppi per cui ogni prodotto di quattro elementi e riordinabile," preprint.

138. Lorenz, M. and Passman, D. S., "Prime ideals in crossed products of finite groups," *Israel J. Math.* **33** (1979), 89–132.

139. Lotharie, M., *Combinatorics on Words*, Addison-Wesley, Reading, Mass., 1983.

140. Makar-Limanov, L., "On free subsemigroups," *S. Forum* **29** (1984), 253–254.

141. Makar-Limanov, L., "On free subsemigroups of skew fields," *Proc. Amer. Math. Soc.* **91** (1984), 189–191.

142. Malcev, A., "On the immersion of an algebraic ring into a field," *Math. Ann.* **113** (1937), 686–691.

143. Malcev, A. I., "Nilpotent semigroups," *Uc. Zap. Ivanovsk. Ped. Inst.* **4** (1953), 107–111. (In Russian.)

144. Matsuda, R., "Notes on torsion free abelian semigroup rings," *Bull. Fac. Sci.*, Ibaraki Univ., Ser. A **20** (1988), 51–59.

145. May, M. K., A von Neumann regular ring that is locally finite but not locally semisimple," *Comm. Algebra* **16** (1988), 103–113.

146. McAlister, D. B., "The category of representations of a completely 0-simple semigroup," *J. Austral. Math. Soc.* **12** (1971), 198–210.

147. McAlister, D. B., "Rings related to completely 0-simple semigroups," *J. Austral. Math. Soc.* **12** (1971), 257–274.

148. McAlister, D. B., "Representations of semigroups by linear transformations," *S. Forum* **2** (1971), 189–263, 283–320.

149. McConnell, J. C. and Robson, J. C., *Noncommutative Noetherian Rings*, Wiley, New York, 1987.

150. McNaughton, R. and Zalcstein, Y., "The Burnside problem for semigroups," *J. Algebra* **34** (1975), 292–299.

151. Mehrvarz, A. A. and Wallace, D. A. R., "On the automorphisms of the group ring of a unique product group," *Proc. Edinburgh Math. Soc.* **2**(30) (1987), 201–205.

152. Menal, P., "The monoid rings that are Bezout domains," *Arch. Math.* **37** (1981), 43–47.

153. Menal, P., "On tensor products of algebras being von Neumann regular or self-injective," *Comm. Algebra* **9** (1981), 691–697.

153a. Mikhalev, A. V. and Zalesskii, A. E., *Group Rings, Modern Problems of Mathematics*, Vol. 2, Moscow, 1973 (in Russian).

154. Munn, W. D., "On semigroup algebras," *Proc. Camb. Philos. Soc.* **51** (1955), 1–15.

155. Munn, W. D., "Semigroups satisfying minimal conditions," *Proc. Glasgow Math. Assoc.* **3** (1957), 145–152.

156. Munn, W. D., "Pseudo-inverses in semigroups," *Proc. Cambridge Philos. Soc.* **57** (1961), 247–250.

157. Munn, W. D., "On the regularity of certain semigroup algebras," *Semigroups*, Academic Press, London, 1980, 207–224.

158. Munn, W. D., "Semiprimitivity of inverse semigroup algebras," *Proc. Roy. Soc. Edinburgh*, Ser. A, **93** (1982), 83–98.

159. Munn, W. D., "On commutative semigroup algebras," *Math. Proc. Cambridge Philos. Soc.* **93** (1983), 237–246.

160. Munn, W. D., "On the Jacobson radical of certain commutative semigroup algebras," *Math. Proc. Cambridge Philos. Soc.* **96** (1984), 15–23.

161. Munn, W. D., "The algebra of a commutative semigroup over a commutative ring with unity," *Proc. Roy. Soc. Edinburgh* **99A** (1985), 387–398.

162. Munn, W. D., "Inverse semigroup algebras," *Group and Semigroup Rings*, North-Holland, New York, 1986, 197–223.

163. Munn, W. D., "Nil ideals in inverse semigroup algebras," *J. London Math. Soc.* **35** (1987), 433–438.

164. Munn, W. D., "Two examples of inverse semigroup algebras," *S. Forum* **35** (1987), 127–134.

165. Munn, W. D., "A class of contracted semigroup rings," *Proc. Roy. Soc. Edinburgh* **107A** (1987), 175–196.

166. Munn, W. D., "The Jacobson radical of a band ring," *Math. Proc. Camb. Phil. Soc.* **105** (1989), 277–283.

167. Nakayama, T., "On Frobeniusean algebras I," *Ann. Math.* **40** (1939), 611–633.

168. Nastasescu, C. and Van Oystaeyen, F., *Graded Ring Theory*, North-Holland, New York, 1982.

169. Nastasescu, C. and Van Oystaeyen, F., *Dimensions of Ring Theory*, Reidel, Dordrecht, 1987.

170. Nordahl, T., "Residual finiteness in permutation varieties of semigroups," *S. Forum* **31** (1985), 33–46.

171. Nordahl, T., "On permutative semigroup algebras," *Algebra Universalis* **25** (1988), 322–333.

172. Nordahl, T., "Separative *PI* semigroups are embeddable into a semilattice of groups," *S. Forum* **38** (1989), 119–122.

173. Okniński, J., "On spectrally finite algebras," *Bull. Acad. Polon. Sci.* **27** (1979), 515–519.

174. Okniński, J., "Spectrally finite and semilocal group rings," *Comm. Algebra* **8** (1980), 533–541.

175. Okniński, J., "Spectrally finite algebras and tensor products," *Comm. Algebra* **10** (1982), 1939–1950.

176. Okniński, J., "Artinian semigroup rings," *Comm. Algebra* **10** (1982), 109–114.

177. Okniński, J., "When is the semigroup ring perfect?" *Proc. Amer. Math. Soc.* **89** (1983), 49–51.

178. Okniński, J., "On regular semigroup rings," *Proc. Roy. Soc. Edinburgh* **99A** (1984) 145–151.

179. Okniński, J., "On self-injective semigroup rings," *Arch. Math.* **43** (1984), 407–411.

180. Okniński, J., "Finiteness conditions for semigroup rings," *Acta. Univ. Carolinae* **25** (1984), 29–32.

181. Okniński, J., "Semilocal semigroup rings," *Glasgow Math. J.* **25** (1984), 37–44.

182. Okniński, J., "Strongly $\pi$-regular matrix semigroups," *Proc. Amer. Math. Soc.* **93** (1985), 215–217.

183. Okniński, J., "On the radical of semigroup algebras satisfying polynomial identities," *Math. Proc. Cambridge Philos. Soc.* **99** (1986), 45–50.

184. Okniński, J., "On semigroup algebras satisfying polynomial identities," *S. Forum* **33** (1986), 87–102.

185. Okniński, J., "Radicals of group and semigroup rings," *Contributions to General Algebra* **4**, Verlag Hölder-Pichler-Tempsky, Vienna, 1987, 125–150.

186. Okniński, J., "On cancellative semigroup rings," *Comm. Algebra* **15** (1987), 1667–1677.

187. Okniński, J., "Semigroup rings as excellent extensions and the regular radical," Simon Stevin **61** (1987), 301–311.

188. Okniński, J., "Noetherian property for semigroup rings," *Ring Theory, Granada, 1986*, Lect. Notes in Math. **1328**, Springer-Verlag, New York, 1988, pp. 209–218.

189. Okniński, J., "A note on the *PI*-property of semigroup rings," *Perspectives in Ring Theory*, Kluwer, Dordrecht, 1988, pp. 275–278.

190. Okniński, J., "Commutative monoid rings with Krull dimension," *Semigroups*, Lect. Notes in Math. **1320**, Springer-Verlag, New York, 1988, pp. 251–259.

191. Okniński, J., "On monomial algebras," *Arch. Math.* **50** (1988), 417–423.

192. Okniński, J. and Putcha, M. S., "Complex representations of matrix semigroups," Trans. Amer. Math. Soc., to appear.

193. Okniński, J. and Putcha, M. S., "Sandwich matrices of linear semigroups," to appear.

193a. Okniński, J. and Putcha, M. S., "*PI* semigroup algebras of linear semigroups," Proc. Amer. Math. Soc., to appear.

194. Okniński, J. and Van Oystaeyen, F., "Cancellative semigroup rings which are Azumaya algebras," *J. Algebra* **117** (1988), 290–296.

195. Okniński, J. and Wauters, P., "Radicals of semigroup rings of commutative semigroups," *Math. Proc. Cambridge Philos. Soc.* **99** (1986). 435–445.

196. Olshanskii, A. Yu., *On a geometric method in the combinatorial group theory*, Proceedings of the International Congress of Mathematicians, Warsaw, 1983, PWN, Warsaw, 1984, pp. 415–424.

197. Ovsyannikov, A. Ya., "On radical semigroup rings," *Mat. Zametki* **37** (1985), 452–459. (In Russian.)

198. Ovsyannikov, A. Ya., "On the radical property of fundamental ideals of semigroup rings," *Issled. Algebr. Systems.*, Sverdlovsk, 1985, 119–127. (In Russian.)

199. Park, J. K., "Skew group rings with Krull dimension," *Math. J. Okayama Univ.* **25** (1983), 75–80.

200. Parker, J. and Gilmer, R., "Nilpotent elements of commutative semigroup rings," *Mich. Math. J.* **22** (1975/76), 97–108.

201. Passman, D. S., *Infinite Group Rings*, Marcel Dekker, New York, 1971.

202. Passman, D. S., "On the semisimplicity of group rings of some locally finite groups," *Pac. J. Math.* **58** (1975), 179–207.

203. Passman, D. S., *The Algebraic Structure of Group Rings*, Wiley-Interscience, New York, 1977; 2nd ed., Robert E. Krieger Publishing, Melbourne, FL, 1985.

204. Passman, D. S., "Group rings of polycyclic groups," *Essays for Phillip Hall*, London Mathematical Society, 1984, pp. 207–256.

205. Passman, D. S., *Infinite Crossed Products*, Academic Press, Boston, Mass., 1989.

206. Patterson, E. M., "On the radicals of rings of row finite matrices," *Proc. Roy. Soc. Edinburgh* **66A** (1961/62), 42–46.

207. Petrich, M., *Introduction to Semigroups*, Merrill, Columbus, 1973.

208. Petrich, M., *Rings and Semigroups*, Lect. Notes in Math. **380**, Springer-Verlag, New York, 1974.

209. Petrich, M., *Lectures in Semigroups*, Wiley, Berlin, 1977.

210. Pierce, R. S., *Associative Algebras*, Springer-Verlag, New York, 1982.

211. Pirillo, G., "On permutation properties for finitely generated semigroups," *Combinatorics*, 1986, Annals of Discrete Mathematics, Vol. **37**, North Holland, New York, 1988, pp. 375–376.

212. Ponizovskii, I. S., "On matrix representations of associative systems," *Mat. Sb.* **38** (1956), 241–260. (In Russian.)

213. Ponizovskii, I. S., "On Frobeniusness of a semigroup algebra of a finite commutative semigroup," *Izv. Akad. Nauk. SSSR*, **32** (1968), 820–836. (In Russian.)

213a. Ponizovskii, I. S., "On quasi-Frobeniusness of a semigroup algebra of a finite regular semigroup," *Sib. Math. J.* 11 (1970), 1312–1320. (In Russian.)

214. Ponizovskii, I. S., "Some remarks on representation theory of semigroups and semigroup algebras," XXIV Herzen. Ctenija, Len. Gos. Ped. Herz. Inst., Leningrad, 1971, 22–23. (In Russian.)

215. Ponizovskii, I. S., "On modules over semigroup rings of completely 0-simple semigroups," *Semigroup Constructions*, Leningrad, 1981, pp. 85–94. (In Russian.)

215a. Ponizovskii, I. S., "On irreducible matrix semigroups," S. Forum 24 (1982), 117–148.

216. Ponizovskii, I. S., "On some classes of semigroup rings," XVIII Vsesojuzn. Algebr. Conf., Theses, Part II, Minsk, 1983, 188–189. (In Russian.)

217. Ponizovskii, I. S., "An example of a semiprimitive semigroup algebra," *S. Forum* **26** (1983), 225–228.

218. Ponizovskii, I. S., "On semigroup rings," *S. Forum* **28** (1984), 143–154.

219. Ponizovskii, I. S., "On semiprimitive and semiprime finite semigroup rings," *Semigroups*, Szeged 1981, Coll. Math. Soc. Janos Bolyai **39**, North-Holland, New York, 1985, pp. 273–285.

220. Ponizovskii, I. S., "Some examples of semigroup algebras of finite representation type," *Analytic Number Theory and Theory of Functions* **8**, pp. 229–238, Zap. Naucn. Sem. LOMI 160, Leningrad, Nauka, 1986. (In Russian.)

221. Ponizovskii, I. S., "Semigroup rings," *S. Forum* **36** (1987), 1–46.

221a. Ponizovskii, I. S. On the semiprimitivity of inverse semigroup algebras and on theorems of Domanov and Munn, S. Forum 40, no. 2 (1990), 181–185.

222. Procesi, C., *Rings with Polynomial Identities*, Marcel Dekker, New York, 1973.

223. Puczyłowski, E. R., "Radicals of polynomial rings in non-commutative indeterminates," *Math. Z.* **182** (1983), 63–67.

224. Puczyłowski, E. R., "Radicals of semigroup algebras of commutative and cancellative semigroups," *Proc. Roy. Soc. Edinburgh* **103A** (1985), 317–323.

225. Putcha, M. S., "On linear algebraic semigroups," *Trans. Amer. Math. Soc.* **259** (1980), 457–469.

226. Putcha, M. S., "Matrix semigroups," *Proc. Amer. Math. Soc.* **88** (1983), 386–390.

227. Putcha, M. S., *Linear Algebraic Monoids*, London Math. Soc. Lecture Note Series **133**, Cambridge University Press, Cambridge, UK, 1988.

228. Putcha, M. S. and Yaqub, A., "Semigroups satisfying permutation identities," *S. Forum* **3** (1971), 68–73.

229. Putcha, M. S. and Yaqub, A., "Rings satisfying monomial identities," *Proc. Amer. Math. Soc.* **32** (1972), 52–56.

230. Quesada, A., "On semiprime twisted semigroup rings," *S. Forum* **25** (1982), 339–344.

231. Quinn, D., "Group-graded rings and duality," *Trans. Amer. Math. Soc.* **292** (1985), 155–167.

232. Renault, G., *Algebre Non-Commutative*, Gauthier-Villars, Paris, 1975.

233. Restivo, A. and Reutenauer, C., "On the Burnside problem for semigroups," *J. Algebra* **89** (1984), 102–104.

234. Restivo, A. and Reutenauer, S., "Rational languages and the Burnside problem," *Theor. Comput. Sci.* **40** (1985), 13–30.

235. Richter, G., "Noetherian semigroup rings with several objects," *Group and Semigroup Rings*, pp. 231–246, North Holland, New York, 1986.

236. Rips, E. and Segev, Y., "Torsion-free group without unique product property," *J. Algebra* **108** (1987), 116–126.

237. Rosenberg, A. and Zelinsky, D., "Tensor products of semiprimary algebras," *Duke Math. J.* **24** (1957), 555–559.

238. Rosenblatt, Y. M., "Invariant measures and growth conditions," *Trans. Amer. Math. Soc.* **193** (1974), 33–53.

239. Rowen, L. H., *Polynomial Identities in Ring Theory*, Academic Press, New York, 1980.

240. Rudin, W. and Schneider, H., "Idempotents in group rings," *Duke Math. J.* **31** (1964), 585–602.

240a. Saito, T., "Note on the minimal conditions for principal ideals in a semigroup," *Math. Japon.* **13** (1968), 95–104.

241. Sato, H., "Semigroup ring construction of Frobenius extensions," *J. Reine Angew. Math.* **324** (1981), 211–220.

242. Schneider, H. and Weissglass, J., "Group rings, semigroup rings and their radicals," *J. Algebra* **5** (1967), 1–15.

243. Sehgal, S. K., *Topics in Group Rings*, Marcel Dekker, New York, 1978.

244. Shapiro, J., "A note on separable integral semigroup rings," *Comm. Algebra* **6** (1978), 1073–1079.

245. Shapiro, J., "The Krull dimension of twisted group rings," *Houston J. Math.* **11** (1985), 121–127.

246. Shehadah, A., "Proper embeddability of inverse semigroups," *Bull. Austral. Math. Soc.* **34** (1986), 383–387.

247. Shevrin, L. N. and Volkov, M. V., "Identities of semigroups," *Izv. Vyss. Uceb. Zaved.* **11** (1985), 3–47. (In Russian.)

248. Shneerson, L. M., "Free subsemigroups of finitely presented semigroups," *Sib. Math. J.* **15** (1974), 450–454. (In Russian.)

249. Shneerson, L. M., "Finitely presented semigroups with nontrivial identities," *Sib. Math. J.* **23** (1982), 160–171. (In Russian.)

250. Shoji, K., "On semigroup rings," *Proceedings of the 9th Symposium on Semigroups and Related Topics*, Naruto Univ., Naruto, 1986, pp. 28–34.

251. Sierpińska, A., "$K$-completeness and its applications to radicals of semigroup rings," *Demonstratio Math.* **13** (1980), 641–645.

252. Slinko, A. M., Shestakov, I. P., Shirshov, A. I., and Zhevlakov, K. A., *Rings That Are Nearly Associative*, Moscow 1978. (In Russian.)

253. Skornjakov, L. A., "On Cohn rings," *Algebra i Logika* **4** (1965), 5–30. (In Russian.)

254. Small, L. W., Stafford, J. T., and Warfield, R. B., "Affine algebras of Gelfand–Kirillov dimension one are *PI*," *Math. Proc. Cambridge Phil. Soc.* **97** (1985), 407–414.

255. Smith, P. F., "On the dimension of group rings," *Proc. London Math. Soc.* **25** (1972), 288–302; Corrigendum: ibid **27** (1973), 766–768.

256. Strojnowski, A., "A note on u.p. groups," *Comm. Algebra* **8** (1980), 231–234.

257. Teply, M. L., "On semigroup algebras and semisimple semilattice sums of rings," *Proc. Edinburgh Math. Soc.* **24** (1981), 87–90.

258. Teply, M. L., Turman, E. G., and Quesada, A., "On semisimple semigroup rings," *Proc. Amer. Math. Soc.* **79** (1980), 157–163.

259. Tomkinson, M. J., *FC-groups*, Pitman, London, 1984.

260. Ufnarovskii, V. A., "A growth criterion for graphs and algebras defined by words," *Mat. Zametki* **31** (1982), 465–472. (In Russian.)

261. Ufnarovskii, V. A., On the use of graphs for computing a basis, growth and Hilbert series of associative algebras, Mat. Sb. **180**, No. 11 (1989), 1548–1560. (In Russian).

262. Van Oystaeyen, F., "Azumaya strongly graded rings and ray classes," *J. Algebra* **103** (1986), 228–240.

263. Van Oystaeyen, F., "On graded *PI* rings and generalized crossed product Azumaya algebras," *Chinese Annals of Math.*, Ser. B. **8** (1987), 13–21.

264. Varricchio, S., "A finiteness condition for finitely generated semigroups," *S. Forum* **38** (1989), 331–335.

265. Wall, E. G., "Finite groups with class preserving outer automorphisms," *J. London Math. Soc.* **22** (1947), 315–320.

266. Wauters, P., "Rings graded by a semilattice—applications to semigroup rings," *Group and Semigroup Rings*, North Holland, New York, 1986, pp. 253–267.

267. Weissglass, J., "Radicals of semigroup rings," *Glasgow Math. J.* **10** (1969), 85–93.

268. Weissglass, J., "Regularity of semigroup rings," *Proc. Amer. Math. Soc.* **25** (1970), 499–503.

269. Weissglass, J., "Semigroup rings and semilattice sums of rings," *Proc. Amer. Math. Soc.* **39** (1973), 471–478.

270. Wenger, R., "Some semigroups having quasi-Frobenius algebras, I," *Proc. London Math. Soc.* **18** (1968), 484–494.

271. Wenger, R., "Self-injective semigroup rings for finite inverse semigroups," *Proc. Amer. Math. Soc.* **20** (1969), 213–216.

272. Wenger, R., "Some semigroups having quasi-Frobenius algebras, II," *Canad. J. Math.* **21** (1969), 615–624.

273. Wenger, R., "Finite commutative semigroups having Frobenius semigroup algebras," *S. Forum* **16** (1978), 427–442.

274. Wong, R. W., "Free ideal monoid rings," *J. Algebra* **53** (1978), 21–35.

275. Woods, S. M., "Existence of Krull dimension in group rings," *J. London Math. Soc.* **9** (1975), 406–410.

276. Zelmanov, E. I., "Semigroup algebras with identities," *Sib. Math. J.* **18** (1977), 787–798. (In Russian.)

277. Zimmermann, W., "($\Sigma$-) algebraic compactness of rings," *J. Pure Appl. Algebra* **23** (1982), 319–328.

# Index

Algebra:
algebraic, 155, 157, 161-163
Azumaya, 313
  local-global property for, 315
fir, 205
Frobenius, 197
graded, 65
  homogeneous component of, 65
  homogeneous subset of, 66
  left noetherian, 77
  radical of, 76
  strongly, 77
graded noetherian, 77
hereditary, 206

[Algebra]
of matrix type, 48
monomial, 39, 293-311
  graph of, 305-310
Munn, 50
  basic ideal of, 51-56
  induced ideal in, 52-55
  modules over, 55-56
  radicals of, 55-56
PI-, 215
right $T$-nilpotent, 170
semifir, 205
semihereditary, 206
semilocal, 155, 159
separable, 313